Geometry Connections

PRENTICE HALL SERIES IN MATHEMATICS FOR MIDDLE SCHOOL TEACHERS

JOHN BEEM *Geometry Connections*
ASMA HARCHARRAS and DORINA MITREA *Calculus Connections*
IRA J. PAPICK *Algebra Connections*
DEBRA A. PERKOWSKI and MICHAEL PERKOWSKI *Data Analysis and Probability Connections*

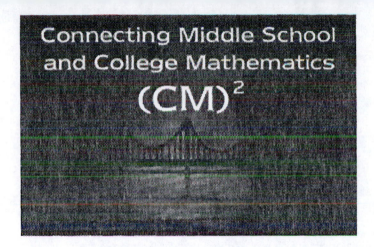

Connecting Middle School
and College Mathematics
(CM)²

Geometry Connections

Mathematics for Middle School Teachers

John K. Beem

Mathematics Department
University of Missouri-Columbia

PEARSON
Prentice
Hall

Upper Saddle River, New Jersey 07458

Beem, John K.
    Geometry connections / John K. Beem.
      p. cm.
    Includes bibliographical references and index.
    ISBN 0-13-144926-5
     1. Geometry—study and teaching (Middle school) I. Title.

QA461.B34 2006
516′.0071′2—dc22

2005048854

Editor in Chief: *Sally Yagan*
Executive Acquisitions Editor: *Petra Recter*
Project Manager: *Michael Bell*
Production Management: *Progressive Publishing Alternatives*
Assistant Managing Editor: *Bayani Mendoza de Leon*
Senior Managing Editor: *Linda Mihatov Behrens*
Executive Managing Editor: *Kathleen Schiaparelli*
Manufacturing Manager: *Alexis Heydt-Long*
Manufacturing Buyer: *Alan Fischer*
Marketing Assistant: *Rebecca Alimena*
Art Director: *Jayne Conte*
Cover Designer: *Bruce Kenselaar*
Art Studio/Formatter: *Laserwords*
Editorial Assistant/Supplement Editor: *Joanne Wendelken*
Cover Image: © *Stockbyte*

©2006 Pearson Education, Inc.
Pearson Prentice Hall
Pearson Education, Inc.
Upper Saddle River, New Jersey 07458

Pearson Prentice Hall™ is a trademark of Pearson Education, Inc.
Development of these materials was supported by a grant from the National Science Foundation
(ESI 0101822).

Printed in the United States of America

10 9 8 7 6 5 4 3 2 1

ISBN: 0-13-144926-5

Pearson Education LTD., *London*
Pearson Education Australia PTY, Limited, *Sydney*
Pearson Education Singapore, Pte. Ltd.
Pearson Education North Asia Ltd., *Hong Kong*
Pearson Education Canada, Ltd., *Toronto*
Pearson Educación de Mexico, S.A. de C.V.
Pearson Education—Japan, *Tokyo*
Pearson Education Malaysia, Pte. Ltd.

Contents

Preface

Improving the quality of mathematics education for middle school students is of critical importance, and increasing opportunities for students to learn important mathematics under the leadership of well-prepared and dedicated teachers is essential. New standards-based curriculum and instruction models, coupled with ongoing professional development and teacher preparation, are foundational to this change.

These sentiments are eloquently articulated in the Glenn Commission Report, *Before It's Too Late: A Report to the Nation from the National Commission on Mathematics and Science Teaching for the 21st Century* (U.S. Department of Education, 2000). In fact, the principal message of the Glenn Commission Report is that America's students must improve their mathematics and science performance if they are to be successful in our rapidly changing technological world. To this end, the report recommends that we greatly intensify our focus on improving the quality of mathematics and science teaching in grades K–12 by bettering the quality of teacher preparation. The report also stresses the need to develop creative plans to attract and retain substantial numbers of future mathematics and science teachers.

Some fifteen years ago, mathematics teachers, mathematics educators, and mathematicians collaborated to develop the architecture for standards-based reform. Their recommendations for the improvement of school mathematics, instruction, and assessment were articulated in three seminal documents published by the National Council of Teachers of Mathematics: *Curriculum and Evaluation Standards for School Mathematics* (1989), *Professional Standards for School Mathematics* (1991), and *Assessment Standards in School Mathematics* (1995); more recently, these three documents were updated and combined into the single book *NCTM Principles and Standards for School Mathematics, a.k.a. PSSM* (2000).

The vision for school mathematics laid out in these three foundational documents was outstanding in spirit and content, yet abstract in practice. Concrete exemplary models reflecting the standards were needed, and implementing the "recommendations" would be unrealizable without significant commitment of resources. Recognizing the opportunity for stimulating improvement in student learning, the National Science Foundation (NSF) made a strong commitment to bring life to the documents' messages and supported several K–12 mathematics curriculum development projects (standards-based curriculum) and other related dissemination and implementation projects.

Standards-based middle school curricula are designed to engage students in a variety of mathematical experiences, including thoughtfully planned explorations that provide and reinforce fundamental skills while illuminating the power and utility of mathematics in our world. These materials integrate central concepts in algebra, geometry, data analysis and probability, and mathematics of change, and focus on important ideas such as proportional reasoning.

The mathematical content of standards-based middle school mathematics materials is challenging and relevant to our technological world. Its effective classroom implementation is dependent upon teachers having strong and appropriate mathematical preparation. *The Connecting Middle School and College Mathematics Project (CM)²* is a three-year (2001–2004) project funded by the National Science Foundation—this project addresses the need for improved teacher qualifications and viable recruitment plans for middle grade mathematics teachers through the development of four foundational mathematics courses, with accompanying support materials and the creation and implementation of effective teacher recruitment models.

The *(CM)²* materials are built upon a framework laid out in the *CBMS Mathematical Education of Teachers Book (MET)* (2001). This report outlines recommendations for the mathematical preparation of middle grade teachers that differ significantly from those for the preparation of elementary school teachers and provides guidance to those developing new programs. Our books are designed to provide middle grade mathematics teachers with a strong mathematical foundation and connect the mathematics they are learning with the mathematics they will be teaching. Their focus is on algebraic and geometric structures, data analysis and probability, and mathematics of change, and they employ standards-based middle grade mathematics curricular materials as a springboard to explore and learn mathematics in more depth. They have been extensively piloted in summer institutes, courses offered at school-based sites, in a variety of professional-development programs, and in semester courses offered at a number of universities throughout the nation.

College students using this book will have learned a fair amount of Euclidean geometry before entering college; consequently, this book is not intended as a first introduction to the subject. Nevertheless, many students will need a substantial review of the subject and will need help in understanding how the content they are learning relates to the mathematics they will eventually be teaching. Throughout the book, the reader will find illustrations of problems and activities from four of the standards-based middle school curricula. These demonstrate the presentation and use of geometry in the middle school. They serve to connect many of the concepts covered in this book to future teaching experiences the college students will eventually encounter and thus provide future teachers with real motivation to learn more mathematical content. Teachers understand they must learn mathematics on a much deeper level than they will be presenting in their future classrooms, but they need to see that what they are learning is related to what they will be teaching.

This book includes a number of Classroom Connections and Classroom Discussions. In both cases, these are activities or discussions that are designed to deepen the connections between the geometry that students are studying now and the geometry they will teach. The **Classroom Connections** are simple activities that teachers might someday use in the middle school classroom. They may be left as part of the readings for the course. The **Classroom Discussions** are explorations that are more central to the book's purpose. These are suggestions that are intended for both the instructor and the students. Some of them may be assigned as readings for the course, and others may be covered as homework assignments. Any of them may be used to shape actual guided discussions in the college classroom. Furthermore, the instructor may wish to assign a few of them as projects to be presented in the classroom by individual students or teams of students.

This book is designed to accommodate both instructors who wish to go slow and spend more time on things such as Classroom Connections and Classroom Discussions and instructors who wish to move more quickly and cover more topics. The former may wish to concentrate on covering the first four chapters. In a course where the chapters are covered more quickly, there should be time to cover all of the first five chapters and most of Chapter 6. A reasonable compromise is to cover Section 1.1 through Section 5.4.

The author would like to thank all of the many faculty members at various schools for piloting preliminary versions of this book and making very helpful suggestions.The author would also like to thank the following reviewers for their helpful suggestions:

Jennifer M. Bay-Williams, Kansas State University; Robert Glasgow, Southwest Baptist University; William James Lewis, University of Nebraska-Lincoln; and Edward Mooney, Illinois State University. Furthermore, the author would like to thank graduate students David Barker, Dustin Foster, and Chris Thornhill for reading over the manuscript and making many valuable suggestions. Also, it is a pleasure to thank Michael Bell and Petra Recter at Pearson/Prentice-Hall for their help in getting this book to appear in print.

REFERENCES FOR THE PREFACE

Conference Board of Mathematical Sciences, *The Mathematical Education of Teachers Book.* Washington, DC: Mathematical Association of America, 2001.

Leitzel, J. (Ed.). *A Call for Change: Recommendations for the Mathematical Preparation of Teachers of Mathematics.* Washington, DC: Mathematical Association of America, 1991.

National Council of Teachers of Mathematics. *Assessment Standards in School Mathematics.* Reston, VA: National Council of Teachers of Mathematics, 1995.

National Council of Teachers of Mathematics. *Curriculum and Evaluation Standards for School Mathematics.* Reston, VA: National Council of Teachers of Mathematics, 1989.

National Council of Teachers of Mathematics. *Principles and Standards for School Mathematics.* Reston, VA: National Council of Teachers of Mathematics, 2000.

National Council of Teachers of Mathematics. *Professional Standards for School Mathematics.* Reston, VA: National Council of Teachers of Mathematics, 1991.

U.S. Department of Education. *Before It's Too Late: A Report to the Nation from the National Commission on Mathematics and Science Teaching for the 21st Century.* John Glenn Commission Chairman. Washington, DC: U.S. Department of Education, 2000.

List of Numbered Figures

List of Classroom Connections

List of Classroom Discussions

Euclid's Geometry

1.1 **EUCLID'S POSTULATES AND COMMON NOTIONS**
1.2 **USING LOGIC**
1.3 **NOTATION AND MEASUREMENT**
1.4 **POLYGONS AND SOLIDS**
1.5 **MEASURING THE EARTH'S RADIUS**

One of the first mathematicians was Thales (*thay leez*) (c. 624 BC–546 BC). He is often credited with being the first person to make extensive use of mathematical arguments that used logical steps. To a large extent he began the tradition of using proofs in mathematics. Many people contributed to mathematics in the years following Thales. One of these was Euclid (born about 325 BC). Among other things, he wrote *The Elements*, which is a collection of thirteen books on mathematics. These books contain more than just geometry, they include several other topics such as ratio and proportion and number theory. Some of the results in these books were due to Euclid, such as his proof that there are an infinite number of primes. On the other hand, most of the mathematics in Euclid's *Elements* represents contributions by others. Euclid took manuscripts of earlier mathematicians and combined them in a very organized fashion.

In this chapter, Euclid's postulates and axioms are given in Section 1.1, and there is a short introduction to logic in Section 1.2. In Section 1.3, some notation is given, and in Section 1.4 there is an introduction to solids. Although the major theme of this book is Euclidean plane geometry, you will also find several results in this book involving solids. At a number of places, spherical geometry is compared and contrasted with Euclidean plane geometry. Section 1.5 describes how Eratosthenes measured the radius of the Earth in about 240 BC. His result was extremely accurate.

Exercises are provided at the end of each section throughout this book. This book also includes Classroom Connections and Classroom Discussions. The

Classroom Connections are simple activities for eventual use in the middle school classroom. Classroom Discussions are explorations that are intended to be more central to the college course. The college instructor may wish to cover Classroom Discussions in various ways. Some might be assigned as readings, others might be covered in classroom discussions, and still others might be assigned as projects that individuals or teams of students will present in the classroom.

1.1   EUCLID'S POSTULATES AND COMMON NOTIONS

Many students find it surprising that Euclid did not prove everything. He started with ten assumptions consisting of five postulates and five common notions. These assumptions are used to prove later results in *The Elements*, but these original ten assumptions are not proven. It is important for teachers to understand and to be able to explain why Euclid did not prove everything. One way to help prepare future mathematics teachers is to have a college classroom discussion such as the one outlined here.

Classroom Discussion 1.1.1

A fundamental concept in mathematics that Euclid evidently understood is that not everything in mathematics is to be proven. In other words, one needs to start with some unproven assumptions and then prove results based on these assumptions. A good classroom discussion can be centered on the question of why one needs to base things on unproven assumptions.   ◆

At the start of Book 1 of Euclid's *Elements*, one finds a list of definitions followed by a list of five **postulates** and then a list of five **common notions**. Although these definitions, postulates, and common notions are not up to modern mathematical standards, we list the postulates and common notions to give an understanding of the historical development of geometry. Also, most middle school and high school treatments of geometry make use of these assumptions. Often Euclid's common notions are called *axioms*. Euclid's original postulates and common notions are listed in Appendix II in translated form. The following are slightly reformulated versions of Euclid's postulates and common notions. You can compare these statements to corresponding statements given in Appendix II.

Postulates:

1. Two distinct points lie on exactly one line.
2. One may extend a line segment indefinitely in each direction.
3. Given any two distinct points, one may construct a circle with one point as center and the segment joining the second point as a radius.
4. Any two right angles are equal in measure to each other.
5. If a straight line falling on two straight lines makes the interior angles on the same side less than two right angles, then the two straight lines, if extended indefinitely, meet on that side on which the angles are less than the two right angles.

Common Notions (Axioms):

1. Things that are equal to the same thing are also equal to each other.
2. If equals are added to equals, the sums are equal.
3. If equals are subtracted from equals, the remainders are equal.
4. Things that coincide with each other are equal to each other.
5. The whole is greater than the part.

Theorems are statements that are proven using postulates, axioms, and previously proven statements. Also, propositions, lemmas, and corollaries are proven statements, just like theorems. The difference between a theorem, a proposition, a lemma, or a corollary is mainly one of semantics. A very important proven statement is generally called a *theorem*. A proven statement of lesser importance might be called a **proposition**. A proven statement of lesser importance that is used to prove a theorem or proposition is considered a **lemma**. A **corollary** is a statement that may be "easily" proven using a preceding theorem.

In today's world, all of Euclid's postulates and common notions would be lumped together and either called *postulates* or *axioms*. Thus, the words *postulate* and *axiom* are used interchangeably in mathematics. However, for historical reasons, when teaching Euclidean geometry, one often makes an artificial distinction and reserves the word *postulate* for Euclid's first five assumptions and reserves the word *axiom* for Euclid's second set of assumptions (i.e., his "common notions"). An **axiom** (or **postulate**) is a statement that is assumed to be true without proof.

EXERCISES 1.1

1. Assume that each line has an infinite number of points. Using Euclid's postulates and common notions, list all possibilities for the number of points in the intersection $\ell_1 \cap \ell_2$ of two (not necessarily distinct) lines ℓ_1 and ℓ_2.
2. Euclid defined a point as that which has no part. Write a paragraph explaining why you think this is either a reasonable definition or an inadequate definition.

1.2 USING LOGIC

The usual statement in mathematics is sometimes known as an "If ..., then ..." sentence. What follows *If* is called the **hypothesis**, and what follows *then* is called the **conclusion**.

EXAMPLE List the hypothesis and conclusion of the following sentence: If a given quadrilateral is a rectangle, then the diagonals of this quadrilateral are of equal length.

Solution The hypothesis is "a given quadrilateral is a rectangle," and the conclusion is "the diagonals of this quadrilateral are of equal length." ∎

An "If ..., then ..." sentence is said to be a **conditional statement** or an **implication**. If P represents the hypothesis and Q represents the conclusion, then the sentence "If P, then Q," may be represented by $P \rightarrow Q$.

$$P \rightarrow Q$$

If P, then Q.

Here, P is the hypothesis, and Q is the conclusion.

The sentence $P \rightarrow Q$ (i.e., "If P, then Q") has a value of F (false) whenever P is T (true) and Q is F (false). In other words, true does not imply false. In all other cases, the sentence $P \rightarrow Q$ has the value T (true).

One often regards a statement P as a variable that either has the value T or the value F. One refers to P as a **truth variable**. Given two statements or truth variables P and Q, then they are regarded as **independent truth variables** if they can independently take on the values of T and F. Thus, a table illustrating the four possibilities can be arranged as follows:

P	Q
T	T
T	F
F	T
F	F

Such a table illustrating possible truth values is called a **truth table**.

The **negation** of P is denoted by $\sim P$ and is read **not** P. The following truth table shows the truth values of $\sim P$ given the truth values of P.

P	$\sim P$
T	F
F	T

In logic and mathematics, the word *and* is a special **logical connective** that yields a truth value of T only when it connects two statements that are both T. In other words, if two statements are joined by *and*, then they must both be true for the entire statement to be true. The symbol for *and* is $\wedge$. Let P and Q be truth variables, and then the sentence $P \wedge Q$ is T if both P and Q are T, and otherwise the value of $P \wedge Q$ is F.

$$P \wedge Q$$

"P and Q" is true only if both P and Q are true.

EXAMPLE Let P be the (true) statement, "The state of New York is on the east coast of the U.S." and let Q be the (true) statement, "The state of North Carolina is on the east coast of the U.S." Find the truth value of $P \land Q$ (i.e., the state of New York is on the east coast of the U.S. and the state of North Carolina is on the east coast of the U.S.).

Solution   The statement $P \land Q$ is true because both P and Q are true.   ■

If at least one of the pair P, Q is false, then the sentence $P \land Q$ is false.

EXAMPLE Let P be the (true) statement, "The state of New York is on the east coast of the U.S.," and let Q be the (false) statement, "The state of California is on the east coast of the U.S." Find the truth value of $P \land Q$ (i.e., the state of New York is on the east coast of the U.S. and the state of California is on the east coast of the U.S.).

Solution   The statement $P \land Q$ is false because at least one of P, Q is false (in this case Q is false).   ■

In logic and mathematics, the word *or* is another special **logical connective**. The truth value of P or Q will be T if either one or both of P, Q is T. The symbol for *or* is $\lor$. If P and Q are truth variables, then the sentence $P \lor Q$ is T if at least one of P or Q is T.

$$P \lor Q$$
"P or Q" is true if at least one of P or Q is true.

EXAMPLE Let P be the statement, "All squares have four sides," and let Q be the statement, "All triangles have a right angle." Investigate the truth value of the sentence "All squares have four sides or all triangles have a right angle." In other words, is the sentence $P \lor Q$ true or false?

Solution   Since all squares do have four sides, it is clear that P is true. On the other hand, since some triangles do not have a right angle, it is clear that Q is false. Since at least one of P or Q is true, it follows that the sentence $P \lor Q$ in this example is true.

Notice that even though the second half (i.e., Q) of the sentence is false, the sentence taken as a whole is true.   ■

It is only when both parts of the pair P, Q are false that the sentence $P \lor Q$ is false.

EXAMPLE Let P be the statement, "All cubes have seven faces," and let Q be the statement, "All pentagons have area greater than 10 square inches." Investigate the truth or falsity of the sentence "All cubes have five faces, or all pentagons have area greater than 6 square inches." In other words, is the sentence $P \lor Q$ true or false?

Solution Since all cubes have only six faces, it is clear that P is false. Furthermore, since some pentagons have an area less than 10 square inches, Q is also false. Since both P and Q are false, it follows that the sentence $P \vee Q$ in this example is false. ∎

The truth table in Figure 1.2.1 gives a comparison of the sentence $P \wedge Q$ with the sentence $P \vee Q$.

Two statements are said to be **logically equivalent** if they always have the same truth value. In other words, they are logically equivalent if they each imply the other. In this case, they are "if and only if" statements. In mathematics one uses **iff** to mean if and only if. One example of logical equivalence is illustrated in the truth table in Figure 1.2.2. In this table, $P \wedge Q$ is shown to always have the same truth value as $Q \wedge P$. Thus, one has $P \wedge Q$ iff $Q \wedge P$. Of course, since $P \wedge Q$ and $Q \wedge P$ are logically equivalent, one may substitute one for the other in any logical expression without changing the value of the expression.

EXAMPLE Use a truth table to show that $\sim(P \vee Q)$ and $(\sim P) \wedge (\sim Q)$ are logically equivalent.

Solution Set up a truth table such as the following with at least one column headed by $\sim(P \vee Q)$ and at least one column headed by $(\sim P) \wedge (\sim Q)$.

P	Q	$P \wedge Q$	$P \vee Q$
T	T	T	T
T	F	F	T
F	T	F	T
F	F	F	F

FIGURE 1.2.1 A Truth Table for $P \wedge Q$ and for $P \vee Q$.

P	Q	$P \wedge Q$	$Q \wedge P$
T	T	T	T
T	F	F	F
F	T	F	F
F	F	F	F

FIGURE 1.2.2 A Truth Table for $P \wedge Q$ and for $Q \wedge P$. Since the columns headed by each of $P \wedge Q$ and $Q \wedge P$ have identical values of T and F on all corresponding rows, it follows that $P \wedge Q$ and $Q \wedge P$ are logically equivalent.

P	Q	$\sim P$	$\sim Q$	$P \vee Q$	$\sim(P \vee Q)$	$(\sim P) \wedge (\sim Q)$
T	T					
T	F					
F	T					
F	F					

Filling in the blanks in the previous table, one obtains the following:

P	Q	$\sim P$	$\sim Q$	$P \vee Q$	$\sim(P \vee Q)$	$(\sim P) \wedge (\sim Q)$
T	T	F	F	T	F	F
T	F	F	T	T	F	F
F	T	T	F	T	F	F
F	F	T	T	F	T	T

This proves that $\sim(P \vee Q)$ and $(\sim P) \wedge (\sim Q)$ are logically equivalent, because the columns below $\sim(P \vee Q)$ and $(\sim P) \wedge (\sim Q)$ are the same. ∎

The **converse** of the original statement $P \rightarrow Q$ is obtained by reversing the roles of the hypothesis and conclusion. The **contrapositive** of the original statement is formed by interchanging the hypothesis and conclusion and by negating each of them. The **inverse** of the original statement $P \rightarrow Q$ is formed by negating each of P and Q.

$P \rightarrow Q$	(original statement)
$Q \rightarrow P$	(converse of original statement)
$\sim Q \rightarrow \sim P$	(contrapositive of original statement)
$\sim P \rightarrow \sim Q$	(inverse of original statement)

EXAMPLE Consider the following original true statement:

$$\text{If } n < 5, \text{then } n < 10.$$

State its converse, contrapositive, and inverse, and in each case state if they are true or false.

Solution Here P is the statement $n < 5$, and Q is the statement that $n < 10$. Of course, the negation of P can be expressed as $n \geq 5$, and the negation of Q can be expressed as $n \geq 10$.

Converse:	If $n < 10$,	then $n < 5$.	(False)
Contrapositive:	If $n \geq 10$,	then $n \geq 5$.	(True)
Inverse:	If $n \geq 5$,	then $n \geq 10$.	(False) ∎

The following truth table shows the values of the expression $P \rightarrow Q$ given the truth values of P and Q. In this table, P and Q are treated as independent variables. Also, note that in the bottom two rows of the table the value of $P \rightarrow Q$ is T independent of the value of Q. In other words, when P is false, the expression $P \rightarrow Q$ is true independent of Q.

P	Q	$P \rightarrow Q$
T	T	T
T	F	F
F	T	T
F	F	T

In general, the truth value of an original sentence of the form $P \rightarrow Q$ is independent of the truth value of its converse. Thus, just because the original sentence is true does not mean the converse is true. However, sometimes one is able to prove both the statement and its converse. In this case, the hypothesis and conclusion are logically equivalent and one combines the two results into a single iff statement. A simple example of a mathematical iff result is the following: The triangle $\triangle ABC$ is equiangular iff the triangle $\triangle ABC$ is equilateral. Of course, P is the statement, "The triangle $\triangle ABC$ is equiangular," and Q is the statement, "The triangle $\triangle ABC$ is equilateral." Here an equiangular triangle is one with all angles of the same measure, and an equilateral triangle is one with all sides of the same length.

Just as with the converse, the truth of $P \rightarrow Q$ does not necessarily imply the truth of the inverse $\sim P \rightarrow \sim Q$. On the other hand, the truth value of the sentence $P \rightarrow Q$ is always the same as the truth value of its contrapositive $\sim Q \rightarrow \sim P$. The original sentence and its contrapositive will either both be T or both F.

Law of the Contrapositive

A sentence of the form $P \rightarrow Q$ is true iff the contrapositive $\sim Q \rightarrow \sim P$ is true.

If you feel that a certain statement P implies a statement Q, then you might state this as a conjecture and try to prove it is true. A **conjecture** is an unproven statement that someone feels may be true. If you try to prove $P \rightarrow Q$ and fail, then it is natural to try and disprove $P \rightarrow Q$. To try and disprove the statement, one tries to find an example such that $P \rightarrow Q$ is false. Such an example is called a **counterexample**. You only need one counterexample

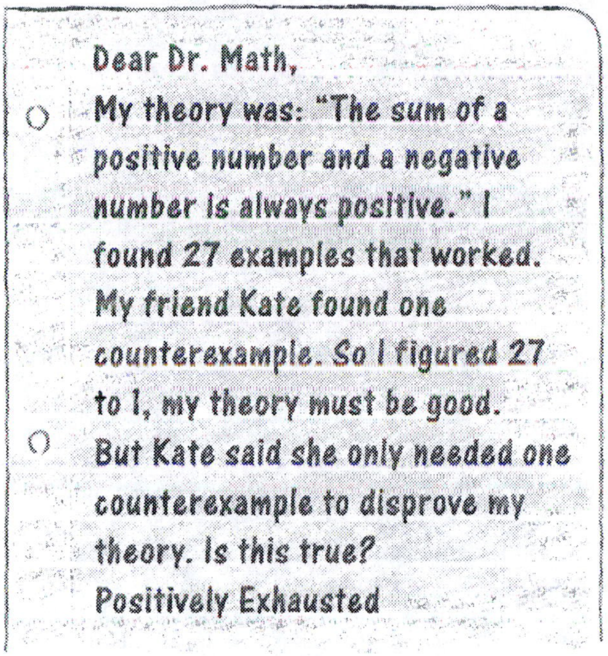

> Dear Dr. Math,
>
> My theory was: "The sum of a positive number and a negative number is always positive." I found 27 examples that worked. My friend Kate found one counterexample. So I figured 27 to 1, my theory must be good. But Kate said she only needed one counterexample to disprove my theory. Is this true?
>
> Positively Exhausted

FIGURE 1.2.3 Students are asked to find and understand counterexamples in grade 7. They are also expected to use logic and to do homework problems that involve writing paragraphs to explain their thinking. Reproduced from page 34 of *Making Mathematical Arguments* in the MathScape grade 7 materials.

to disprove $P \rightarrow Q$. Figure 1.2.3 is part of a page from a seventh-grade homework assignment from MathScape where the term *counterexample* is used. The seventh-grade student is to be "Dr. Math" and is to write out an answer to the question posed.

As the Dr. Math question suggests, it is difficult for some students to understand that one counterexample is all it takes to disprove a conjecture. The Dr. Math question can be used as the starting point for a class discussion of counterexamples and of the strengths and pitfalls of looking for patterns in mathematics. Many students are fearful of searching for counterexamples since, in general, there is no formula for finding them. It is helpful to point out that although there are certainly cases where counterexamples are difficult to find, it is often the case that there are simple counterexamples to a conjecture. Hence it is always wise to at least look at some simple possibilities for counterexamples if you think that a mathematical statement may be invalid.

EXERCISES 1.2

1. Given that Germany is in Europe and that Japan is in Asia, state the truth value of each of the following sentences.
   a. Germany is in Europe, and Japan is not in Asia.
   b. Germany is in Europe, or Japan is not in Asia.

    c. Germany is not in Europe, and Japan is in Asia.

    d. Germany is not in Europe, or Japan is in Asia.

2. Answer the question asked of Dr. Math by Positively Exhausted in Figure 1.2.3. You are to be Dr. Math, and write your answer in paragraph form.

3. State the converse of the following true statement: If $n = 30$, then $n \neq 40$. Also, state if the converse is true or false.

4. Fill in the following truth table to illustrate that $\sim(\sim P)$ and P always have the same truth value. This will show that $\sim(\sim P)$ and P are logically equivalent to each other.

P	$\sim P$	$\sim(\sim P)$
T		
F		

5. Consider the following sentence, which is true in Euclidean geometry: "If $\angle C$ in $\triangle ABC$ is a right angle, then $m(\angle A) < 90°$."

    a. State the converse. Is the converse true?

    b. State the contrapositive. Is the contrapositive true?

    c. State the inverse. Is the inverse true?

6. Consider the following sentence, which is true in Euclidean geometry: "If the quadrilateral $ABCD$ is a square, then all of its interior angles have measure $90°$."

    a. State the converse. Is the converse true?

    b. State the contrapositive. Is the contrapositive true?

    c. State the inverse. Is the inverse true?

7. Give an example of an "If ..., then ..." sentence about triangles that is true and such that its converse is also true.

8. Give an example of an "If ..., then ..." sentence about triangles that is true but is such that its converse is false. (Note that the converse needs only to fail once.)

Classroom Discussion 1.2.1

Students often have difficulty understanding how a sentence $P \rightarrow Q$ can be true and yet the converse $Q \rightarrow P$ can be false. Some future teachers may have memorized this fact from a high school geometry course yet be unable to explain why the given sentence and its converse are logically independent. It is worthwhile to take time in the course to make sure future teachers understand and can explain this logical independence. A possible approach is to make this a homework assignment. One can assign students to write a paragraph or two describing how they might explain to middle school or high school students that the truth value of the original sentence $P \rightarrow Q$ and its converse $Q \rightarrow P$ are logically independent. ◆

Classroom Discussion 1.2.2

Another aspect of logic that students often have difficulty understanding is **proof by contradiction**. Students need to do more than just memorize that $H \rightarrow C$ may be proven by assuming that H is true and C is false and then finding a contradiction. This is another case where it is worthwhile to take time in the course to ensure that future middle school mathematics teachers understand the underlying idea and can explain things to a class they teach someday. ◆

1.3 NOTATION AND MEASUREMENT

Line Segments and Distances. Given two distinct points A and B, the unique **line** containing both of them is denoted by $\overleftrightarrow{AB}$, and the **line segment** with A and B as endpoints is denoted by $\overline{AB}$. A point on the segment $\overline{AB}$ that is not an endpoint is said to be an **interior point** of the segment. The length of the segment $\overline{AB}$ is the distance between A and B and is denoted by AB. A **circle**, C, is the set of points at some fixed positive distance, r, from a **center** point, O (i.e., $C = \{P | OP = r\}$). Lengths may be measured in many different units. One approach is to set up a standard by taking some given line segment as a "unit" length. Then, for example, segments twice as long have length 2 units. Units used in everyday life include English units such as inches and metric units such as centimeters.

Some Equivalent Distance Measures (Approximate)	
English	**Metric**
1 inch	2.54 cm (centimeter)
1 foot	30.48 cm
.3937 inches	1 cm
39.37 inches	1 m (meter)
1 mile (= 5,280 feet)	1.6093 km (kilometer)
.6214 mile	1 km

Units of length may easily be converted from one system to another.

EXAMPLE Convert a length of 65 inches to meters.

Solution From the previous table, one has

$$1 \text{ meter} \approx 39.37 \text{ inches},$$

where the symbol $\approx$ is used to denote "approximately equal." Dividing both sides of the previous equation by 39.37 in. yields

$$\frac{1 \text{ meter}}{39.37 \text{ in.}} \approx 1$$

Hence, multiplying 65 inches by the fraction $\frac{1 \text{ meter}}{39.37 \text{ in.}}$ is (approximately) like multiplying 65 inches by 1. Of course, multiplication by 1 does not change the quantity.

Thus,

$$65 \text{ in.} = (65 \text{ in.}) \cdot (1) \approx (65 \text{ in.}) \cdot \left(\frac{1 \text{ meter}}{39.37 \text{ in.}}\right) \approx 1.65 \text{ meters.}$$

Notice that the units of inches in the product $(65 \text{ in.}) \cdot \left(\frac{1 \text{ meter}}{39.37 \text{ in.}}\right)$ cancel each other. Also notice that the answer is only approximate for two reasons. First, 1 meter is only approximately equal to 39.37 inches. Second, the result for the fraction $\frac{65}{39.37}$ has been rounded to 1.65. ∎

Rays and Angles.   The closed half line with endpoint A and containing the point B will be called a **ray** and will be denoted by $\overrightarrow{AB}$.

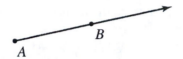

Given two rays $\overrightarrow{AB}$ and $\overrightarrow{AC}$, both starting at the point A, the angle $\angle BAC$ will be the union of the two rays $\angle BAC = \overrightarrow{AB} \cup \overrightarrow{AC}$.

$$\boxed{\angle BAC = \overrightarrow{AB} \cup \overrightarrow{AC}}$$

If the two rays do not lie on the same line (i.e., $\overleftrightarrow{AB} \neq \overleftrightarrow{AC}$), then the **interior** of the angle is defined to be the set of points P, such that P is a point in the interior of a segment $\overline{RS}$ with one endpoint R on $\overrightarrow{AB}$ with $R \neq A$ and one endpoint S on $\overrightarrow{AC}$ with $S \neq A$.

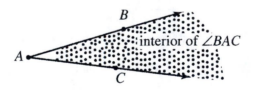

When the two rays point in exactly opposite directions, then one has $\overleftrightarrow{AB} = \overrightarrow{AB} \cup \overrightarrow{AC}$, and the angle is usually called a **straight angle**. Angles may be measured using either **degrees** or **radians**. The measure of angle $\angle BAC$ is denoted by $m(\angle BAC)$. A right angle has a measure of $90°$ or, equivalently, $\pi/2$ radians. A straight angle has measure $180°$ or, equivalently, π radians. To measure an angle $\angle BAC$ in radians, one takes a circle of some radius r about the vertex A and lets S denote the length of the arc of the circle subtended by $\angle BAC$. Using θ to denote $m(\angle BAC)$ in radians, one has $\theta = \frac{S}{r}$, which can also be expressed by $S = r\theta$, see Figure 1.3.1.

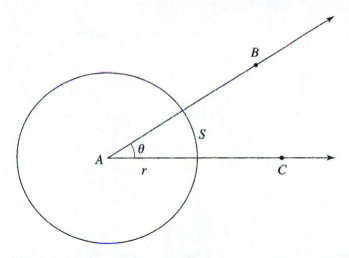

FIGURE 1.3.1 The angle $\angle BAC$ cuts off an arc of length S from the circle with radius r centered at A. In radians, the measure of angle $\angle BAC$ is given by $\theta = \frac{S}{r}$.

Notation	Meaning
$\overleftrightarrow{AB}$	Line containing both A and B
$\overline{AB}$	Line segment with endpoints A and B
AB	Length of segment $\overline{AB}$ (AB = distance from A to B)
$\overrightarrow{AB}$	Ray starting at A and containing B
$\angle BAC$	$\overrightarrow{AB} \cup \overrightarrow{AC}$
$m(\angle BAC)$	Measure of the angle $\angle BAC$

Coordinates. In the 1600s, a coordinate approach to geometry was developed by Pierre de Fermat (*fehr mah*) (1601–1665) and Rene Descartes (*day kart*) (1596–1659). Let R^1 represent the real numbers, and let $R^2 = \{(x,y)|x, y \in R^1\}$ be the *xy*-plane.

$$R^1 \leftrightarrow \text{real numbers}$$
$$R^2 \leftrightarrow xy\text{-plane}$$

The Euclidean plane is said to have R^2 as a **coordinate representation**. One generally refers to R^2 as the *xy*-plane. A point in this model is an ordered pair (x,y) of real numbers, and a straight line in this model is a set of the following general form:

$$\{(x,y)|Ax + By + C = 0, where\ at\ least\ one\ of\ A\ or\ B\ is\ not\ zero\}.$$

> **Lines in the *xy*-plane**
>
> $Ax + By + C = 0$ (general form)
> $y = mx + b$ (slope-intercept form)

The following classroom connection gives one approach a middle school teacher might use to introduce coordinates and graphing to students. Remember, the Classroom Connections are written to give future teachers some ideas about how they might eventually teach things they are learning. Future teachers should at least read the Classroom Connections and give some thought to how they might someday present things in classes they teach.

Classroom Connection 1.3.1

One approach to introducing middle school students to coordinates and graphing is to have them draw several different figures in the *xy*-plane. One first introduces students to coordinates and then has them locate and draw many different individual points. Later one gives them three points that are not collinear and have them sketch the triangle with these points as vertices. After students become more familiar with coordinates, one can have them graph equations by making *xy*-tables then graphing a number of points with *x* and *y* values satisfying the equation. In particular, graphing several equations of the form $Ax + By + C = 0$ and/or $y = mx + b$ will make clear to students that equations like these represent straight lines. If one has sufficient class time in a given semester, then one can develop a discovery lesson where students are asked to discover and describe what kinds of graphical changes correspond to changes in the *m* and the *b* in equations of the form $y = mx + b$. This kind of lesson gives students the opportunity to discover for themselves the importance of the concepts of intercept and slope. ◆

Since two points determine a unique line, it is no surprise that given the coordinates of two distinct points in the *xy*-plane, one can determine the equation of the line. One method of doing this is illustrated in the following example.

EXAMPLE   Find the equation of the line containing both $(1, 2)$ and $(3, 7)$ in general form (i.e., $Ax + By + C = 0$) and in slope-intercept form (i.e., $y = mx + b$).

Solution   There are several techniques for finding the equation of a line through two points. One method is to substitute the coordinates of the two points into the equation $Ax + By + C = 0$ and then solve simultaneously and let one nonzero coefficient be some convenient number such as 1.

$$A \cdot 1 + B \cdot 2 + C = 0 \quad (\text{using } x = 1 \text{ and } y = 2)$$

$$A \cdot 3 + B \cdot 7 + C = 0 \quad (\text{using } x = 3 \text{ and } y = 7)$$

Multiplying the top equation by -3 and leaving the lower equation unchanged except for putting the coefficients in front of the letters yields

$$-3 \cdot A - 6 \cdot B - 3 \cdot C = 0$$

$$3 \cdot A + 7 \cdot B + C = 0.$$

Adding these two equations, one obtains $B - 2 \cdot C = 0$. Hence

$$B = 2 \cdot C.$$

Putting this result into the equation $A \cdot 1 + B \cdot 2 + C = 0$ yields

$$A = -5 \cdot C.$$

Using $C = 1$, one obtains $A = -5$, $B = 2$. Thus, the general form is given by

$$-5x + 2y + 1 = 0.$$

Of course, any nonzero multiple of this equation is also a correct answer. The slope-intercept form is obtained from the previous equation by solving for y, which yields

$$y = \left(\frac{5}{2}\right)x - \frac{1}{2}. \qquad \blacksquare$$

The formula for distance between points in the xy-plane is based on the Pythagorean theorem, which we cover in more detail in Chapter 3. This famous theorem states that in a right triangle, the sum of the squares of the lengths of the two legs is equal to the square of the length of the hypotenuse. Thus if $\triangle ABC$ has a right angle at vertex C, and if the lengths of sides of the triangle are denoted by the lowercase letter of the letter corresponding to the opposite vertex, then $a^2 + b^2 = c^2$.

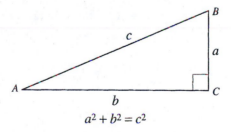

$$a^2 + b^2 = c^2$$

Classroom Connection 1.3.2

After students have learned the Pythagorean theorem, this theorem can be used to justify the usual formula for distance in the xy-plane. Of course, one generally denotes the vertical axis as the y-axis and denotes the horizontal axis as the x-axis. Let two points $A = (x_1, y_1)$ and $B = (x_2, y_2)$ be given and assume they have different first and second coordinates. Then construct $C = (x_2, y_1)$ as shown next where, for purposes of illustration, the case $0 < x_1 < x_2$ and $0 < y_1 < y_2$ is considered.

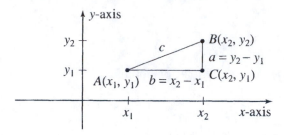

Notice that $\triangle ABC$ is a right triangle with right angle at the vertex C. Furthermore, $a = y_2 - y_1$, $b = x_2 - x_1$, and $c = AB$ is the distance from A to B. Thus, $a^2 + b^2 = c^2$ becomes $(y_2 - y_1)^2 + (x_2 - x_1)^2 = c^2$, which yields the distance $c = AB$ from A to B and is given by

$$c = \sqrt{(y_2 - y_1)^2 + (x_2 - x_1)^2}$$
$$= \sqrt{(x_1 - x_2)^2 + (y_1 - y_2)^2}.$$

After the previous discussion, there are several questions one might ask a middle or high school class. For example, Does the final formula still hold when the two points lie on a common vertical line or lie on a common horizontal line? One can also ask, Does the formula make sense when the points A and B are the same point (i.e., $x_1 = x_2$ and $y_1 = y_2$)? Of course, one can also ask: Why does the formula $\sqrt{(y_2 - y_1)^2 + (x_2 - x_1)^2} = \sqrt{(x_1 - x_2)^2 + (y_1 - y_2)^2}$ hold? ◆

As seen in the previous Classroom Connection the **Euclidean distance** in the xy-plane between the points $P = (x_1, y_1)$ and $Q = (x_2, y_2)$ is given by $\sqrt{(x_1 - x_2)^2 + (y_1 - y_2)^2}$ and may be denoted by either PQ or $d_E(P, Q)$.

Euclidean Distance in the xy-plane

$$PQ = \sqrt{(x_1 - x_2)^2 + (y_1 - y_2)^2}$$

Alternatively,

$$d_E(P, Q) = d_E\left[(x_1, y_1), (x_2, y_2)\right] = \sqrt{(x_1 - x_2)^2 + (y_1 - y_2)^2}$$

For three-dimensional Euclidean space, one has a corresponding coordinate representation of $R^3 = \{(x, y, z) | x, y, z \in R^1\}$.

$$R^3 \leftrightarrow xyz\text{-space}$$

The Euclidean distance in xyz-space between the points $S = (x_1, y_1, z_1)$ and $T = (x_2, y_2, z_2)$ is given by $\sqrt{(x_1 - x_2)^2 + (y_1 - y_2)^2 + (z_1 - z_2)^2}$ and may be denoted by either ST or $d_E(S, T)$.

> **Euclidean Distance in *xyz*-space**
>
> $$ST = \sqrt{(x_1 - x_2)^2 + (y_1 - y_2)^2 + (z_1 - z_2)^2}$$
>
> Alternatively,
>
> $$d_E(S, T) = d_E\left[(x_1, y_1, z_1), (x_2, y_2, z_2)\right]$$
> $$= \sqrt{(x_1 - x_2)^2 + (y_1 - y_2)^2 + (z_1 - z_2)^2}$$

EXERCISES 1.3

1. Find the approximate height in feet of someone who is 1.7 meters tall and explain your reasoning.
2. In a, b, and c, convert the measurements given in degrees to radians. In d, e, and f, convert the measurements given in radians to degrees. Also, explain your reasoning in parts b and e.
   a. $360°$
   b. $45°$
   c. $1°$
   d. π radians
   e. $\frac{\pi}{3}$ radians
   f. 1 radian
3. In the *xy*-plane, find the distance from the point $(0, 0)$ to the point $(3, 3)$ using any technique.

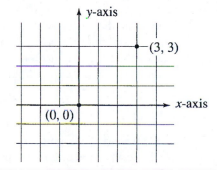

4. In the *xy*-plane, find the distance from the point $(1, 3)$ to the point $(4, 7)$ using (a) the Pythagorean theorem and (b) the distance formula.

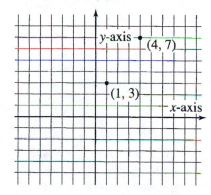

5. A first line in the xy-plane is given by $y = 2x - 1$ and a second line is given by $y = x + 1$.

   a. Find the coordinates of the point P of intersection of these two lines.

   b. Find the distance from the origin $(0, 0)$ to the point P.

6. The **perimeter** of a triangle is the sum of lengths of its three sides. Find the perimeter of the triangle with vertices at the three points $(0, 1)$, $(2, 1)$, and $(0, 3)$. What strategies did you use to find the lengths of each side?

7. In the xy-plane, let P be the point with coordinates $(1, 2)$ and let Q be the point with coordinates $(2, 4)$.

   a. Find the equation of the line through both P and Q in the form $Ax + By + C = 0$.

   b. Find the equation of the line through both P and Q in the form $y = mx + b$.

   c. Graph the line through P and Q.

   d. Find where the line through P and Q intersects the x-axis.

8. Find the Euclidean distance between the points $(0, 0, 0)$ and $(1, 1, 1)$ in xyz-space.

9. Find the perimeter of the triangle lying in xyz-space with vertices at the points $(7, 2, 3)$, $(7, 14, 3)$, and $(12, 14, 3)$.

Classroom Connection 1.3.3

In problem number 3 of Exercise Set 1.3, you found that the distance between the points $(0, 0)$ and $(3, 3)$ in R^2 is given by $3\sqrt{2}$, which is approximately equal to 4 1/4. This can be illustrated in middle school, high school, or college classrooms using a yard stick. Have the students take the origin $(0, 0)$ as a corner of the classroom, and let the x-axis and y-axis be the lines formed by the intersection of the walls and the floor. Find the point A on the floor corresponding to 3 feet along the x-axis and 3 feet along the y-axis. Then have a student measure the distance from the corner to the point A on the floor. It should be about 4 1/4 feet (i.e., about 4 feet 3 inches). ◆

Classroom Connection 1.3.4

In problem number 8 of Exercise Set 1.3, you found that the distance between the points $(0, 0, 0)$ and $(1, 1, 1)$ in xyz-space is given by $\sqrt{3}$, which is approximately equal to 1.7321, which in turn is close to 1 3/4. This can be illustrated in the middle school or high school classroom using a yard stick. Have the students take the origin $(0, 0, 0)$ as a corner of the classroom and let the z-axis be the line formed by the intersection of the two walls. Let the x-axis and y-axis be the lines formed by the intersection of the walls and the floor. Find the point A on the floor corresponding to 1 foot along the x-axis and 1 foot along the y-axis. Now have a student find the point B 1 foot directly above this point A. Another student can then measure the distance from the corner to the point B. It should be about 1 3/4 feet (i.e., about 1 foot and 9 inches). ◆

Classroom Discussion 1.3.1

Many students wonder why they should learn about radians when degrees seem to be so much easier. A good classroom discussion could center on the role of units in mathematics in general and especially in geometry. A discussion of radians vs.

degrees is a particularly good idea since many students feel degrees are more natural. They are often very surprised to learn that radians represent a more natural measure of angles from a mathematical standpoint than do degrees. One question that brings home the mathematical artificiality of degrees is the following: How many degrees do you think there would be in one straight angle if it took 100 days for the Earth to circle halfway around the sun rather than a little more than 182.5? ◆

1.4 POLYGONS AND SOLIDS

A triangle is a special case of a more general plane figure known as a *polygon*. Polygons are now being covered in middle school by at least some curricula (compare Figure 1.4.1).

Polygons.   A **polygon** $(P_1P_2\ldots P_n)$ is a union of distinct line segments $(\overline{P_1P_2} \cup \overline{P_2P_3} \cup \ldots \cup \overline{P_{n-1}P_n} \cup \overline{P_nP_1})$ lying in a plane with no three consecutive vertices lying on a common line and no two sides meeting except at a vertex and only when the sides are in consecutive order. Of course, the **sides** are the segments $\overline{P_iP_{i+1}}$. The **vertices** are the points $P_1, P_2, \ldots, P_n$, and one sets $P_{n+1} = P_1$. The perimeter of a polygon with vertices $P_1, P_2, \ldots, P_n$ is the sum of the lengths of the sides. In other words, the perimeter p is given by $p = \sum_{i=1}^{n} P_iP_{i+1}$.

These are polygons.   These are *not* polygons.

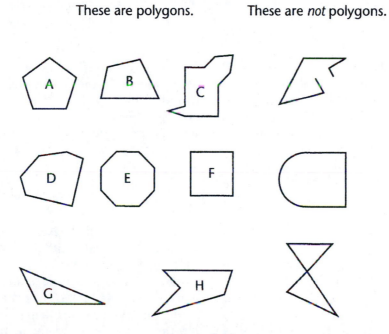

FIGURE 1.4.1 Note that polygons are closed curves with each side lying on a straight line and that two different sides cannot "cross" each other. Reproduced from page 81 of Book 1 in the Math Thematics grade 6 materials.

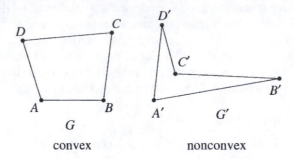

convex   nonconvex

A polygon is **convex** if it is contained by the sides and interior of each of its vertex angles—$\angle P_{i-1}P_iP_{i+1}$. All triangles are convex, but polygons with four or more sides need not be convex. Shown here is a convex quadrilateral ($G = ABCD$) and a nonconvex quadrilateral ($G' = A'B'C'D'$). Notice that the quadrilateral G' cannot be convex since it is not contained in the sides and in the interior of the angle $\angle B'$ (i.e., $\angle A'B'C'$) formed by three consecutive vertices.

Classroom Connection 1.4.1

Figure 1.4.1 shows examples of polygons and shows figures that are not polygons. To get middle school students to better understand polygons, make up a list of figures with some figures being polygons and some that are not polygons. Use figures in Figure 1.4.1 to give some ideas on how to generate such a list. Have the students identify which figures are polygons. Also, have them explain in writing why the nonpolygons fail to be polygons.   ◆

Polyhedrons.   Many things about polyhedrons are now taught in middle school mathematics courses. In particular, it is common to teach about certain well-known polyhedrons encountered in everyday life. Here a **polyhedron** is the surface of a solid that has flat faces made up of polygons. For example, a cube has faces made of squares and is an example of a more general figure known as a "box," which has faces made up of rectangles (boxes are more precisely known as **right rectangular prisms**). It is worth noting that both *polyhedrons* and **polyhedra** are commonly used as the plural of the word *polyhedron*. The volume enclosed by a polyhedron is said to be the volume of the polyhedron.

Prisms.   A **prism** is a polyhedron with two faces, called **bases**, that are congruent polygons and lie in parallel planes. The other faces, called either **lateral faces** or **vertical faces**, are required to be parallelograms. When the lateral faces are rectangles, the prism is called a **right prism**. Otherwise, the prism is called an **oblique prism**. Examples of prisms are illustrated in Figures 1.4.2 and 1.4.3. The reader is cautioned that in a number of books, the definition of *prism* requires that all lateral faces be rectangles. Thus, in these books, all prisms are what in this monograph are called *right prisms*. In any case, **prisms are named according to the polygonal shape of their bases**. Thus, a triangular prism is one where the bases are triangles. The **height** of a prism is the (perpendicular) distance between the parallel planes containing the bases.

oblique prism A prism whose vertical faces are not all rectangles.

prism A three-dimensional shape with a top and bottom that are congruent polygons and faces that are parallelograms.

rectangular prism A prism with a top and bottom that are congruent rectangles.

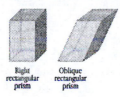

Right rectangular prism   Oblique rectangular prism

FIGURE 1.4.2 Prisms may be right or oblique. Reproduced from page 79 of *Filling and Wrapping* in the Connected Mathematics Project (CMP) grade 7 materials.

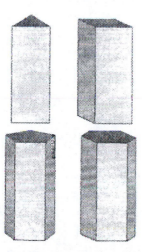

FIGURE 1.4.3 The lateral faces of a right prism are rectangles. The lateral faces are also known as *vertical faces*. Reproduced from page 80 of *Filling and Wrapping* in the Connected Mathematics Project grade 7 materials.

Pyramids.   A **pyramid** is a polyhedron formed by taking a polygon in one plane, selecting a point off the plane of the polygon, and then joining the vertices of the polygon to the point selected off the plane. The point off the plane is called the **(top) vertex,** and the **height** of the pyramid is the (perpendicular) distance from the top vertex to the plane of the polygon. The polygon is called the **base,** and the **lateral surface** of the pyramid consists of triangles with one vertex equal to the top vertex of the pyramid. Much like prisms, pyramids are named by the shape of their base. The well-known pyramids in Egypt are **square pyramids,** and the top vertex is directly above the center of the square base.

Hexahedrons.   A **hexahedron** (hexa = six; hedron = face[s]) is a polyhedron with six faces, such as a box (i.e., a right rectangular prism) or a pyramid with a base that is a pentagon (i.e., a pentagonal pyramid). Since a cube has six faces, it is a special case of a hexahedron. If one is given a plain unmarked cube, then all of the faces look the same and all of the vertices look the same. Of course, each face of a cube is a square, and thus each face is an example of a regular polygon. Clearly, each vertex of a cube is also the vertex of three faces. If one fixes a vertex of the cube, then the cube may be rotated onto itself while holding the given vertex fixed such that any one of the faces having that vertex may be rotated to any other of the faces having that same vertex. Also, given any two vertices of the cube, the cube may be rotated onto itself such that the first vertex moves to the

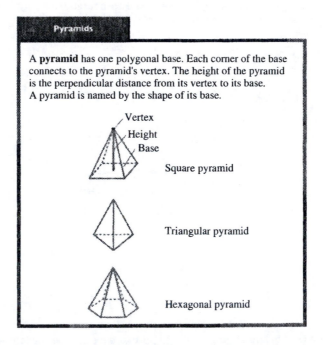

FIGURE 1.4.4 Note that the base of a pyramid need not be a square. These examples show the top vertex located over a central point of the base polygon. In general, the top vertex need not be located over the center of the base polygon. Reproduced from page 26 of *Shapes and Space* in the MathScape grade 8 materials.

second. Because the cube has the aforementioned properties, it is called a *regular polyhedron*.

Regular Polyhedrons. A polyhedron is **regular** if, given any first vertex and face at that first vertex along with any second vertex and face at that second vertex, there is a rotation, taking the polyhedron onto itself and taking the first face at the first vertex to the second face at the second vertex. Thus, a regular polyhedron may be rotated onto itself to take any vertex to any other vertex. Also, fixing any vertex of a regular polyhedron, one may rotate any face at that vertex to any other face at that vertex. In general, the faces of a regular polyhedron must be regular polygons that are all congruent to each other. For a regular polyhedron, all of the faces "look the same," and also all of the vertices "look the same."

Platonic Solids. The **Platonic solids** are the regular polyhedrons. There are known to be only five types of regular polyhedrons: **tetrahedron, cube, octahedron, dodecahedron**, and **icosahedron**. The numbers of faces of these are 4, 6, 8, 12, and 20, respectively. Of course, *tetra, octa, dodeca*, and *icosa* refer to the numbers 4, 8, 12, and 20, respectively.

Plato (427–347 BC) was a famous Greek philosopher who founded an academy. He felt that serious thinkers should be schooled in geometry, and he put an inscription over the entrance to his academy that indicated that only geometers need enter. The five Platonic solids were known at the time of Plato. Plato used these solids in his writings, and over time they became associated with his name. Plato listed the four elements as fire, earth, air, and water, which he identified with the tetrahedron, cube, octahedron, and icosahedron, respectively. He identified the universe with the dodecahedron.

Spheres. A **sphere** is the set of points in three-dimensional Euclidean space at a given (positive) distance from a given point called the **center**. Thus, if the center of the sphere is a point O, and the radius is the positive number r, then the sphere S is the set of points P, such that the distance OP equals r (i.e., $S = \{P \mid OP = r\}$). A **great circle** on the sphere S is the intersection of S with a plane through the center O of S. In **spherical geometry**, the great circles

Platonic Solids

Regular Tetrahedron. A regular tetrahedron has four faces, and each face is an equilateral triangle

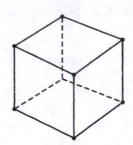

Regular Hexahedron (Cube). A regular hexahedron is a cube. There are six faces, and each face is a square

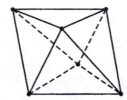

Regular Octahedron. A regular octahedron has eight faces, and each face is an equilateral triangle

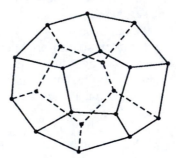

Regular Dodecahedron. A regular dodecahedron has twelve faces, and each face is a regular pentagon

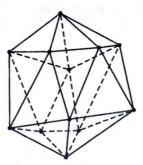

Regular Icosahedron. A regular icosahedron has twenty faces, and each face is an equilateral triangle

are referred to as **geodesics**, and they play the role in spherical geometry that straight lines play in Euclidean geometry. If S represents the earth, then the equator is one example of a great circle. One may study spheres **synthetically** (i.e., without using coordinates) or **analytically** (i.e., using coordinates). When using coordinates, for simplicity it is often useful to choose the center of the sphere to be the origin. If the center is chosen as the origin $(0, 0, 0)$ of R^3 and the radius of the sphere is some positive number r, then the equation of the sphere is $x^2 + y^2 + z^2 = r^2$. More generally, the sphere with center (a, b, c) and the radius r in R^3 is given by the equation $(x - a)^2 + (y - b)^2 + (z - c)^2 = r^2$.

sphere of radius r and center (a, b, c)

$$(x - a)^2 + (y - b)^2 + (z - c)^2 = r^2$$

Euler Number.

A square pyramid, as shown in the first example of Figure 1.4.4, has five faces (the base and four triangles for its lateral surface), eight edges (four on the base and four on the lateral surfaces), and five vertices (the top vertex and four vertices for the square base). Using F, E, V to denote the number of faces, edges, and vertices,

Faces, Vertices, and Edges

In the previous section, you discovered a relationship between the numbers of faces, vertices, and edges of a polyhedron. This relationship is explained in Euler's Formula:

| Number of Faces | + | Number of Vertices | − | Number of Edges | = 2 |

It is named for the Swiss mathematician Leonhard Euler. The relationship is usually written the following way:

$$F + V - E = 2$$

1. Does Euler's Formula work for the polyhedra pictured below? Explain.

a.

F =
V =
E =
F + V − E =

b.

F =
V =
E =
F + V − E =

c.

F =
V =
E =
F + V − E =

FIGURE 1.4.5 The formula due to Euler relates the number of faces (F), edges (E), and vertices (V) of a polyhedron that may be deformed to a sphere. Although middle school textbooks usually write it as $F + V − E = 2$, mathematicians generally prefer to use $F − E + V = 2$, which represents the sum with alternating $\pm$ signs of two-dimensional parts, one-dimensional parts, and zero-dimensional parts. Reproduced from page 29 of *Packages and Polygons* in the Mathematics in Context (MiC) grade 7/8 materials.

respectively, one has $F = 5$, $E = 8$, and $V = 5$ for a square pyramid. Computing $F - E + V$ (which, of course, can also be written as $F + V - E$), one finds $F - E + V = 5 - 8 + 5 = 2$. Note that the alternating sum $F - E + V$ corresponds to (two-dimensional parts) − (one-dimensional parts) + (zero-dimensional parts). Leonhard Euler (*oy luhr*) (1707–1783) found that the alternating sum $F - E + V$ was a number that only depended on the **topological** type of the surface. Here two surfaces are said to have the same topological type if one may be continuously deformed into the other. Thus, if one thinks of a cube's surface as made of rubber, one may deform it into a sphere. The deformation is continuous and allows stretching, bending, and shrinking but does not allow tearing. Likewise, a square pyramid may be deformed into a sphere with stretching, bending, and shrinking but without tearing. Two surfaces that have the same topological type will yield the same value for $F - E + V$. Thus, the same number (i.e., 2) will be obtained for a cube as for a square pyramid. On the other hand, a polyhedron that can be deformed into a doughnut (i.e., *torus*) shape will have a different value for the quantity $F - E + V$. The number $F - E + V$ is called the **Euler number** of the surface. This number is also known as the *Euler characteristic* of the surface. Euler, a Swiss mathematician, continued to do mathematics even after he went blind; the Swiss are very proud of producing such a great mathematician. He is shown in Figure 1.4.5.

EXERCISES 1.4

1. Find the perimeter of the quadrilateral with vertices at $P_1 = (0,0)$, $P_2 = (3, 4)$, $P_3 = (3, 5)$, and $P_4 = (0,5)$.
2. Explain why a circle is not a polygon.
3. Fill in the following table for the five Platonic solids.

Polyhedron	Types of Faces	F (# faces)	E (# edges)	V (# vertices)	F − E + V
Tetrahedron	Equilateral triangles				
Cube	Squares				
Octahedron	Equilateral triangles				
Dodecahedron	Regular pentagons				
Icosahedron	Equilateral triangles				

4. Fill in the table with values of F, E, and V for the three polyhedrons in Figure 1.4.5. Do all three of these polyhedrons satisfy $F - E + V = 2$?
5. Check that $F - E + V = 2$ for a pentagonal prism.
6. Explain why a circular cylinder is not a polyhedron.

Classroom Connection 1.4.2

One good middle school or high school project is to have students make models of the five Platonic solids. Nets (also known as *flat patterns*) for the Platonic solids are given in Section 6 of Chapter 2. Construction paper and tape are the only materials necessary for these constructions. ◆

Classroom Connection 1.4.3

The Euler number for polyhedrons that are topologically like spheres can be illustrated using a block of cheese. Have students form teams of three to five and give each of them a block of cheese. Use cheese that does not crumble too easily. Using plastic knives, the students can construct new polyhedra and check that one still has the relation $F - E + V = 2$. Students should cut in ways that yield shapes with surfaces that are still polygons—thus they should cut away pieces with flat surfaces. They should also not make "holes" in the blocks of cheese. This activity was devised by a team of college students studying to become middle school mathematics teachers. ◆

Classroom Discussion 1.4.1

Middle school students only need to know that the Euler number (i.e., Euler characteristic) is 2 for a polyhedron that is topologically equivalent to a sphere. On the other hand, middle school mathematics teachers should know that surfaces such as the torus (i.e., a "doughnut" shape) have a different Euler number. In particular, a good college classroom discussion might center on the justification that the torus has an Euler number of zero. To work up to this, first note that a simple way to "build" several different polyhedrons in the classroom is to paste faces of cubes together. For example, one can paste two cubes together as illustrated.

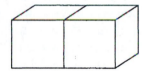

The two faces pasted together are deleted from the figure, the two sets of four edges along the pasting are identified to get a single set of four edges around the "middle" of the figure, and the two sets of four vertices along the pasting are identified to get a single set of four vertices around the "middle" of the figure. Thus, for the new composite figure, one has $F = 10, E = 20$, and $V = 12$, which yields the expected $F - E + V = 2$. At this point it becomes easy to see that one gets this same value $F - E + V = 2$ if one pastes the cubes into a straight line of seven cubes. If one changes the line of seven cubes into a "U" shape with three cubes in a line and two cubes on either side at right angles to the line of three cubes, again one has $F - E + V = 2$. By adding one more cube to get a 3 by 3 "square," with a missing cube in the middle, one obtains a polyhedron that is topologically a torus. However, this last cube is increasing F by 2 and E by 4 while leaving V the same. The net result is that for the polyhedron that is topologically a torus, one finds zero for the sum $F - E + V$. ◆

Classroom Discussion 1.4.2

In the previous example, one discovered that $F - E + V = 0$ for a polyhedron that is topologically like a torus. The previous exploration can be generalized to a polyhedron that is like a doughnut with exactly n holes. For example, for $n = 2$, one can use thirteen cubes to make a polyhedron with two holes. One just adds five cubes to the above 3 by 3 "square," with a missing cube in the middle, to get a 3 by 5 "rectangle" that has two missing cubes that represent the two holes. Using cubes, one can have students discover what the value of $F - E + V$ is for various numbers of holes. Let n represent the number of holes. After doing $n = 2, 3$, and 4, students should be able to guess the general formula for $F - E + V$ given any value of n. ◆

1.5 MEASURING THE EARTH'S RADIUS

It is interesting to note that the ancient Greeks knew the Earth was round and were interested in measuring its radius. In about 240 BC, a Greek mathematician and scientist named Eratosthenes (*ehr uh tahs thuh neez*) (c. 276–194 BC) measured the Earth's radius very accurately. The basic mathematical tool that Eratosthenes used was the theorem that states that if one has a transversal to two parallel lines, then corresponding angles are congruent. This is illustrated in Figure 1.5.1. In fact, two lines are parallel iff each transversal forms corresponding angles that are congruent. This result was known to Euclid and is found in Euclid's *Elements*. It will be considered along with other results having to do with parallel lines and transversals in the next chapter.

Eratosthenes assumed that light travels in straight lines and that the sun was far enough away that light rays were very close to parallel. He also knew that on the day of the summer solstice, the sun was directly overhead in Syene, a city in southern Egypt that is now known as Aswan. When the sun was directly overhead in Syene, Eratosthenes measured it to be at an angle of 7.2° (about .12566 radians) from straight overhead in Alexandria, Egypt. Using this angle and the north-south distance between Alexandria and Syene, Eratosthenes calculated the Earth's radius, see Figure 1.5.2.

Since a sphere's surface is curved, it is clear that the shortest curve on its surface joining two points cannot be a straight line. In fact, the shortest curve on a sphere joining two points on a sphere's surface will lie on a great circle. For the Earth, the equator is one example of a great circle. An arc of a great circle that is less than half the circumference of the circle (i.e., an arc that is less than a semicircle) will minimize the distance between its endpoints. Consequently, planes fly great-circle routes between cities in order to minimize the distance traveled.

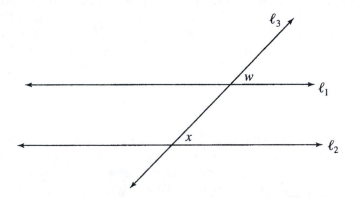

FIGURE 1.5.1 In this figure, $\angle w$ and $\angle x$ are called *corresponding angles*. The two lines ℓ_1 and ℓ_2 are parallel iff $m(\angle w) = m(\angle x)$.

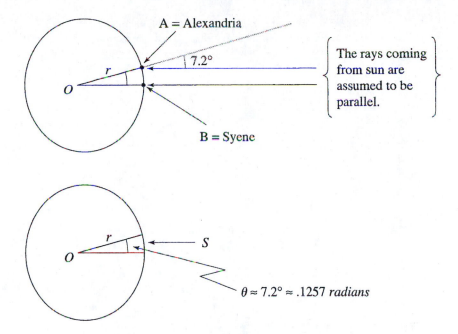

The rays coming from sun are assumed to be parallel.

FIGURE 1.5.2 A cross section of the Earth appears as a circle of radius r. Eratosthenes measured the angle $\angle AOB$ as $\frac{1}{50}$ of a complete (i.e., $360°$) angle. Thus, he found $m(\angle AOB) = \frac{360°}{50} = 7.2° = \frac{(7.2)\cdot\pi \text{ radians}}{180} \approx .1257$ radians. If the north-south distance from Alexandria to Syene is S, then the radius of the Earth is found using $S = r\theta$. This yields $r = \frac{S}{\theta} \approx \frac{S}{.1257}$. Using 490 miles for the north-south distance between Alexandria and Syene, one obtains $r \approx \frac{490 \text{ mi.}}{.1257} \approx 3,900$ miles, which is not far from the correct value of the Earth's radius at the equator. Of course, Eratosthenes did not use miles for his length estimates—he used the stadia.

EXERCISES 1.5

1. Assuming the Earth's radius is 4,000 miles, what is the Earth's radius in kilometers? Round your answer.

2. Do the exercise indicated in Figure 1.5.3. In other words, assuming that the Earth's radius is approximately 4,000 miles, write a brief paragraph outlining what this fact tells you about Earth's measurements, including diameter and circumference. Be sure to explain how you found your results.

3. Assuming the Earth is a sphere, find the distance around the Earth at the equator measured in both miles and kilometers. Round your answers.

4. A person travels directly south along the Earth's surface from the North Pole until reaching the equator. Assuming the Earth's radius is 4,000 miles, find the distance traveled, measured in miles.

5. Assuming Mars is a sphere with a diameter of 6,790 kilometers, find the distance around Mars at its equator in kilometers. Round your answer.

6. Assume the Earth's radius is 4,000 miles and that the Earth is a perfect sphere. Let one rope be snugly placed around the Earth at the equator, and let a second rope be placed around the Earth's equator exactly 1 foot above the Earth's surface. How much longer is the second rope than the first rope?

The radius of Earth is approximately 4,000 miles. Write a brief paragraph outlining what this fact tells you about Earth's measurements, including diameter and circumference. Also provide an explanation of how you found your results.

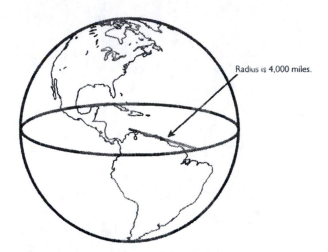

Radius is 4,000 miles.

FIGURE 1.5.3 Notice that seventh-grade students are asked to answer questions about the Earth in paragraph form and are asked to explain their reasoning. Reproduced from page 12 of *Getting in Shape* in the MathScape grade 7 materials.

Classroom Connection 1.5.1

The easy way to calculate the Earth's surface area is to use the formula $S = 4\pi r^2$. This says the surface area of a sphere or radius r is four times the area πr^2 of a circle of the same radius r. To illustrate this formula in the middle or high school classroom, bring in oranges and have the students make four circles on paper with the radius of each circle being the same radius as the orange. Then peel the oranges and flatten the peelings onto the four circles. In theory, the peeling should just fill the four circles. The activity works reasonably well but is a bit messy. The next three classroom connections form a collection of related activities that can be combined with this activity. ◆

Classroom Connection 1.5.2

It is known that one cannot make an accurate flat map of the Earth. If one has a relatively small area of say 50 miles by 50 miles on the Earth's surface, then one can make a fairly good flat map to describe the area. This flat map will not be truly accurate but will be of reasonable accuracy. On the other hand, when one has a large area on the Earth, say the size of Asia, then any flat map will have major distortions. To "physically illustrate" to middle or high school students that flat maps are not accurate, note that in the previous "orange peel" exploration one finds that "large" pieces of orange peel will split when they are flattened out. ◆

Classroom Connection 1.5.3

Use oranges to demonstrate that the shortest curve joining two points on a sphere lies on a great circle. Have middle or high school students take an orange and fix two points. It is best if they are not diametrically opposed points. Have them take a string with one end at each of the fixed points and pull the string tight. This yields the shortest curve on the orange's surface from one point to the other. Have them mark the curve on the orange's surface along the string, and then, using a plastic knife, have them cut the orange into two pieces with flat surfaces such that the resulting edges of the orange pieces contain the marked curve. The resulting two pieces of the orange will be hemispheres. In other words, the knife will cut through the sphere's center, and the marked curve will lie on a great circle. ◆

Classroom Connection 1.5.4

In spherical geometry, the word parallel has a special meaning which is different from its meaning in plane Euclidean geometry. Given a sphere with one great circle designated as the equator, then a **parallel** is a circle on the sphere that comes from intersecting the sphere with a plane parallel to the equator. For example, on the Earth, the forty-fifth northern parallel is a circle that is in the Northern Hemisphere; points on this parallel are halfway from the equator to the North Pole. To demonstrate to middle or high school students that a parallel (different from the equator) is not a great circle, have them mark a point on an orange as the North Pole and prepare a length of string that does not reach to the "equator" of the orange. Hold one end of the string at the "North Pole" and use this string to mark a "parallel" consisting of a circle with all of its points at an equal distance from the "North Pole." Then cut the orange such that this "parallel" is the edge. The flat open face will not contain the orange's center. ◆

CHAPTER 1 REVIEW

This chapter listed the fundamental assumptions of Euclid and contained an introduction to the basics of logic. It also introduced some notational conventions, terms, and aspects of measurement. Euclid's program was to make a few assumptions and then use logic and reasoning to deduce important consequences.

Since logic is at the foundations of mathematics, it is important to be familiar with some of the results, terms, and symbolism used in logic. Four logical connectives introduced in this chapter are *or*, *and*, *negation*, and *implication*. These are denoted by the symbols $\vee$, $\wedge$, $\sim$, and $\rightarrow$, respectively. Always keep in mind that both the words *or* and *and* have precise logical meanings in mathematics. The typical result in mathematics is stated in the following sentence form: If P, then Q (i.e., $P \rightarrow Q$). Here P is the hypothesis and Q is the conclusion. If one reverses the positions of P and Q, then one obtains its converse (i.e., $Q \rightarrow P$). The original sentence and its converse are logically independent in that they may have the same truth values or opposite truth values. On the other hand, a sentence $P \rightarrow Q$ always has the same truth value as its contrapositive (i.e., $\sim Q \rightarrow \sim P$), and thus the original is logically equivalent to its contrapositive.

Polygons are an important class of curves in the plane. They have straight line segments for sides, are closed, and the sides do not cross each other. The corresponding three-dimensional surfaces are known as *polyhedrons*. Each face of a polyhedron lies in a plane and is bounded by a polygon. Given a polyhedron, let F denote the number of faces, let E denote the number of edges, and let V denote the number of vertices. Then $F - E + V$ is called the *Euler number* (or *Euler characteristic*) of the polyhedron. The Euler number is 2 for any polyhedron that is topologically equivalent to a sphere (i.e., can be continuously deformed into a sphere). Thus, one has $F - E + V = 2$ for the Platonic solids, prisms, pyramids, and many other polyhedrons.

Some of the power of Euclidean geometry is illustrated by Eratosthenes' measurement of the Earth's radius in about 240 BC. This would not have been possible without the use of Euclidean geometry. Since Euclidean geometry is a very excellent approximation to the geometry in the world of ordinary experience, it serves as a basis for much of engineering, science, and technology. The most widely used measurement system in the world is the metric system, and the second most widely used is the English system. Thus, it makes sense to be able to use metric measures, such as centimeters, meters, and kilometers, as well as English measures such as feet, yards, and miles. Angular measurements are most often given in degrees, but it also important to understand radian measure.

Selected formulas and notational conventions:

P or *Q*:	$P \vee Q$
P and *Q*:	$P \wedge Q$
negation of *P*:	$\sim P$
If *P*, then *Q*.	$P \rightarrow Q$
radian measure:	$\theta = \dfrac{S}{r}$
Perimeter of polygon:	$p = \sum\limits_{i=1}^{n} P_i P_{i+1}$ (= sum of lengths of sides)

The Euler number for a polyhedron that is topologically a sphere:

$$F - E + V = 2$$

Euclidean distance in the *xy*-plane:

$$d_E([x_1, y_1], [x_2, y_2]) = \sqrt{(x_1 - x_2)^2 + (y_1 - y_2)^2}$$

Euclidean distance in *xyz*-space:

$$d_E[(x_1, y_1, z_1), (x_2, y_2, z_2)] = \sqrt{(x_1 - x_2)^2 + (y_1 - y_2)^2 + (z_1 - z_2)^2}$$

CHAPTER 1 REVIEW EXERCISES

1. Consider the following sentence, which is true in Euclidean geometry: "If the quadrilateral $ABCD$ is a square, then all of its sides have the same length."
   a. State the converse. Is the converse true?
   b. State the contrapositive. Is the contrapositive true?
   c. State the inverse. Is the inverse true?

2. Fill in the following truth table to illustrate that $\sim[Q \wedge (\sim P)]$ and $(\sim Q) \vee P$ are logically equivalent.

P	Q	$\sim P$	$\sim Q$	$Q \wedge (\sim P)$	$\sim(Q \wedge (\sim P))$	$(\sim Q) \vee P$
T	T					
T	F					
F	T					
F	F					

3. Find the distance between the two points $(1, 1)$ and $(2, 2)$ in the xy-plane.

4. Find the distance between the two points $A = (2, 3)$ and $B = (6, 6)$ in the xy-plane.

5. Find the distance between the two points $C = (5, 5, 5)$ and $D = (17, 10, 5)$ in xyz-space.

6. Find the perimeter of the triangle with vertices at the three points $(0, 0)$, $(6, 0)$, and $(0, 8)$.

7. Find the perimeter of the quadrilateral with vertices at the points $(1, 1)$, $(5, 1)$, $(6, 2)$, and $(2, 2)$.

8. In the xy-plane let P be the point with coordinates $(7, 9)$ and let Q be the point with coordinates $(11, -3)$. Find the equation of the line through both P and Q in the general form $Ax + By + C = 0$.

9. Given a hexagonal pyramid, find the number F of faces, the number E of edges, and the number V of vertices. Also, find the value of $F - E + V$.

10. Assuming the moon is a sphere of radius 1,080 miles, find the length of a great circle on the moon.

11. Assume that the sun is fixed and that the Earth moves in a circle of radius 93 million miles about the sun. Assuming that a year is 365.25 days long, find how far the Earth travels in one day.

12. Assume the Earth is a sphere with a radius of 4,000 miles and that the North Star (i.e., Polaris) lies directly above the North Pole. Thus, at the North Pole, the North Star makes an angle of $0°$ with the vertical upward direction. If a person is $1,000\,\pi$ miles ($\approx 3,140$ miles) south of the North Pole, find the angle that the North Star will make with the vertical upward direction for that person. Explain your reasoning.

RELATED READING FOR CHAPTER 1

Asimov, I. *Asimov's Biographical Encyclopedia of Science and Engineering*. New York: Equinox, 1964.

Bell, E. T. *Men of Mathematics*. New York: Simon and Schuster, 1965.

Billstein, R., and J. Williamson. *Math Thematics: Book 1*. Evanston, IL: McDougal Littell, 1999.

Clemens, C. H., and M. A. Clemens. *Geometry for the Classroom*. New York: Springer-Verlag, 1991.

Heath, Sir T. *Euclid: The Thirteen Books of the Elements, Vol. 1*. 2nd ed. New York: Dover Publications, 1956.

James, G., and R. James. *Mathematics Dictionary*. Princeton, NJ: D. Van Nostrand, 1968.

Jensen, G. R. *Arithmetic for Teachers*. Providence, RI: American Math. Soc., 2003.

Kleiman, G. "Getting in Shape." In *MathScape: Seeing and Thinking Mathematically*. Grade 7 materials. New York: Glencoe/McGraw-Hill, 1999.

Kleiman, G. "Making Mathematical Arguments." In *MathScape: Seeing and Thinking Mathematically*. Grade 7 materials. New York: Glencoe/McGraw-Hill, 1999.

Kleiman, G. "Shapes and Space." In *MathScape: Seeing and Thinking Mathematically*. Grade 8 materials. New York: Glencoe/McGraw-Hill, 1999.

Kline, M. *Mathematical Thought from Ancient to Modern Times*. New York: Oxford University Press, 1972.

Lappan, G., et al. *Filling and Wrapping*. Connected Mathematics Project. Needham, MA: Pearson/Prentice Hall, 2004.

Romberg, T. A., et al. *Packages and Polygons*. Mathematics in Context. Chicago: Encyclopedia Britannica Educational Corporation, 1998.

Serra, M. *Discovering Geometry: An Inductive Approach*. Berkeley, CA: Key Curriculum Press, 1993.

Wenninger, M. J. *Polyhedron Models*. New York: Cambridge University Press, 1971.

Congruent Figures, Areas, and Volumes

Many of the most important concepts in Euclidean geometry are introduced in this chapter. Congruent triangles are studied in Section 2.1. Two triangles are defined to be congruent if their corresponding sides and corresponding angles are congruent. Thus, the definition requires that six parts (three sides and three angles) of the first triangle match the corresponding six parts of the second triangle. In Section 2.2, we investigate the many implications of Euclid's fifth postulate, the **parallel postulate**. Section 2.3 introduces parallelograms, rectangles, and trapezoids. In Section 2.4, we consider the area of plane figures, in particular, we study the areas of triangles, rectangles, parallelograms, and trapezoids. We also discuss **Cavalieri's principle**. In Section 2.5, we study circles, and in Section 2.6 we look at volume and surface area.

2.1 CONGRUENT TRIANGLES

If two plane figures are congruent, then one may be placed exactly on top of the other. Thus, if one cuts two congruent triangles out of paper, then either triangle may be placed exactly on top of the other with each point of the bottom triangle covered by a point of the top triangle and with each point of the top triangle right over a point of the bottom triangle. (Of course, one may draw triangles on sheets of tracing paper or on clear transparencies instead of cutting them out.) In general, congruent figures have the same shape and size. Introduce middle school students to congruence by having them cut figures or use tracing paper to compare paper figures, as in the following Classroom Connection.

Classroom Connection 2.1.1

When introducing congruence to middle school students, one may hand out a sheet to each student with a number of figures such as line segments, triangles, quadrilaterals, circles, etc. Some of the figures shown should be congruent to others, and it is convenient to have all figures labeled. Have the students cut out the figures, use sheets of tracing paper or use transparencies to check which figures are congruent to which other figures. This can be made a discovery lesson with the students encouraged to discover what key properties that congruent figures of a certain type have in common. For example, congruent line segments have the same length, and congruent circles have the same radius. Students should be able to discover that congruent triangles always have corresponding sides of equal length and corresponding angles of equal measure. Of course, a similar statement holds for congruent quadrilaterals. By the time this activity is over, all of the students should understand that congruent polygons always have corresponding sides of equal length and corresponding angles of equal measure.   ◆

Another activity that one can use to encourage students to begin thinking about measurements and relationships in geometry is outlined in the following Classroom Connection. This activity gives students experience in searching for patterns and, in particular, allows them to discover the scalene inequality (Theorem 2.1.6) as a conjecture before learning it as a rigorous result.

Classroom Connection 2.1.2

Have middle or high school students make several triangles and then, using protractors and rulers, measure the angles and sides. One can list the results for several triangles on the board and ask the students to see if they can find some relationships between lengths of sides and measures of angles. The class should be able to come to the conjecture that in a given triangle the largest side is opposite the largest angle and that the smallest side is across from the smallest angle. After making this conjecture, some students may feel this result is "so obvious" that it does not require proof. How would you explain to such students that this result is really a little less obvious than it looks?   ◆

Two segments, $\overline{AB}$ and $\overline{CD}$, are **congruent** (denoted by $\overline{AB} \cong \overline{CD}$) if and only if they have the same length. Likewise, two angles, $\angle A$ and $\angle B$, are congruent if and only if they have the same measure.

$$
\begin{array}{lll}
\overline{AB} \cong \overline{CD} & \text{iff} & AB = CD \\
\angle A \cong \angle B & \text{iff} & m(\angle A) = m(\angle B)
\end{array}
$$

The **midpoint** of the segment $\overline{AB}$ is the point $D \in \overline{AB}$ such that $AD = BD$. A set of points that all lie on a single line is said to be a **collinear** set. A set of points that are not all contained in a single line is said to be a **noncollinear** set. Since there is always a line joining two distinct points, it follows that any noncollinear set must have at least three points. A **triangle** is defined to be the union of the three line segments joining three noncollinear points. Thus, if the three points A, B, C fail to be collinear,

$$\triangle ABC = \overline{AB} \cup \overline{AC} \cup \overline{BC}.$$

Definition 2.1.1 Two triangles are congruent if there is a correspondence between them such that corresponding sides are congruent and corresponding angles are congruent.

Two triangles, $\triangle ABC$ and $\triangle DEF$ are congruent with the sides and vertices in the indicated order, written $\triangle ABC \cong \triangle DEF$, if all pairs of corresponding sides are congruent and all pairs of corresponding angles are congruent. Thus, $\triangle ABC \cong \triangle DEF$ means all six of the following: $\overline{AB} \cong \overline{DE}, \overline{AC} \cong \overline{DF}, \overline{BC} \cong \overline{EF}$, $\angle A \cong \angle D, \angle B \cong \angle E$, and $\angle C \cong \angle F$.

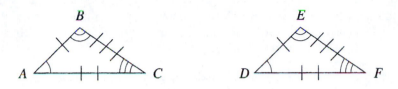

$\triangle ABC \cong \triangle DEF$		
1. $\overline{AB} \cong \overline{DE}$	2. $\overline{AC} \cong \overline{DF}$	3. $\overline{BC} \cong \overline{EF}$
4. $\angle A \cong \angle D$	5. $\angle B \cong \angle E$	6. $\angle C \cong \angle F$

Notice that the order of the vertices is very important when using the symbol $\cong$. Thus, one only writes $\triangle ABC \cong \triangle DEF$ when one has congruence of sides and angles in the indicated order (i.e., $A \leftrightarrow D, B \leftrightarrow E$, and $C \leftrightarrow F$). Congruence of triangles is an example of an equivalence relation.

Equivalence Relation.   A relation $\sim$ on a set S is said to be an **equivalence relation** if it satisfies the reflexive, symmetric, and transitive laws. Congruence of segments, angles, and triangles are all examples of equivalence relations. Other equivalence relations include equality of real numbers and similarity of figures.

Equivalence Relation		
1. Reflexive Law	$a \sim a$	for all $a \in S$.
2. Symmetric Law	$a \sim b$ implies $b \sim a$	for all $a, b \in S$.
3. Transitive Law	$a \sim b$ and $b \sim c$ imply $a \sim c$	for all $a, b, c \in S$.

An example of a relation on the set of real numbers that is not an equivalence relation would be the relation $a \sim b$ if $|a - b| \le 1$. This relation is reflexive and symmetric. However, this relation fails to be transitive since $a = 0, b = 1$, and $c = 2$ yields a case where $a \sim b$ and $b \sim c$ does *not* imply $a \sim c$.

Relations may be defined given any set S. Of course, if a relation satisfies all three laws (i.e., if the relation is reflexive, symmetric, and transitive), then it is an equivalence relation. If one or more of the three laws fails, then it is not an equivalence relation.

EXAMPLE Let S be a set of four students, Jane, Pedro, Bob, and Louise. Jane and Pedro are 12 years old, Bob is 11 years old, and Louise is 13 years old. Let each student be equivalent to themselves and also to all other students in S with the same age. This is an equivalence relation since it is reflexive, symmetric, and transitive. Find all of the students in S who are equivalent to Jane and all who are equivalent to Louise. Use $\sim$ to denote this equivalence relation.

Solution Since Jane is herself, one has Jane $\sim$ Jane. Since Pedro has the same age as Jane, one has Jane $\sim$ Pedro. Since Jane is not the same age as Bob and is not the same age as Louise, it follows that Jane is not equivalent to either of these two. In the case of Louise, one has only Louise $\sim$ Louise, since the other three students have ages different from Louise. ■

EXAMPLE Let S be the set of all real numbers. Let $a \sim b$ mean that $a^4 = b^4$. Show this is an equivalence relation.

Solution One must show that this relation $\sim$ is reflexive, symmetric, and transitive. In order to show that this relation is reflexive, one must show that $a \sim a$ for all a in S. Notice that if a is any real number, then one always has $a^4 = a^4$. Thus, $a \sim a$ for all elements of S. In order to show that the relation is symmetric, one must show that $a \sim b$ implies $b \sim a$. Notice that if $a \sim b$, then a and b are any real numbers with $a^4 = b^4$. However, this last equation implies that $b^4 = a^4$, which means that $b \sim a$. Hence the relation $\sim$ is symmetric. To show that the relation is transitive, one must show that $a \sim b$ and $b \sim c$ taken together imply that $a \sim c$. Notice that $a \sim b$ and $b \sim c$ mean $a^4 = b^4$ and $b^4 = c^4$, respectively. However, these last two equations clearly imply $a^4 = c^4$, which yields $a \sim c$, as desired. ■

Often a set may have more than one relation defined on it.

EXAMPLE Let the set S be two sets of identical twins from different families $S = \{Jane, Betty, Bill, Joe\}$. Here *Jane* and *Betty* are the first set of identical twins, and *Bill* and *Joe* are the second set of identical twins in a second family. Consider the following two relations: (1) $a \sim_1 b$ means a and b are twins to each other, and (2) $a \sim_2 b$ means both a and b belong to the same family. Which of the reflexive, symmetric, and transitive laws do these relations satisfy? Are either of them equivalence relations?

Solution Note that $\sim_1$ fails to be reflexive since each individual fails to be a twin of himself or herself. On the other hand, each of the four people in S is related to himself or herself using $\sim_2$ since each is in the same family as himself or herself. Thus, $\sim_2$ is reflexive. Note that $\sim_1$ is symmetric, because if a and b are twins to each other, then clearly b and a are twins to each other. Also, $\sim_2$ is symmetric, because if a and b belong to the same family, then b and a also belong to the same family. The relation

$\sim_1$ fails to be transitive because, for example, *Jane* $\sim_1$ *Betty* and *Betty* $\sim_1$ *Jane* does not imply *Jane* $\sim_1$ *Jane*. On the other hand, $\sim_2$ does satisfy the transitive relation, because if a and b are in the same family and b and c are in the same family, then a and c are in the same family. Note that $\sim_1$ is not an equivalence relation since it only satisfies the symmetric law. Finally, note that $\sim_2$ is an equivalence relation since it satisfies all of the three laws.   ∎

Triangles may be classified according to the relative lengths of their sides and/or their angles. A **scalene triangle** is one with all sides of different lengths. If a triangle has two or more sides of the same length, then it is an **isosceles triangle**. A triangle with all three sides of the same length is an **equilateral triangle**. Note that equilateral triangles are also isosceles triangles. A triangle with all angles of equal measure is called an **equiangular triangle**. One of the exercises in this section will be to prove that a triangle is equilateral iff it is equiangular.

One of Euclid's mistakes was in thinking he had a valid proof of SAS. In fact, one must assume SAS is an axiom or else assume some equivalent axiom. The SAS axiom states that if two triangles have two pairs of corresponding sides that are congruent, and the corresponding pair of included angles are congruent, then the triangles are congruent. This axiom is illustrated in Figure 2.1.1.

SAS axiom: If two triangles, $\triangle ABC$ and $\triangle DEF$, satisfy $\angle A \cong \angle D$, $\overline{AB} \cong \overline{DE}$, and $\overline{AC} \cong \overline{DF}$, then $\triangle ABC \cong \triangle DEF$.

Classroom Connection 2.1.3

A good activity for middle or high school students is to have them draw diagrams to illustrate very clearly that AAA and SSA (=ASS) do not, in general, imply congruence. A related question is: What can one say in the special case when the A in SSA is a right angle? Another related question is: What can one say if AAA holds? The first related question becomes much easier to answer after students have encountered the Pythagorean theorem and the answer is known as the HL theorem. The AAA question will be answered in the next chapter where similarity is studied. It should be noted that for triangles in Euclidean geometry, AAA is equivalent to AA.   ◆

In a thorough treatment of Euclidean geometry, one would proceed more or less like Euclid did around 300 BC and carefully build up the theorems of Euclidean

FIGURE 2.1.1 The SAS axiom states that side-angle-side implies congruent triangles. Hence $\angle A \cong \angle D$, $\overline{AB} \cong \overline{DE}$, and $\overline{AC} \cong \overline{DF}$ imply $\triangle ABC \cong \triangle DEF$. It is very important that the corresponding pair of angles be between the corresponding pairs of sides since SSA (=ASS) does not imply congruence.

geometry. This is an excellent approach in many respects, but it is also a very time-consuming approach. This book is not intended to be a comprehensive treatment of Euclidean geometry, and hence a number of things will be assumed without going over the rigorous proofs. It will be assumed that students are familiar with angular measure and such things as vertical angles being equal in measure. Among other things, ASA, SSS, and AAS (= SAA) will all be stated without proofs. These theorems are illustrated in Figures 2.1.2, 2.1.3, and 2.1.4. However, there will still be a fair number of proofs given so that students get to see much of the underlying structure of Euclidean geometry.

Theorem 2.1.1 (ASA Theorem). *If two triangles, $\triangle ABC$ and $\triangle DEF$, satisfy $\angle A \cong \angle D$, $\overline{AB} \cong \overline{DE}$, and $\angle B \cong \angle E$, then $\triangle ABC \cong \triangle DEF$.*

Theorem 2.1.2 (SSS Theorem). *If two triangles, $\triangle ABC$ and $\triangle DEF$, satisfy $\overline{AB} \cong \overline{DE}$, $\overline{AC} \cong \overline{DF}$, and $\overline{BC} \cong \overline{EF}$, then $\triangle ABC \cong \triangle DEF$.*

Theorem 2.1.3 (AAS Theorem). *If two triangles, $\triangle ABC$ and $\triangle DEF$, satisfy $\angle A \cong \angle D$, $\angle B \cong \angle E$, and $\overline{BC} \cong \overline{EF}$, then $\triangle ABC \cong \triangle DEF$.*

The **isosceles triangle theorem** is very useful and was established early in Book 1 of Euclid's *Elements*. In proving this result, a given triangle will be shown to be congruent to itself with the vertices listed in different orders. Recall that the order in which the vertices of a triangle are listed makes an important difference.

FIGURE 2.1.2 The ASA theorem states that angle-side-angle implies congruent triangles. Hence, $\angle A \cong \angle D$, $\overline{AB} \cong \overline{DE}$, and $\angle B \cong \angle E$ imply $\triangle ABC \cong \triangle DEF$.

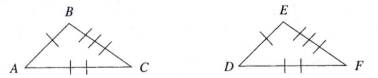

FIGURE 2.1.3 The SSS theorem states that side-side-side implies congruent triangles. Hence, $\overline{AB} \cong \overline{DE}$, $\overline{AC} \cong \overline{DF}$, and $\overline{BC} \cong \overline{EF}$ imply $\triangle ABC \cong \triangle DEF$.

FIGURE 2.1.4 The AAS theorem states that angle-angle-side implies congruent triangles. Hence $\angle A \cong \angle D$, $\angle B \cong \angle E$, and $\overline{BC} \cong \overline{EF}$ imply $\triangle ABC \cong \triangle DEF$.

Theorem 2.1.4 (Isosceles Triangle Theorem). *Two sides of a triangle are congruent if and only if the angles opposite these sides are congruent. Thus, in* $\triangle ABC$*, one has* $\overline{AB} \cong \overline{AC}$ *iff* $\angle B \cong \angle C$.

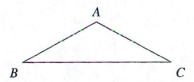

Proof Part 1 ($\Rightarrow$). Using $\overline{AB} \cong \overline{AC}$, show that $\angle B \cong \angle C$. Notice that $\angle A \cong \angle A$ and $\overline{AB} \cong \overline{AC}$ yield that $\triangle ABC \cong \triangle ACB$ by the SAS axiom. Hence, since corresponding parts of congruent figures (CPCF) are congruent, the angle at the second listed vertex of $\triangle ABC$ (i.e., $\angle B$) must be congruent to the angle at the second listed vertex of the triangle $\triangle ACB$ (i.e., $\angle C$). Thus, $\angle B \cong \angle C$, as desired for the proof of Part 1.

Part 2 ($\Leftarrow$). Using $\angle B \cong \angle C$, show that $\overline{AB} \cong \overline{AC}$. Notice that $\overline{BC} \cong \overline{CB}$ and $\angle B \cong \angle C$ yield $\triangle ABC \cong \triangle ACB$ using the ASA theorem. Hence, the side joining the first two listed vertices of $\triangle ABC$ must be congruent to the side joining the first two listed vertices of $\triangle ACB$ by CPCF. Thus, $\overline{AB} \cong \overline{AC}$, as desired.

Let $\triangle ABC$ be given with $D \in \overleftrightarrow{AB}$, such that B lies strictly between A and D as shown here.

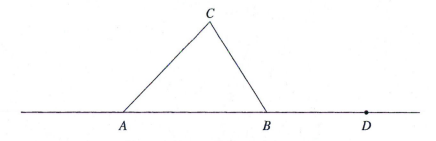

The angle $\angle DBC$ is said to be an **exterior angle** of the triangle. Of course, $\angle DBC$ is supplementary to the angle $\angle B$ (which also may be denoted by $\angle CBA$), and $\angle DBC$ may be larger, smaller, or equal to the angle $\angle B$. The two interior angles $\angle A$ and $\angle C$ are said to be the **remote interior angles** to the exterior angle $\angle DBC$.

Classroom Connection 2.1.4

One can have middle or high school students make several triangles (some with obtuse angles) and then measure the exterior angles and compare them with the remote interior angles. This exploration can be done either by using paper and a protractor or by using the Geometer's Sketchpad (GSP). Later in this chapter we show that an exterior angle of a triangle in the Euclidean plane has measure equal to that of the sum of the two remote interior angles. The next theorem is a weaker version of this. It only states that an exterior angle is greater than either of the two remote interior angles. ◆

Theorem 2.1.5 (Exterior Angle Inequality). *An exterior angle of a triangle is larger than either of the two remote interior angles.*

Proof. Given $\triangle ABC$ with exterior angle $\angle CBD$ as shown here, one wishes to prove that both $m(\angle C) < m(\angle CBD)$ and $m(\angle A) < m(\angle CBD)$. However, it suffices to show that $m(\angle C)$ is less than $m(\angle CBD)$ since $\angle CBD \cong \angle ABE$. The same reasoning that can be used to show $m(\angle C) < m(\angle CBD)$ can also be used to show that $m(\angle A) < m(\angle ABE) = m(\angle CBD)$.

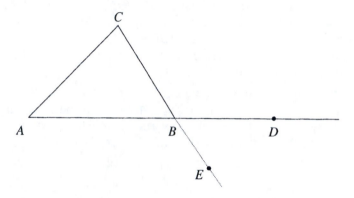

Let F be the midpoint of the segment $\overline{BC}$ and extend $\overline{AF}$ beyond F to obtain the point G, such that $AF = FG$ as shown here.

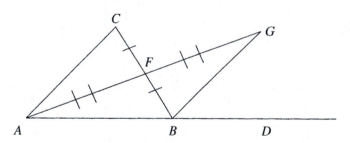

Since F is the midpoint of $\overline{BC}$, it follows that $BF = CF$. Because vertical angles are congruent, one obtains $\angle CFA \cong \angle GFB$. It follows that $\triangle AFC \cong \triangle GFB$ using SAS. Thus, $\angle C \cong \angle FBG$ by CPCF. Since $m(\angle FBG) < m(\angle FBD) = m(\angle CBD)$, it follows that $m(\angle C) < m(\angle CBD)$, as desired.

The next theorem uses the exterior angle inequality and the isosceles triangle theorem to prove that if two sides are unequal in a triangle, then the angle across from the larger of the two sides is greater than the angle across from the shorter of the two sides. This theorem is covered in the Mathematics in Context grade 7/8 materials, see Figure 2.1.5.

The largest angle of a triangle lies opposite the longest side.

The second largest angle of a triangle lies opposite the second longest side.

The smallest angle of a triangle lies opposite the shortest side.

Isosceles triangles have two equal sides and two equal angles.
Equilateral triangles have three equal sides and three equal angles.

FIGURE 2.1.5 Reproduced from page 26 of *Triangles and Beyond* in the Mathematics in Context grade 7/8 materials.

Theorem 2.1.6 (Scalene Inequality). *Given two unequal sides in a triangle, the angle opposite the larger side is greater than the angle opposite the smaller side.*

   Proof. Given $\triangle ABC$ with $BC > AC$, show that $m(\angle A) > m(\angle B)$.

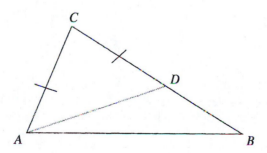

Since $BC > AC$, there is a point $D \in \overline{BC}$ with $AC = DC$. The exterior angle inequality yields $m(\angle CDA) > m(\angle B)$. Using the isosceles triangle theorem and the fact that the whole is bigger than the part, one finds $m(\angle A) > m(\angle CAD) = m(\angle CDA) > m(\angle B)$. Hence, $m(\angle A) > m(\angle B)$, as desired.

EXAMPLE The triangle $\triangle ABC$ has $m(\angle A) = 35°$, $m(\angle B) = 95°$, and $m(\angle C) = 50°$. How do the lengths of the three sides compare to each other?

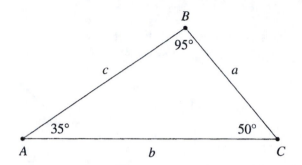

Solution Since $m(\angle B) > m(\angle C) > m(\angle A)$, it follows that $b > c > a$ (i.e., $AC > AB > BC$). ∎

EXAMPLE The triangle $\triangle DEF$ in the xy-plane has vertices at the points $D = (2,0)$, $E = (10,0)$, and $F = (5,4)$. How do the measures of the three angles compare to one another?

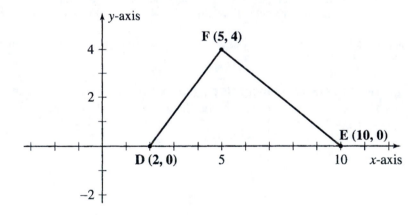

Solution Using the distance formula in the xy-plane, it is easy to find that the lengths of this triangles sides are $DE = 8$, $EF = \sqrt{41} \approx 6.4$, and $FD = 5$. Hence, $DE > EF > FD$, which implies using the scalene inequality that $m(\angle F) > m(\angle D) > m(\angle E)$. ∎

Recall the **midpoint** of the segment $\overline{AB}$ is the point $D \in \overline{AB}$, such that $AD = BD$. The **perpendicular bisector** of the segment $\overline{AB}$ is the line ℓ, which passes through the midpoint D and is also perpendicular to the segment $\overline{AB}$.

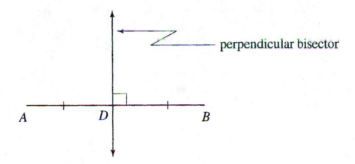

perpendicular bisector

Classroom Connection 2.1.5

Have middle or high school students make a perpendicular bisector of a segment, $\overline{AB}$, using the GSP or protractors and rulers. Have them select several points, $P_1, P_2, P_3, \ldots$, on the perpendicular bisector. For each point P_i, have them (1) measure the distance AP_i and then (2) measure the distance BP_i. They should find that the distance from A to P_i is equal to the distance from B to P_i for each i. A possible follow-up question is to ask your students if what they have learned can be used to give a compass and straight edge construction of the perpendicular bisector. One of the exercises of this section is to show that the perpendicular bisector of the segment $\overline{AB}$ is exactly equal to the set of all points P, such that $AP = BP$. In other words, the set of points that are equidistant from two fixed points in the Euclidean plane is the perpendicular bisector of the segment joining the two points. What do you think the set of points equidistant from two fixed points in Euclidean three-dimensional space looks like? ◆

EXERCISES 2.1

1. The triangle $\triangle ABC$ shown here has sides of lengths 16, 17, and 18 units. List the angles in order from smallest to largest. Explain your reasoning.

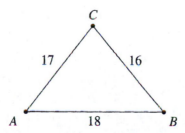

2. The triangle $\triangle LMN$ in the following diagram has angles of $65°, 58°$, and $57°$. List the sides in order from shortest to longest. Explain your reasoning.

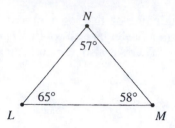

3. The triangle $\triangle ABC$ in the xy-plane has vertices at the points $A = (1,0)$, $B = (8,0)$, and $C = (4,1)$. List the angles in order from smallest to largest. Explain your reasoning.

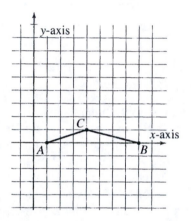

4. The triangle $\triangle DEF$ has $DE = DF = 3$ and $EF = 4$.
   a. Which two angles have the same measure? Explain your reasoning.
   b. Is the third angle larger or smaller than the two angles that have equal measure? Explain your reasoning.
5. The isosceles triangle $\triangle GHK$ in the xy-plane has vertices at the points $G = (0,2)$, $H = (1,8)$, and $K = (0,14)$.
   a. Which two sides have the same length?
   b. Which two angles have the same measure?
   c. Is the third angle larger or smaller than the two angles that have equal measure?
6. Let $\triangle ABC$ be isosceles with $AC = BC$ and let D be the midpoint of the segment $\overline{AB}$ as shown here. Prove $\triangle CDB \cong \triangle CDA$, and prove that $\angle CDA$ is a right angle.

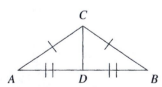

7. Let G be the midpoint of the segment $\overline{EF}$ and let segment $\overline{HG}$ be perpendicular to segment $\overline{EF}$ as shown here.
   a. Prove that $\triangle HGE \cong \triangle HGF$.
   b. What does your answer to part a imply about the lengths of $\overline{HE}$ and $\overline{HF}$?

c. What special type of triangle is $\triangle HEF$?

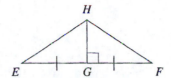

8. The **equidistant locus** of points A and B consists of the points that are at an equal distance from A and B. Thus, the equidistant locus is the set of all points P, such that $AP = BP$. Recall the **perpendicular bisector** of a segment $\overline{AB}$ is the line that contains the midpoint of the segment $\overline{AB}$ and that is perpendicular to this segment. Prove that the equidistant locus of A and B consists of the set of points on the perpendicular bisector of $\overline{AB}$.

9. Let the set S under consideration be given by the whole numbers from 1 to 5 (i.e., $S = \{1,2,3,4,5\}$). Which of the reflexive, symmetric, and transitive laws are valid for the relation: $a \sim b$ means $a < b$?

10. Let the set S under consideration be given by the whole numbers from 1 to 5 (i.e., $S = \{1,2,3,4,5\}$). Which of the reflexive, symmetric, and transitive laws are valid for the relation: $a \sim b$ means $a + b \leq 10$?

11. Let the set S under consideration be the set of real numbers (i.e., $S = R^1$). Which of the reflexive, symmetric, and transitive laws are valid for the following two relations?
    a. $a \sim b$ means $|a| = |b|$.
    b. $a \sim b$ means $a^3 = b^3$.

12. The tennis team at Blank College has seven members. Three of them have blue eyes, and four of them have brown eyes. Let the set S under consideration be the members of the tennis team. Let $a \sim b$ mean both a and b have the same color eyes. Assume that Jack has blue eyes and Jane has brown eyes.
    a. Is Jack $\sim$ Jane valid or not valid?
    b. Is Jane $\sim$ Jack valid or not valid?
    c. Is this relation reflexive?
    d. Is this relation symmetric?
    e. Is this relation transitive?
    f. Is this an equivalence relation?

13. Prove that a triangle in the Euclidean plane can have at most one right angle.

Classroom Discussion 2.1.1

A key method used in mathematics is generalization. If one is given a theorem and one finds a second theorem that includes the first as a special case, then the second theorem is called a generalization of the first. For example, the Pythagorean theorem states that if $\triangle ABC$ has a right angle at vertex C, then $c^2 = a^2 + b^2$. There are many proofs of this theorem, and we give one of these in Chapter 3. An important generalization of the Pythagorean theorem is the law of cosines, which we will also cover in the next chapter. The law of cosines applies to any triangle $\triangle ABC$ and can be stated as $c^2 = a^2 + b^2 - 2ab\cos(C)$. Note that in the special case where $\angle C$ is a right angle, one has $\cos(C) = 0$, and the formula for the law of cosines reduces to the formula for the Pythagorean theorem. Have students discuss what the word *generalization* means in mathematics and then have the students find a generalization of the result that a triangle in Euclidean geometry can have at most one right angle. ◆

Classroom Discussion 2.1.2

One possible conjecture is that the scalene inequality generalizes to all polygons with an odd number of sides. Conjecture: Given two unequal angles in a polygon with an odd number of sides, the angle opposite the larger side is greater than the angle opposite the smaller side. As a possible classroom discussion, try and determine if this conjecture is correct. One might first ask why the conjecture is stated for polygons, which have an odd number of sides. Of course, a resolution of the conjecture should either be a proof that the conjecture is correct or a counterexample to show it is false. If students have difficulty finding a counterexample to the conjecture, ask them to draw a number of (nonregular) pentagons. ◆

2.2 PARALLEL LINES

One problem that arises with teaching parallelism to very young students has to do with the way eyesight can mislead people. If one has a pair of railroad tracks that go straight on a flat plain, the tracks appear to converge in the far distance. Of course, the distance between the two parallel rails stays the same. However, things at farther distances appear to be smaller. For example, things at twice the distance appear to have half the diameter.

This section is devoted to the parallel postulate (i.e., Euclid's fifth postulate) and its consequences. One of the most important consequences of the parallel postulate is that a triangle in the Euclidean plane always has the sum of the measures of its angles equal to exactly $180°$. This result is proven in Theorem 2.2.4. In fact, the result that all triangles have an angle sum equal to $180°$ is equivalent to the parallel postulate. Figure 2.2.1 illustrates that triangles in Euclidean geometry have an angle sum of $180°$ and that given any three angles, one can construct a triangle with these three angles if and only if their measures add up to exactly $180°$.

The sum of the measures of the three angles of any triangle is $180°$.

If the measures of three angles total $180°$, then a triangle can be made with these angles.

If the measures of three angles do not total $180°$, then a triangle cannot be made with these angles.

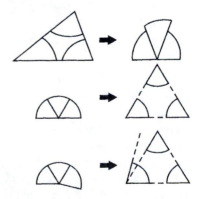

FIGURE 2.2.1 The angles of a triangle add up to $180°$. Also, if three angles add up in measure to $180°$, then one may construct a triangle having these angles. Reproduced from page 20 of *Triangles and Beyond* in the Mathematics in Context grade 7/8 materials.

Classroom Connection 2.2.1

There is an activity that can easily be done in the middle or high school classroom with paper and scissors to illustrate that the angle sum of a triangle is a straight angle. Have students cut out a triangle and then cut off a piece near each angle. Then have them lay the angles together to form a straight angle. For example, assume one has a triangle such as $\triangle ABC$ shown in Figure 2.2.1. Cut off pieces near the three angles and fit them together as shown at the top of Figure 2.2.1 to get a straight angle.   ◆

Recall that two lines in the Euclidean plane are defined to be parallel if they don't intersect (i.e., if their intersection is the empty set). **Playfair's postulate** states that given a point, P, off a line, ℓ, in the Euclidean plane, there is exactly one line containing P that is parallel to ℓ. Playfair's postulate may be shown to be equivalent to Euclid's fifth postulate and thus may be substituted for Euclid's fifth postulate whenever it is convenient.

Transversal.   Given two distinct lines, ℓ_1 and ℓ_2, then a third line, ℓ_3, is said to be a **transversal** if (1) ℓ_3 has a nonempty intersection with each of ℓ_1 and ℓ_2, and if (2) $\ell_1 \cap \ell_2 \notin \ell_3$. Given a pair of lines with a transversal, there are several pairs of angles that are useful to consider when investigating the possibility that ℓ_1 and ℓ_2 may be parallel. There are **alternate interior angles**, which are on alternate sides of ℓ_3 and are between ℓ_1 and ℓ_2. There are **corresponding angles**, which are on the same side of ℓ_3 where one angle is between ℓ_1 and ℓ_2 and one is not, nor is it adjacent to the first angle. There **are interior angles on the same side of** ℓ_3, which are both between ℓ_1 and ℓ_2. Figure 2.2.2 illustrates such pairs.

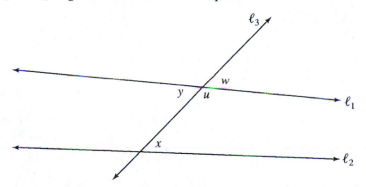

FIGURE 2.2.2 Angles x and y are alternate interior angles, while angles x and w are corresponding angles. Angles x and u are interior angles on the same side of ℓ_3. The lines ℓ_1 and ℓ_2 need not be parallel.

Theorem 2.2.1 (Alternate Interior Angle Criteria). *A transversal ℓ_3 to ℓ_1 and ℓ_2 has a pair of alternate interior angles of equal measure iff ℓ_1 and ℓ_2 are parallel.*

> ***Proof.*** **Part 1 ($\Rightarrow$).** Assuming $m(\angle x) = m(\angle y)$ in Figure 2.2.2, one must show that ℓ_1 and ℓ_2 are parallel (i.e., one must prove $\ell_1 \cap \ell_2 = \emptyset$). In order to prove this by contradiction, assume that ℓ_1 and ℓ_2 have point P in common. Then P is either on the same side of ℓ_3 as $\angle x$ or on the same side as $\angle y$. Assume without loss of generality that P is on the same side as $\angle x$. Then the diagram is as shown on the following page.

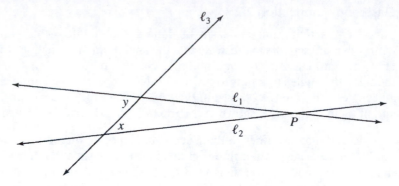

However, this is impossible since $\angle x$ is a remote interior angle to the exterior angle $\angle y$, and thus $m(\angle x) = m(\angle y)$ yields a contradiction to the exterior angle inequality.

Part 2 ($\Leftarrow$). Assuming that ℓ_1 and ℓ_2 are parallel, one must show that $m(\angle x) = m(\angle y)$. Let $Q = \ell_3 \cap \ell_1$ and let ℓ_4 be the line through Q such that ℓ_3 forms equal alternate interior angles with ℓ_1 and ℓ_4. In other words, let ℓ_3 and ℓ_4 make an angle $\angle x'$ equal in measure to $\angle x$ as indicated in the following diagram.

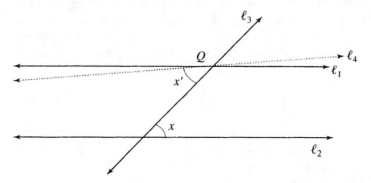

According to Part 1 of this proof, ℓ_4 must be parallel to ℓ_2. Playfair's postulate now guarantees that $\ell_4 = \ell_1$ since there is only one parallel line to ℓ_2 through the point Q. Hence, $\angle x'$ is the same as $\angle y$, and one has $m(\angle x) = m(\angle x') = m(\angle y)$, as desired. ∎

EXAMPLE Assume that lines ℓ_1 and ℓ_2 in the following diagram are parallel. Find the measures of $\angle x, \angle y,$ and $\angle u$.

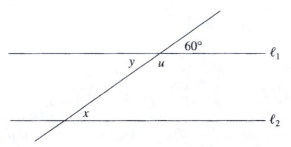

Solution   Note that $m(\angle y) = 60°$ since vertical angles are congruent. Also, $\angle y$ and $\angle u$ are supplementary since they form a linear pair (i.e., taken together they yield the straight angle corresponding to line ℓ_1). Hence, $m(\angle u) = 180^0 - m(\angle y) = 120°$. Finally, $m(\angle x) = m(\angle y) = 60°$ using the alternate interior angle criteria. ■

Using Theorem 2.2.1, it is not too difficult to establish the following two corollaries.

Corollary 2.2.2 (Corresponding Angle Criteria). *A transversal ℓ_3 to ℓ_1 and ℓ_2 has a pair of corresponding angles of equal measure iff ℓ_1 and ℓ_2 are parallel.*

Corollary 2.2.3 (Same-Side Interior Angle Criteria). *A transversal ℓ_3 to ℓ_1 and ℓ_2 has a pair of interior angles on the same side of ℓ_3 that are supplementary (i.e., have measures that sum to a straight angle) iff ℓ_1 and ℓ_2 are parallel.*

Figure 2.2.3 illustrates Theorem 2.2.1, Corollary 2.2.2, and Corollary 2.2.3.

Theorem 2.2.4 (Angle Sum Theorem). *The sum of the measures of a triangle's angles is equal to a straight angle.*

Proof.   Consider $\triangle ABC$ in Figure 2.2.4 with angles $\angle x, \angle y$, and $\angle z$ at respective vertices A, B, and C. Let line ℓ be the unique line parallel to the line $\overleftrightarrow{BC}$ passing through vertex A. Using the alternate interior angle criteria (Theorem 2.2.1), one finds $m(\angle y) = m(\angle y')$ and $m(\angle z) = m(\angle z')$. The result now follows using $m(\angle x) + m(\angle y) + m(\angle z) = m(\angle x) + m(\angle y') + m(\angle z') = 180°$.

EXAMPLE   Let $\triangle ABC$ be isosceles with $AB = AC$. Find $m(\angle B)$ and $m(\angle C)$ given that $m(\angle A) = 40°$.

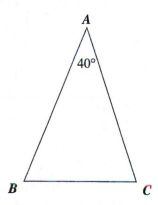

Solution   Notice that $m(\angle B) = m(\angle C)$ by the isosceles triangle theorem. Using this and the fact that the angle sum of the triangle is $180°$, one obtains $m(\angle A) + m(\angle B) + m(\angle C) = 180°$.
    Hence,
$$40° + m(\angle B) + m(\angle B) = 180°.$$

Solving for $m(\angle B)$, one finds $m(\angle B) = 70°$. Thus, $m(\angle C) = m(\angle B) = 70°$. ■

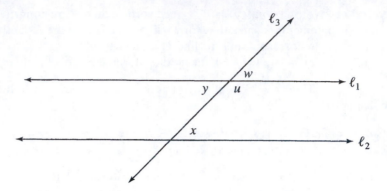

FIGURE 2.2.3 The equations given in a, b, and c are equivalent to one another, and each is equivalent to the statement that ℓ_1 and ℓ_2 are parallel. a. $m(\angle x) = m(\angle y)$, b. $m(\angle x) = m(\angle w)$, and c. $m(\angle x) + m(\angle u) = 180°$.

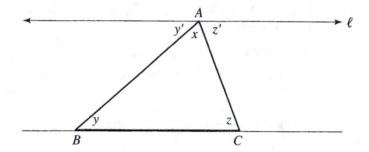

FIGURE 2.2.4 The angle sum of a triangle is one straight angle (i.e., 180°). Thus, $m(\angle x) + m(\angle y) + m(\angle z) = m(\angle x) + m(\angle y') + m(\angle z') = 180°$.

Recall that the exterior angle inequality from the previous section states that a triangle's exterior angle is strictly larger than each of the two remote interior angles. The proof of this result that was given did not use the parallel postulate (i.e., Euclid's fifth postulate). Euclid put off using the parallel postulate for as long as possible, evidently he wanted to do as much geometry as he could without using the parallel postulate. He established the exterior angle inequality early on and made extensive use of it. However, after he started using the parallel postulate, he was able to establish the stronger result that a triangle's exterior angle is equal in measure to the sum of the measures of the two remote interior angles. This stronger result is known as the **exterior angle equality** and, in general, is not valid in non-Euclidean geometries where one does not have a parallel postulate. For spherical geometry, one uses great circles as the "lines" of the geometry. Classroom Discussion 2.2.1 at the end of this section involves discovering that the angle sum of a spherical triangle is not 180°. In fact, the angle sum of a spherical triangle is always larger than 180°, and the following exterior angle equality result is false for all spherical triangles. Thus, an understood assumption in Corollary 2.2.5 is that one is working in Euclidean geometry.

Corollary 2.2.5 (Exterior Angle Equality). *The measure of a triangle's exterior angle is equal to the sum of the measures of the two remote interior angles.*

> **Proof.** In the following figure, $m(\angle x) + m(\angle y) + m(\angle z) = 180°$ by the angle sum theorem.

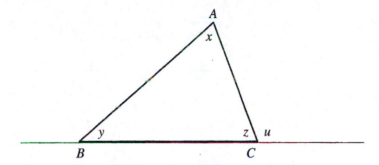

> Also, $m(\angle u) + m(\angle z) = 180°$ since the two angles yield a straight angle and are thus supplementary. By subtracting the previous two equations and using basic algebra, it follows that $m(\angle u) = m(\angle x) + m(\angle y)$, as desired.

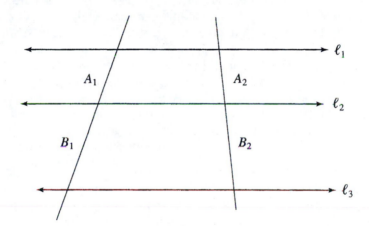

An important property of parallels is that when one has three parallel lines, then they cut transversals in proportional ratios. In particular, assume that in this diagram the parallel lines ℓ_1, ℓ_2, and ℓ_3 intersect the transversals shown forming segments of lengths A_1, A_2, B_1, and B_2. Then the following equality holds $\frac{A_1}{B_1} = \frac{A_2}{B_2}$. One also has $\frac{A_1}{A_2} = \frac{B_1}{B_2}$ and $\frac{A_1+B_1}{B_1} = \frac{A_2+B_2}{B_2}$, etc.

EXAMPLE Let the three lines ℓ_1, ℓ_2, and ℓ_3 in the following diagram be parallel. Assume they cut one transversal to yield segments that are 8 units and 4 units long, as shown. Let another transversal yield segments that are x units and $x + 3$ units long, as also shown. Find x.

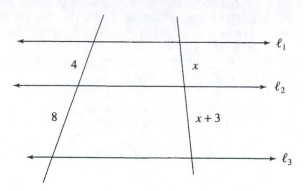

Solution Set up the proportion $\frac{4}{8} = \frac{x}{x+3}$ and cross multiply to obtain $4 \cdot (x + 3) = 8 \cdot (x)$. Solving for x yields $x = 3$ units. ∎

Interiors and Exteriors of Simple Closed Curves. A curve in the Euclidean plane is said to be **closed** if it is constructed as starting and ending at the same point. Here the word *curve* includes figures that may have straight parts such as polygons. Circles, ellipses, and polygons are closed, but a single line segment is not closed. Also, parabolas fail to be closed. A curve is said to be **simple** if it never crosses itself. Thus, circles, ellipses, and triangles are simple, but a curve shaped like a figure eight fails to be simple. All polygons are simple closed curves. An important result in mathematics, known as the **Jordan curve theorem**, states that any simple closed curve in a Euclidean plane separates the plane into exactly two components. In other words, if one takes a Euclidean plane and subtracts a simple closed curve, then the remainder of the plane has two pieces, and one cannot construct a curve going from one piece to the other since one would have to cross the deleted curve. Of course, one of the pieces of the plane left after deleting the simple closed curve is called the *interior of the curve*, and the other is called the *exterior*. Since polygons are simple closed curves, they have interiors and exteriors.

Congruence of Polygons. Recall that two triangles are defined to be congruent if their corresponding sides are congruent and if their corresponding angles are congruent. This definition may be extended to polygons.

Definition 2.2.1 Two polygons are congruent if there is a correspondence between them such that their corresponding sides are congruent and their corresponding angles are congruent.

Let G_1 and G_2 be polygons with the same number of vertices, assume there is a correspondence between their vertices such that each pair of vertices joined by a side of G_1 corresponds to a pair of vertices joined by a side of G_2. Here a correspondence between the vertices of G_1 and G_2 is a $1-1$ onto function, taking the vertices of G_1 onto the vertices of G_2. According to Definition 2.2.1, this correspondence yields congruent polygons if each pair of corresponding sides are congruent and each pair of corresponding angles are congruent.

The symbol for congruence of polygons is the same as for triangles. Thus, one writes either $G_1 \cong G_2$ or $P_1 P_2 \ldots P_n \cong Q_1 Q_2 \ldots Q_n$.

$$P_1 P_2 \ldots P_n \cong Q_1 Q_2 \ldots Q_n$$
$$\angle P_i \cong \angle Q_i \quad \text{and} \quad \overline{P_i P_{i+1}} \cong \overline{Q_i Q_{i+1}} \quad \text{for all } i = 1, 2, \ldots, n$$

Just as with congruence of triangles, the order in which the vertices are listed is important. Two polygons may be congruent with one correspondence of vertices and yet fail to be congruent with a different correspondence of vertices.

Classroom Connection 2.2.2

A discovery lesson for middle and high school students could start by having them measure the angle sum of several polygons. It is best to have them work on convex polygons. The students should be able to discern a pattern that suggests a relationship between the number of sides of the polygon and the polygon's angle sum. After the students discover a pattern, ask how they can justify the pattern. If students need an added hint, suggest that they think of some way to use the knowledge that a triangle always has an angle sum of 180° in Euclidean geometry. In general, a polygon with n sides may be subdivided into $(n - 2)$ triangles. It follows that the sum S of the "interior" angles is $S = (n - 2) \cdot 180°$. For convex polygons, there is no real problem knowing how to measure the "interior" angles. Figure 2.2.5 illustrates that for nonconvex polygons, one needs to interpret some "interior" angles as having measure more than 180°. ◆

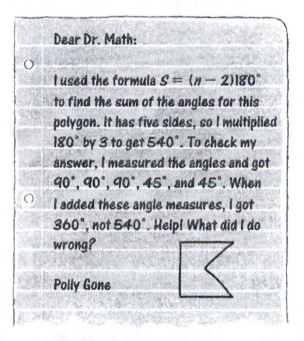

Dear Dr. Math:

I used the formula $S = (n - 2)180°$ to find the sum of the angles for this polygon. It has five sides, so I multiplied 180° by 3 to get 540°. To check my answer, I measured the angles and got 90°, 90°, 90°, 45°, and 45°. When I added these angle measures, I got 360°, not 540°. Help! What did I do wrong?

Polly Gone

FIGURE 2.2.5 For convex polygons, the formula $S = (n - 2) \cdot 180°$ is fairly easy to apply, but for nonconvex polygons, one must assume some "interior" angles are larger than 180°. Reproduced from page 39 of *Getting in Shape* in the MathScape grade 7 materials.

EXERCISES 2.2

1. Assume ℓ_1 and ℓ_2 are parallel in the following diagram and that $m(\angle s) = 120°$. Find the measures of the angles $\angle x, \angle y, \angle u, \angle w, \angle k,$ and $\angle h$.

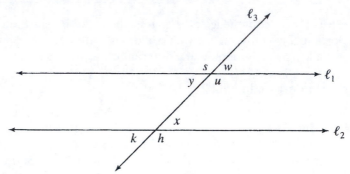

2. Answer the question asked of Dr. Math by Polly Gone in Figure 2.2.5. You are to be Dr. Math, write your answer in paragraph form.

3. Find the sum of the measures of the interior angles of each of the following polygons.
   a. A quadrilateral
   b. A pentagon
   c. A hexagon

4. A regular polygon has all sides of equal length and all angles of equal measure. Find the measure of each interior angle of the following polygons.
   a. A regular pentagon
   b. A regular hexagon
   c. A regular octagon

5. The sum of the measures of a polygon's interior angles is $900°$. How many sides does this polygon have?

6. The angle measurements shown in the figure have are given in degrees. Find the measures of angles $\angle t, \angle u, \angle v, \angle w, \angle x, \angle y,$ and $\angle z$.

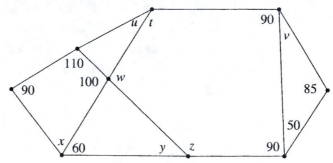

7. A regular polygon has eighteen sides. Find the measure of each of its interior angles.

8. Let two lines ℓ_1 and ℓ_2 be parallel, and let ℓ_3 be a transversal that is perpendicular to the first line ℓ_1. Prove that ℓ_3 must then also be perpendicular to ℓ_2.

9. Let two lines ℓ_1 and ℓ_2 be parallel with $A, C \in \ell_1, B, D \in \ell_2$. Assuming both $\overline{AB}$ and $\overline{CD}$ are perpendicular to one of the parallel lines, hence also to both of them, prove that $\overline{AB} \cong \overline{CD}$.

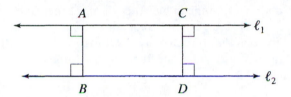

Classroom Discussion 2.2.1

Recall that a great circle on a sphere comes from intersecting the sphere's surface with a plane through the sphere's center. Since the shortest curve joining two points on a sphere's surface will lie along an arc of a great circle, it follows that great circles play the role of straight lines on the sphere. Hence, **spherical triangles** have three sides, and each side lies on a different great circle. Start a class discussion by asking students if they think the angle sum theorem of Euclidean plane geometry is also valid for geometry on a sphere. If students have difficulty with this exploration, then suggest thinking of the Earth and looking at spherical triangles with one vertex at the North Pole and the other two vertices at the equator. These spherical triangles will always have at least two right angles since each great circle joining the North Pole to a point on the equator will meet the equator in a right angle. ◆

2.3 QUADRILATERALS

A **quadrilateral** is a polygon with exactly four sides. A quadrilateral with both pairs of opposite sides being parallel is a **parallelogram**. A **rectangle** is a parallelogram with a right angle. A parallelogram with all four sides of the same length is a **rhombus**. A **square** is a rhombus that is also a rectangle. A **trapezoid** is a quadrilateral with at least one pair of opposite sides being parallel. Notice that parallelograms are also trapezoids. A **proper trapezoid** is one that is not a parallelogram. In a trapezoid, two opposite sides that are parallel are called *bases*. When the trapezoid is also a parallelogram, then either pair of opposite sides may be considered bases. Proper trapezoids have only one pair of opposite sides that are the bases. The reader should be cautioned that some books, especially on the precollege level, do not include parallelograms as trapezoids. In such books, the only figures called *trapezoids* are proper trapezoids. A diagram illustrating the relationships between quadrilaterals, trapezoids, parallelograms, rectangles, squares, and rhombi is given in Figure 2.3.1.

   Many of the results on quadrilaterals can be illustrated with class activities. The following Classroom Discussion is an activity that one can use in the college classroom to demonstrate that results on quadrilaterals can often be illustrated in interesting ways. Future teachers may find this activity interesting and may wish to someday use it in a classroom they will be teaching.

Classroom Discussion 2.3.1

A physical demonstration that the diagonals of a rhombus are perpendicular can easily be done in any classroom. Have students take four sticks (or geo-strips or poly-strips) of equal length. Make holes near the end of each stick and fit them loosely together to get a movable figure. Since all four sides are of the same length, the figure remains a rhombus. For different positions, use rubber bands for the two diagonals and measure the angles they form at their intersection. Since a rhombus has diagonals that are perpendicular, one should always find 90° angles. Holding the rhombus such that one diagonal is horizontal and then moving the top vertex vertically up will help illustrate that the two diagonals of a rhombus are always perpendicular. ◆

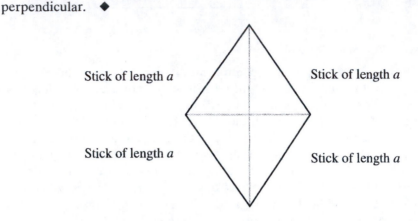

Stick of length a Stick of length a

Stick of length a Stick of length a

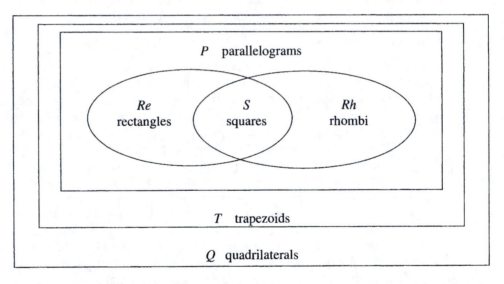

FIGURE 2.3.1 Working in the Euclidean plane, let Q be the collection of all quadrilaterals, T the collection of all trapezoids, P the collection of all parallelograms, Re the collection of all rectangles, Rh the collection of rhombi, and S the collection of all squares. Then $S \subset Re \subset P \subset T \subset Q$ and $S \subset Rh \subset P \subset T \subset Q$. Furthermore, $S = Re \cap Rh$. Notice that the various areas shown here for the different types of quadrilaterals do not really represent the relative abundance of the various types. For example, a quadrilateral "chosen at random" is very unlikely to be a trapezoid.

Theorem 2.3.1. *A quadrilateral is a parallelogram iff opposite sides are of equal length.*

    ***Proof* Part 1 (⟹).** Assume that $ABCD$ is a parallelogram with $\overleftrightarrow{AB}$ parallel to $\overleftrightarrow{DC}$ and $\overleftrightarrow{AD}$ parallel to $\overleftrightarrow{BC}$. Draw diagonal $\overline{AC}$.

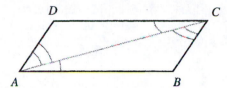

Applying the alternate interior angle criteria first to $\overleftrightarrow{AC}$ (considered as a transversal to the pair of parallel lines $\overleftrightarrow{AB}$, $\overleftrightarrow{DC}$) and then to the pair of parallel lines $\overleftrightarrow{AD}$, $\overleftrightarrow{BC}$, one finds first that $\angle CAB \cong \angle ACD$ and then that $\angle CAD \cong \angle ACB$. Using these pairs of congruent angles and $\overline{AC} \cong \overline{AC}$, one obtains $\triangle ACB \cong \triangle CAD$ by ASA. It follows that $\overline{AB} \cong \overline{DC}$ and $\overline{AD} \cong \overline{BC}$ by CPCF (corresponding parts of congruent figures). Hence, parallelograms have opposite sides that are congruent, and Part 1 has been established.

Part 2 ($\Leftarrow$). Assume that the quadrilateral $ABCD$ has congruent opposite sides. Thus, $\overline{AB} \cong \overline{DC}$ and $\overline{AD} \cong \overline{BC}$.

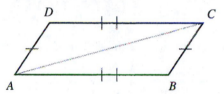

Using these congruent pairs and $\overline{AC} \cong \overline{AC}$, one obtains $\triangle ACB \cong \triangle CAD$ by SSS. It follows that $\angle CAB \cong \angle ACD$ and $\angle CAD \cong \angle ACB$ by CPCF. The alternate interior angle criteria now yields that $\overleftrightarrow{AB}$ and $\overleftrightarrow{DC}$ are parallel and that $\overleftrightarrow{AD}$ and $\overleftrightarrow{BC}$ are parallel. Thus, a quadrilateral with congruent opposite sides is a parallelogram, and Part 2 has been established.

The next three results are often very useful when working with parallelograms.

Theorem 2.3.2. *The diagonals of a parallelogram bisect each other.*

Theorem 2.3.3. *A quadrilateral is a parallelogram iff each pair of consecutive angles are supplementary.*

Corollary 2.3.4. *All four angles of a rectangle are right angles.*

EXAMPLE Let $ABCD$ be a parallelogram with $m(\angle A) = 40°$ as shown here. Find $m(\angle B), m(\angle C)$, and $m(\angle D)$.

Solution   Since $\angle A$ and $\angle B$ are supplementary, one has $m(\angle A) + m(\angle B) = 180°$. Thus, $m(\angle B) = 180° - 40° = 140°$. Also, $m(\angle C) = m(\angle A) = 40°$ since these two angles are opposite. Likewise, $m(\angle D) = m(\angle B) = 140°$, since $\angle D$ and $\angle B$ are opposite angles.   ∎

EXERCISES 2.3

1. The quadrilateral *PQRS* shown here has $m(\angle P) = m(\angle R) = 60°$ and $m(\angle Q) = 120°$.
   a. Find $m(\angle S)$.
   b. Is this quadrilateral a parallelogram? Why or why not?

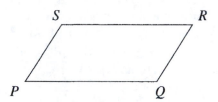

2. The quadrilateral *EFGH* shown here has $m(\angle E) = m(\angle G) = 60°$ and $m(\angle F) = 115°$.
   a. Find $m(\angle H)$
   b. Is this quadrilateral a parallelogram? Why or why not?

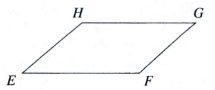

3. The parallelogram *JKLM* has $m(\angle J) = 50°$. Find the measures of the other three angles.

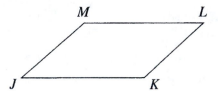

4. The parallelogram *ABCD* has $AB = 5$ and $BC = 9$.
   a. Find *AD* and explain how you arrived at your answer.
   b. Find *CD* and explain how you arrived at your answer.

5. Prove that in a parallelogram the angles at opposite vertices are congruent.

6. Prove that all four angles of a rectangle are right angles (i.e., prove Corollary 2.3.4). Don't forget that the definition of a rectangle only requires that it be a parallelogram with at least one right angle.

2.4   AREAS OF FIGURES

Students in middle school and high school often have difficulty understanding area. One particular problem is that they often confuse perimeter and area. Even when they understand the intrinsic difference between these concepts, they often feel that if you know one of these quantities, then the other is determined. One approach a teacher can use with middle school students is illustrated in Figure 2.4.1. Sixth-grade students are asked to give a written explanation of how two different rectangles with the same perimeter can have different areas.

Classroom Connection 2.4.1

Give middle or high school students sixteen squares that are 1 unit by 1 unit and ask them to construct as many different rectangles as possible of area sixteen square units using these squares. Have them record the perimeter in each case. Which of the figures will have the minimum perimeter? Ask them to guess which rectangle would have the least perimeter if they did this problem with n^2 unit squares instead of sixteen unit squares.   ◆

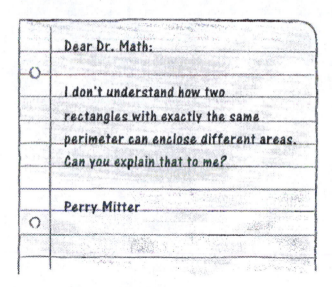

FIGURE 2.4.1 Sixth-grade students are expected to give written answers explaining things that often trouble middle school or even high school students. One of the most common difficulties encountered with students is the tendency to confuse the length of the perimeter of a plane figure with the area enclosed by the perimeter. Reproduced from page 41 of *Designing Spaces* in the MathScape grade 6 materials.

Area is one way to measure the size of certain sets lying in a Euclidean plane. If set A consists of two disjoint sets, B and C, in the Euclidean plane, having areas $Area(B)$ and $Area(C)$, then the area of A is the sum of the areas of B and C. In other words, if $A = B \cup C$ with $B \cap C = \emptyset$, then $Area(A) = Area(B) + Area(C)$. When B and C have areas and C is a subset of B, the area of the set $B - C$ (i.e., the area of the set of points in B that are not in C) is given by $Area(B) - Area(C)$. More generally, when B and C have areas but are not disjoint (i.e., $B \cap C \neq \emptyset$), then the area of $B \cap C$ is included once in $Area(B)$ and once in $Area(C)$. Hence, $Area(A) = Area(B) + Area(C) - Area(B \cap C)$. This is often stated as the following:

$$Area(B \cup C) = Area(B) + Area(C) - Area(B \cap C)$$

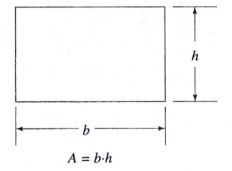

One very important property of area is that if two figures are congruent and one has area, then the second one has the same area. Also, the area of a point is zero, and the area of a line is zero.

Areas of Rectangles.   For **rectangles**, the formula for area is very simple. It is just base, b, times height, h.

$$A = b \cdot h$$

Classroom Connection 2.4.2

The formula $A = b \cdot h$ for the area of a rectangle may be demonstrated to very young students first learning about area by using unit squares for (small) integer values of b and h. For example, using twelve unit squares, one can have middle school students build 2×6 and 3×4 rectangles and observe that the formula holds for these cases. This can be made into a discovery lesson by asking them to build some rectangles using squares of 1 unit by 1 unit and to try to devise a way of finding (i.e., formula for calculating) the area of rectangles.   ◆

Areas of Parallelograms.   The formula for the area of a parallelogram is $A = b \cdot h$, where the height must be measured perpendicular to the base.

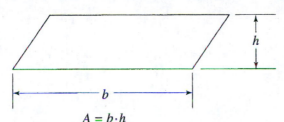

$$A = b \cdot h$$

There are several ways to verify the formula for the area of a parallelogram. One way is to "square off the ends" so as to embed the parallelogram in a rectangle that has a base of length $b + k$ and height h. The diagram is shown here.

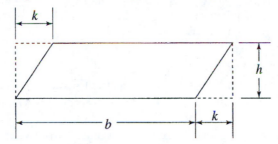

Since the rectangle has base $b + k$ and height h, the rectangle has area given by $A_{rectangle} = (b + k) \cdot h = b \cdot h + k \cdot h$. On the other hand, the two end triangles are congruent to each other. They are both right triangles with one leg of length h and one leg of length k. These two end triangles can be pasted together to obtain a rectangle as shown here with base of length k and height of length h.

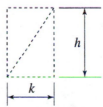

Thus, the two end triangles taken together have an area of $A_{ends} = k \cdot h$. It follows that the area A of the parallelogram must be

$$A = A_{rectangle} - A_{ends} = (b + k) \cdot h - k \cdot h$$
$$= (b \cdot h + k \cdot h) - k \cdot h = b \cdot h,$$

which is the desired formula.

Notice that the area of a parallelogram of base b and height h is the same as that of a rectangle of base b and height h (i.e., $A = b \cdot h$).

Another way of thinking about obtaining a rectangle from a parallelogram is illustrated here. Think of the parallelogram as being "straightened up" to obtain a rectangle. Here, line segments that are parallel to the base are moved holding their length and direction fixed to fill in the rectangle. For example, the dotted line segment of length b shown in the parallelogram is moved to a corresponding dotted line segment of the same length b in the rectangle.

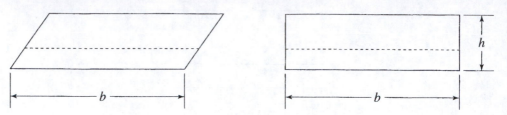

There is an underlying principle due to Cavalieri (*ca val eyar e*) (1598–1647) that implies when one "straightens up" a figure by moving parallel slices in the same direction, the area (for plane figures) or volume (for solids) remains the same. For plane figures, areas of the figures are the same, but in general the perimeters may be different. For solids, volumes of the solids are the same, but surface areas may be different. Cavalieri's principle is illustrated for solids in Figure 2.6.1.

Areas of Triangles. In order to obtain the formula for the area of a triangle, take a given triangle and duplicate it. Then position the two triangles such that the original triangle and duplicate triangle form a parallelogram.

The parallelogram in Figure 2.4.2 is made up of two triangles that are congruent to the original, and thus the original triangle must have half the area of the parallelogram. Hence, the formula for the area of a triangle is $A = \frac{1}{2}b \cdot h$.

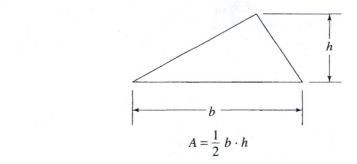

$$A = \frac{1}{2} b \cdot h$$

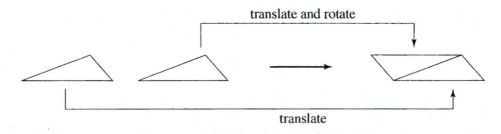

FIGURE 2.4.2 Duplicate the first triangle to get the second. Then position the two triangles such that the original triangle and duplicate triangle form a parallelogram with twice the area of the original triangle. The top triangle should be rotated by 180° and translated.

An alternative formula for the area of a triangle is known as Heron's formula. Let triangle $\triangle ABC$ have sides of lengths a, b, and c, as shown here.

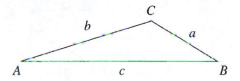

The **semi-perimeter** s of $\triangle ABC$ is defined to be $s = \frac{1}{2} \cdot (\textit{perimeter}) = \frac{1}{2} \cdot (a + b + c)$. **Heron's formula** for the area of the triangle is

$$A = \sqrt{s \cdot (s - a) \cdot (s - b) \cdot (s - c)}.$$

EXAMPLE   Let $\triangle ABC$ have $AB = 5$ cm, $BC = 6$ cm, and $CA = 7$ cm. Find the area of this triangle.

Solution   Notice that $a = BC = 6$ cm, $b = CA = 7$ cm, and $c = AB = 5$ cm. Thus, $s = \left(\frac{1}{2}\right)(a + b + c) = \left(\frac{1}{2}\right)(6 + 7 + 5) = 9$ cm. Hence,

$$A = \sqrt{s \cdot (s - a) \cdot (s - b) \cdot (s - c)} = \sqrt{9(9 - 6)(9 - 7)(9 - 5)}$$

$$= 6\sqrt{6} \text{ cm}^2 \approx 14.7 \text{ cm}^2 \qquad \blacksquare$$

Areas of General Figures.   For a polygon, one can subdivide the area into triangles and/or other figures with known area formulas. Then if one has sufficient information, compute the area of the polygon by adding the areas of the figures in the subdivision. Compare Figure 2.4.3. For more general figures, the situation can be more complicated. In integral calculus, one learns techniques used to find the areas of very general figures. The basic idea is to take a sequence of approximations to the area of a given figure and find the limiting value of the approximations. One generally uses rectangles to approximate the area of a figure and then takes the limit as the rectangles get very small in both dimensions or, in certain cases, just very thin. To find approximations to the area of an irregularly shaped figure, take a grid and, after placing the grid over the figure, count the squares that lie over the figure. In general, one gets better approximations by using finer grids. Figure 2.4.4 shows a problem for sixth-grade students from *Book 1* of the *Math Thematics* materials in which the students are asked to use the grid approach to approximate the area of Madagascar.

Geoboards.   A geoboard can be a useful tool in teaching areas. A geoboard has pegs spaced uniformly in a grid pattern, and one selects polygonal areas using rubber bands. Think of a geoboard as a model for the set of all the lattice points of the xy-plane where both coordinates are integers.

Find the area of each polygon.

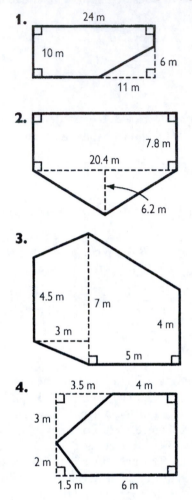

FIGURE 2.4.3 Notice that the interior of each polygon can be split up into familiar figures such as triangles, rectangles, trapezoids, etc. Reproduced from page 42 of *From The Ground Up* in the MathScape grade 7 materials.

An important result for calculating areas on a grid such as a geoboard is Pick's theorem. In 1900, George Pick provided a formula for the area of a polygonal region with vertices at integer lattice points in the xy-plane. If one lets the number of lattice points interior to the polygonal region be denoted by I and lets the number of lattice points on the boundary be denoted by B, then **Pick's theorem** says the formula for the area is $A = I + (1/2) \cdot B - 1$. An example illustrating this theorem is shown in Figure 2.4.5.

$$A = I + (1/2) \cdot B - 1 \text{ (Pick's theorem)}$$

Area of an Irregularly Shaped Figure

You can estimate areas using grid squares.

First Trace the outline of the figure.

Next Place the tracing on a piece of centimeter graph paper.

Then Count the complete grid squares. Add on the area from the partially filled grid squares. You can use a fraction to estimate a part of a grid square or combine parts to form whole squares.

34. a. On the map, the area of Madagascar is about how many square centimeters? Explain how you made your estimate.

b. An area on the map that is 1 cm by 1 cm represents how many square kilometers? Estimate the actual area of Madagascar.

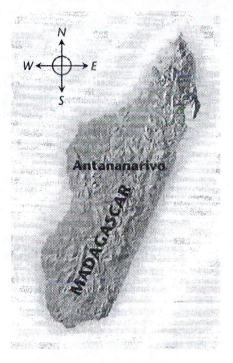

1 cm is about 200 km.

FIGURE 2.4.4 As you can see, sixth-grade students are taught to estimate irregularly shaped areas using grids. Reproduced from page 464 of *Book 1* in the *Math Thematics* grade 6 materials.

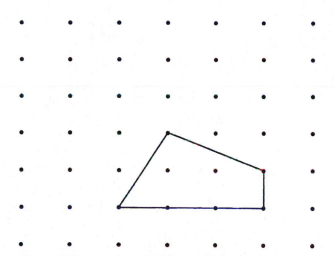

FIGURE 2.4.5 This geoboard illustrates a rubber band with two interior pegs and six boundary pegs. Thus, $I = 2$ and $B = 6$. Hence, by Pick's theorem, the area of the indicated region is $A = I + (1/2) \cdot B - 1 = 2 + (1/2) \cdot (6) - 1 = 4$ square units.

EXERCISES 2.4

1. A rubber band on a geoboard encloses an area of 5 square units and there are 8 boundary pegs. How many interior pegs are there?

2. Answer the question asked of Dr. Math by Perry Mitter in Figure 2.4.1. You are to be Dr. Math, write your answer in paragraph form.

3. The parallelogram $EFGH$ in the xy-plane has vertices given by $E = (0,0), F = (4,4), G = (4,14)$, and $H = (0,10)$.
   a. Graph this parallelogram.
   b. Find the area of this parallelogram.

4. Given a triangle, a segment from one vertex to the middle of the opposite side is called a **median**. Prove that a median divides a triangle into two triangles with equal areas.

5. The following trapezoid $WXYZ$ has bases with lengths of 2.5 inches and 3.5 inches The height of the trapezoid is 1 inch.
   a. Find the area of the triangle $\triangle WXY$.
   b. Find the area of the triangle $\triangle WYZ$.
   c. Find the area of the trapezoid $WXYZ$.

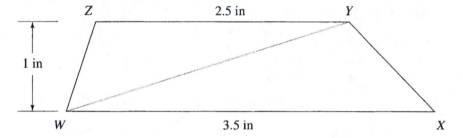

6. Find the area of each of the four polygons shown in Figure 2.4.3.

7. Let a triangle have sides of lengths 9 cm, 10 cm, and 11 cm. Find the area of this triangle.

8. Find the area of the triangle $\triangle ABC$ in the xy-plane with vertices at $A = (-1,2), B = (3,2)$, and $C = (-20, -1)$.

9. Find the area of the trapezoid $ABCD$ with vertices at $A = (1,1), B = (4,1), C = (4,4)$, and $D = (1,6)$.

10. Prove that if a given trapezoid has height h and bases of lengths b_1 and b_2, then the area A of the trapezoid is given by the formula $A = \frac{1}{2}(b_1 + b_2) \cdot h$.

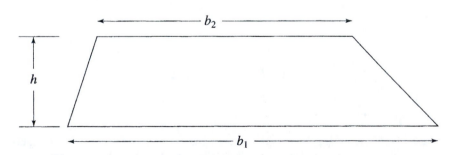

Classroom Connection 2.4.3

Pick's theorem is illustrated in Figure 2.4.5. Have students verify this particular case and/or several other cases by subdividing the interior of a given polygon into more familiar figures such as triangles, etc. In fact, the interior of a polygon may always be subdivided into triangles. Thus, given sufficient information about a polygon, one may find the area of a given polygon using the formula for the area of a triangle. ◆

Classroom Connection 2.4.4

Middle and high school students often find it tedious to learn the formulas for areas of triangles, squares, rectangles, parallelograms, and trapezoids. Furthermore, they often find it difficult to keep straight which formula goes with which type of figure. They often fail to see any unity in the various formulas. In particular, they often have most difficulty with the formula for the area of a trapezoid—$A = \frac{1}{2}(b_1 + b_2) \cdot h$—and don't see how it is related to anything else they have learned. A classic teaching strategy is to emphasize that a trapezoid can be divided into two triangles having height h and having bases of b_1 and b_2, respectively. Another strategy is to ask students to discover how one might start with the formula for the area of a trapezoid and then derive the formulas for areas of squares, rectangles, and parallelograms. The formulas for these figures are easy to obtain from that of a trapezoid. By learning how various formulas are related, students often see more unity in mathematics. They may even begin to see how they might develop new formulas and relations on their own. ◆

2.5   CIRCLES

Understanding circles and their various parts is extremely important in our society. Among other things, middle school students should become familiar with parts of circles such as center, radius, and diameter. Figure 2.5.1 illustrates some of what seventh-grade students should know about circles.

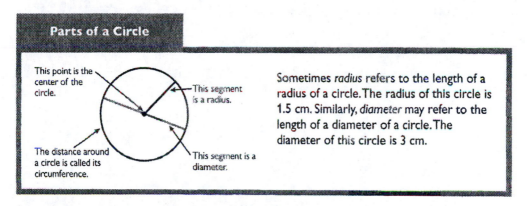

FIGURE 2.5.1 Students in seventh grade are expected to know and understand the important parts of circles. Reproduced from page 26 of *Getting in Shape* in the MathScape grade 7 materials.

A circle is defined to be the set of points at some fixed (positive) distance from a given point. The fixed distance is the **radius**, and the given point is the **center**. The word *radius* is also used to denote a segment with an endpoint at the center and the other endpoint at a point on the circle. The interior of the circle consists of the points whose distances from the center are less than the circle's radius. Note that technically the circle is just the "edge" and not the interior. The word *diameter* is used to denote a line segment with both endpoints on the circle, such that this line segment contains the circle's center. Also, *diameter* is used to denote the length of such a segment. Using r to denote the radius (considered as the length of a radius segment) and d to denote the diameter (considered as the length of a diameter segment), one has $d = 2r$.

Tangents and Secants. A **tangent line** to a circle is a line that has exactly one point in common with the circle. One should be warned that for more general curves, the word *tangent* must be defined in a different way, which involves concepts from calculus. If a line has exactly two points in common with a circle, then it is called a **secant**. A secant may or may not pass through the circle's center. A line segment joining two points of the circle is called a **chord**. Figure 2.5.2 illustrates tanget line, secant line, and chord.

Central and Inscribed Angles. A **central angle** is an angle with a vertex at the circle's center. An **inscribed angle** is one with a vertex on the circle and with sides pointing into the circle. Given a central angle, the arc of the circle with which the sides and interior of the angle intersect has **angular measure**, the same as the measure of the central angle. Notice that a circle's arc has two different measures. One is the arc's angular measure and the other is the arc's length. Central angle, inscribed angle, and angular measure are illustrated in Figure 2.5.3.

Radian Measure. As mentioned previously, angles can be measured with degrees or with radian measure. Find the radian measure θ of an angle by first taking a circle of some radius, r, centered at the vertex and measuring the arc length, S, of the intercepted arc. Then the angle measure in radians is $\theta = \dfrac{S}{r} = \dfrac{\text{length of intercepted arc}}{\text{radius of the circle}}$.

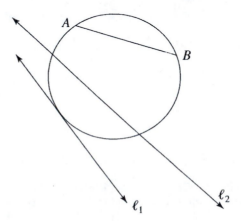

FIGURE 2.5.2 Line ℓ_1 is a tangent line, line ℓ_2 is a secant line, and segment $\overline{AB}$ is a chord.

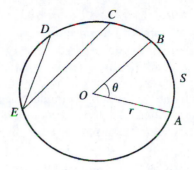

FIGURE 2.5.3 The circle with center O has radius r, and the shorter arc $\overparen{AB}$ has length S. The **central angle** $\angle BOA$ has **radian** measure $\theta = \frac{S}{r}$. The **angular measure** $m(\overparen{AB})$ of arc $\overparen{AB}$ is the same as the measure of the central angle subtended by the arc [i.e., $m(\overparen{AB}) = m(\angle BOA)$]. Thus, using radian measure, one has $m(\overparen{AB}) = \theta = \frac{S}{r}$. The angle $\angle CED$ is an **inscribed angle**.

Classroom Connection 2.5.1

Have middle or high school students measure several inscribed angles and ask them to compare those to the angular measure of the arc intercepted by the inscribed angle. They should look for a simple relation. The exact relation is described by the inscribed angle theorem, which is stated as Theorem 2.5.1. This activity is especially convincing when it is done on the Geometer's Sketchpad, but it can be done reasonably well with protractor and straight edge. For a related activity, see the next Classroom Connection. ◆

    If an inscribed angle has one side passing through the circle's center, then the diagram will look like the following illustration.

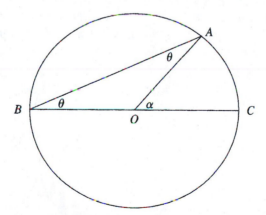

    Here the inscribed angle $\angle ABC$ has measure θ and intercepts an arc $\overparen{AC}$ with angular measure α, which is also the measure of the angle $\angle AOC$. Since $\overline{OA}$ and $\overline{OB}$ are each radii of the same circle, they must be congruent. Thus, the isosceles triangle theorem yields that $m(\angle OAB)$ and $m(\angle ABC)$ are equal. Let $\theta =$

$m(\angle OAB) = m(\angle ABC)$. Using the exterior angle equality, one finds $\alpha = \theta + \theta$, which yields $\theta = \frac{\alpha}{2} = \frac{1}{2} m(\widehat{AC})$. Thus, the measure of the inscribed angle is one-half of the angular measure of the intercepted arc.

Of course, most inscribed angles fail to have either side pass through the center of the circle. One can investigate these inscribed angles by using an additional inscribed angle with one side passing through the center of the circle. One finds that the measure of an inscribed angle is always one-half the measure of the intercepted arc. This result is known as the inscribed angle theorem and will be stated without proof.

Theorem 2.5.1 (Inscribed Angle Theorem). *The measure of an inscribed angle is one-half the angular measure of the arc it intercepts. One of the most important numbers in mathematics is π.*

Definition 2.5.1 The number π is defined to be the ratio of the circumference of a circle with its diameter. In other words, $\pi = \frac{C}{d}$.

$$\pi \approx 3.14159 \quad \text{(approximate)}$$

Using $d = 2r$, it follows that the circumference C of a circle of radius r is given by $C = 2\pi r$. The area of a circle of radius r is $A = \pi r^2$.

$$\text{Circle}$$
$$C = 2\pi r \quad A = \pi r^2$$

Areas of Sectors of Circles. An area of a wedge cut out by a circle's central angle is illustrated in Figure 2.5.4. The central angle $\angle COB$ has radian measure equal to θ. Thus, the arc subtended by this central angle has length $S = r\theta$. Turning $360°$ about the center O would be 2π radians; thus, the ratio $\frac{\theta}{2\pi}$ is the ratio of

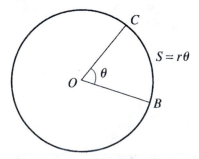

FIGURE 2.5.4 The central angle $\angle COB$ has radian measure equal to θ. The area of the sector (i.e., "wedge") cut from the circle by the central angle $\angle COB$ is $A = \frac{\theta \cdot r^2}{2}$, where the angle θ is measured in radians.

measure of the central angle to one complete turn about the center. The area of the *sector* (i.e., wedge) cut from the circle by the central angle $\angle COB$ must be the entire area of the circle $\pi \cdot r^2$ times the ratio $\frac{\theta}{2\pi}$. Hence, the area of the sector is

$$A = (\pi \cdot r^2) \cdot \frac{\theta}{2 \cdot \pi} = \frac{\theta \cdot r^2}{2},$$ where the angle θ is measured in radians.

EXAMPLE A circle of radius 12 cm has a given central angle of $10°$. Find the area of the sector cut from the circle by this central angle.

Solution Calculating the central angle in radians, one finds

$$\theta = (10°)\left(\frac{2\pi \text{ radians}}{360 \text{ degrees}} \right) = \frac{\pi}{18} \text{ radians.}$$

Hence, the area is given by

$$A = \left(\frac{1}{2} \right)(\theta)(r)^2 = \left(\frac{1}{2} \right)\left(\frac{\pi}{18} \right)(12 \text{ cm})^2 = 4\pi \text{ cm}^2.$$ ∎

EXERCISES 2.5

1. A circle with a radius of 1 unit has a given central angle of $30°$. Find the length of the arc that this angle intercepts.

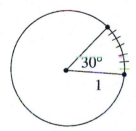

2. A circle with a radius of 4 units contains an arc $\overset{\frown}{AB}$, which has length 2π. Find the measure of the central angle subtended by this arc in (a) radians and (b) degrees.

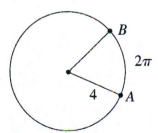

3. A circle of radius of 6 units has a given central angle of $60°$. Find the area of the sector that this central angle cuts from the circle.

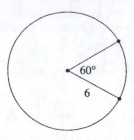

4. A certain circle has a radius of r units. A central angle of $30°$ cuts a sector with an area of 12π square units from this circle. Find the radius r.

5. Prove that two chords of a given circle have the same length iff the central angles they subtend are congruent. In other words, using circle O in the following diagram, prove that $AB = CD$ iff $\theta_1 = \theta_2$.

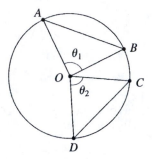

Classroom Connection 2.5.2

As a follow-up activity to that described in the previous Classroom Connection, have middle or high school students choose a given circle and a point P interior to a circle. Also, have them choose segments $\overline{AB}$ and $\overline{CD}$, which are chords containing P, as shown in Figure 2.5.5. Have the students measure the angles $\angle APC$ and $\angle BPD$

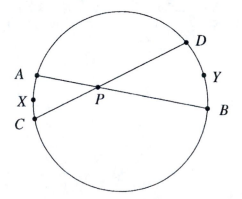

FIGURE 2.5.5 If P is an interior point of a circle, then $m(\angle APC) = m(\angle BPD) = \frac{1}{2} \times [m(\widehat{AXC}) + m(\widehat{BYD})]$.

and find the angular measures of $\overset{\frown}{AXC}$ and $\overset{\frown}{BYD}$. Here the arc $\overset{\frown}{AXC}$ is the arc joining A to C, which includes the point X. Ask the students if they can find any relationships between the four measures they have found. If necessary, remind them of the inscribed angle theorem and point out that when the arc $\overset{\frown}{AXC}$ is very small, one expects that $m(\angle APC)$ [which equals $m(\angle BPD)$] will be approximately equal to one-half the angular measure of $\overset{\frown}{BYD}$. They should be able to discover that $m(\angle APC) = m(\angle BPD) = \frac{1}{2}[m(\overset{\frown}{AXC}) + m(\overset{\frown}{BYD})]$. ◆

2.6 VOLUMES AND SURFACE AREAS

Students have difficulty with volumes because humans see two-dimensional images since the retina is essentially a two-dimensional surface. The human brain must take two-dimensional images and interpret what they mean in terms of a three-dimensional world.

In this section, we study three-dimensional objects, in particular, volume and surface area. Since middle school students often confuse area and perimeter in studying plane figures, it is no surprise that they often confuse volume and surface area when studying three-dimensional figures. One activity a teacher can use to give middle school students an initial feeling for these two concepts is outlined in the following Classroom Connection.

Classroom Connection 2.6.1

An informative activity for middle school students is to make eight identical cubes using paper. Then make a larger cube with each edge having length equal to twice the length of an edge of the first eight cubes. You may then compare surface areas and volumes fairly easily. In particular, the eight smaller cubes fill up the larger cube, and the sum of the surface areas of four of the smaller cubes is equal to the surface area of the larger cube. In this activity, one demonstrates that when a cube's edges are doubled, the volume is multiplied by 2^3, and the surface area is multiplied by 2^2. After doing this activity, ask your students what they think would happen to the volume and surface area of a unit cube that had all of its edges tripled in length. ◆

Boxes.   The simplest solid is a cube with each edge of 1-unit length. The volume of such a cube is 1 cubic unit, and since each face is a square with an area of 1 square unit, it follows that the surface area of such a cube is 6 square units—one square unit coming from each of the six faces. If a cube C has each edge of length ℓ, then the volume is ℓ^3 and the surface area is $6 \cdot \ell^2$. Of course, the units for the volume will be in cubic units of length, and the surface area will be in square units of length. A box (i.e., right rectangular prism) with dimensions of ℓ, h, and w (for length, height, and width, respectively) has volume $V = \ell \cdot h \cdot w$. The surface area S of the box will equal the sum of the areas of the six faces. In fact, $S = 2 \cdot \ell \cdot h + 2 \cdot \ell \cdot w + 2 \cdot h \cdot w$.

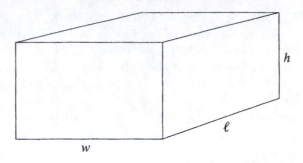

Box

Right Rectangular Prism

$$V = \ell \cdot h \cdot w$$

$$S = 2 \cdot \ell \cdot h + 2 \cdot \ell \cdot w + 2 \cdot h \cdot w$$

Nets, also known as **flat patterns**, are flat diagrams that can be folded to form solid figures. They are especially helpful in illustrating volume and surface area of many common polyhedrons. Figure 2.6.2 illustrates three homework problems from the Math Thematics materials involving nets.

Figures 2.6.3 and 2.6.4 show more examples of nets.

The nets for the Platonic solids are given here.

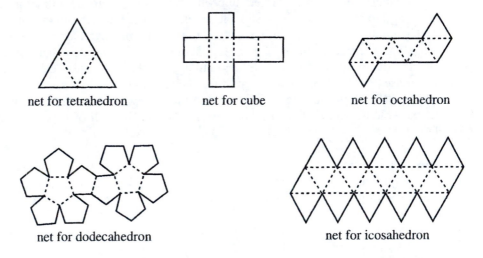

net for tetrahedron   net for cube   net for octahedron

net for dodecahedron   net for icosahedron

Prisms.   If one has a right prism with a base of area B and a height h, then the volume is $V = B \cdot h$. If one takes an oblique prism, with area of base B and height h, then, using Cavalieri's principle, one finds it has the same volume as a right prism with base B and height h (i.e., $V = B \cdot h$). Cavalieri's principle for solids is illustrated in Figure 2.6.1.

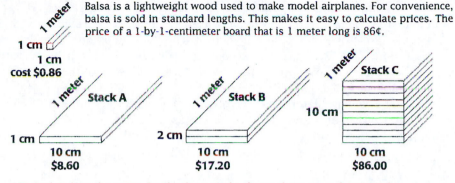

Balsa is a lightweight wood used to make model airplanes. For convenience, balsa is sold in standard lengths. This makes it easy to calculate prices. The price of a 1-by-1-centimeter board that is 1 meter long is 86¢.

7. Explain how the prices for the three stacks shown above were determined.

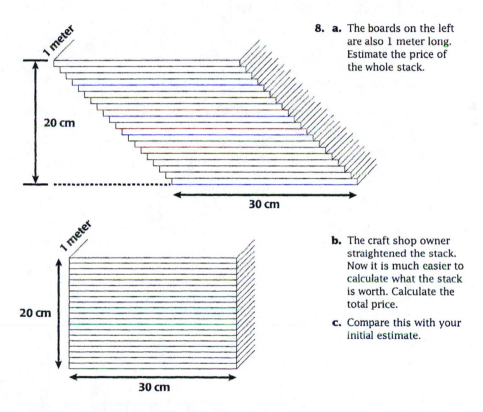

8. a. The boards on the left are also 1 meter long. Estimate the price of the whole stack.

b. The craft shop owner straightened the stack. Now it is much easier to calculate what the stack is worth. Calculate the total price.

c. Compare this with your initial estimate.

FIGURE 2.6.1 *Cavalieri's principle* for three-dimensional objects is illustrated in this figure. When the boards are "straightened up," we see that the volume is equal to that of a box. Reproduced from page 19 of *Reallotment* in the Mathematics in Context grade 6/7 materials.

Classroom Connection 2.6.2

Using construction paper, have middle or high school students make the prism in problem 1 in Figure 2.6.5. Fill the model with rice and measure using a graduated cylinder. Compare to calculated volume. ◆

An interesting comparison can be made of the volumes enclosed by (1) a right circular cylinder of radius r and height $2r$, (2) a sphere of radius r, and (3) a right circular cone of radius r and height $2r$. These volumes are in the ratio $3 : 2 : 1$. This ratio of $3 : 2 : 1$ can be demonstrated in the classroom by activities found in the Filling and Wrapping module of the Connected Mathematics Project (see Figures 2.6.6 and 2.6.7).

The volume, lateral surface area, and total surface area of a right circular cylinder of height h and radius of base r, as shown in Figure 2.6.8, are $V = \pi r^2 h$, $A_{\text{lateral}} = 2\pi rh$, and $A_{\text{total}} = 2\pi rh + 2\pi r^2$, respectively. A net like the one in Figure 2.6.3 can be used to justify the formulas for A_{lateral} and A_{total}.

EXAMPLE   A can in the shape of a right circular cylinder has a height of 15 cm and a radius of base equal to 6 cm. Find the volume enclosed by this can in cubic centimeters. Also, find the lateral and total surface areas in square centimeters.

Solution   The volume is given by

$$V = \pi r^2 h = \pi(6 \text{ cm})^2(15 \text{ cm}) = 540\pi \text{ cm}^3 \approx 1{,}696 \text{ cm}^3.$$

The lateral surface area is given by

$$A_{\text{lateral}} = 2\pi rh = 2\pi(6 \text{ cm})(15 \text{ cm}) = 180\pi \text{ cm}^2 \approx 565 \text{ cm}^2.$$

The total surface area is given by

$$A_{\text{total}} = 2\pi rh + 2\pi r^2 = 2\pi(6 \text{ cm})(15 \text{ cm}) + 2\pi(6 \text{ cm})^2$$

$$= 252\pi \text{ cm}^2 \approx 792 \text{ cm}^2. \qquad \blacksquare$$

Choose the letter of the space figure that can be formed with each net.

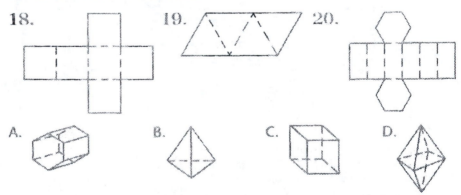

18.   19.   20.

A.   B.   C.   D.

21. Name space figures A and C shown with Exercises 18–20.

22. Is space figure D shown above a prism? Why or why not?

FIGURE 2.6.2 Nets (i.e., flat patterns) can help students get a better understanding of both volume and surface area. Reproduced from page 475 of *Book 1* in the Math Thematics grade 6 materials.

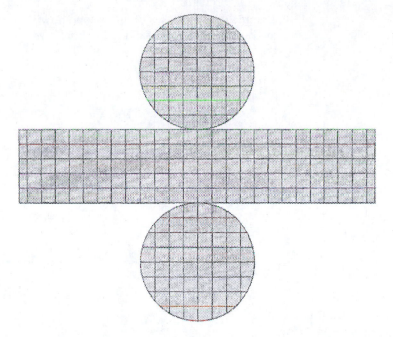

FIGURE 2.6.3 Nets (i.e., flat patterns) can help students get a better understanding of both volume and surface area. They are particularly useful in helping students understand how surface area differs from volume. Reproduced from page 39 of *Filling and Wrapping* in the Connected Mathematics Project grade 7 materials.

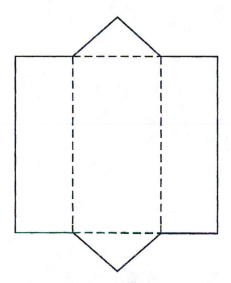

FIGURE 2.6.4 Students are asked to predict the type of solid that results from this net. Reproduced from page 476 of *Book 1* in the Math Thematics grade 6 materials.

Assume that the bases of all prisms shown are regular polygons. Find the volume and surface area of each prism.

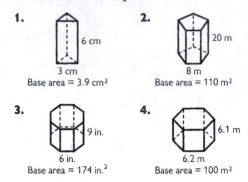

1.

6 cm

3 cm
Base area = 3.9 cm²

2.

20 m

8 m
Base area = 110 m²

3.

9 in.

6 in.
Base area = 174 in.²

4.

6.1 m

6.2 m
Base area = 100 m²

FIGURE 2.6.5 Reproduced from page 39 of *Shapes and Space* in the MathScape grade 8 materials.

Classroom Discussion 2.6.1

One activity that illustrates that surface area is not related in any simple fashion to volume can be done with a sheet of standard-sized paper. Take a sheet of 8.5 in by 11 in paper and roll it in two different directions to get two different cylinders. One will be 11 inches high and the other will be 8.5 inches high. Each cylinder has the same lateral surface area of $A_{\text{lateral}} = 8.5 \text{ in} \times 11 \text{ in} = 93.5 \text{ in}^2$. Ask your students which cylinder has the most volume. It is quite likely that many students will say they enclose the same volume, a few will feel that, if there is a difference in volume, then it must be the taller cylinder that has the most volume. To illustrate that the cylinders have different volumes, close off the bottoms of the two cylinders and fill them with rice. One cylinder will hold more rice than the other. Typically, many students find it quite strange that the short cylinder has more volume. In fact, if the height h of a sheet of paper is greater than the base b, then the shorter of the two cylinders formed by rolling the sheet one way and then the other way always has more volume (compare Figure 2.6.9). ◆

The volume of both cones and pyramids is given by one-third the area of the base times the height.

$$\boxed{\begin{array}{c} \text{Cones and Pyramids} \\ V = \frac{1}{3} \cdot B \cdot h = \frac{1}{3} \cdot (\text{area of base}) \cdot (\text{height}) \end{array}}$$

For a right circular cone of height h and radius of base r, as shown in Figure 2.6.10, the volume is given by $V = \frac{1}{3}\pi r^2 h$ which is exactly $\frac{1}{3}$ the volume of a right circular cylinder of the same height and radius of base. See Figure 2.6.10 for formulas for lateral surface area and total surface area of such cones.

Problem 5.1

- Using modeling dough, make a sphere with a diameter between 2 inches and 3.5 inches.

- Using a strip of transparent plastic, make a cylinder with an open top and bottom that fits snugly around your sphere. Trim the height of the cylinder to match the height of the sphere. Tape the cylinder together so that it remains rigid.

- Now, flatten the sphere so that it fits snugly in the bottom of the cylinder. Mark the height of the flattened sphere on the cylinder.

A. Measure and record the height of the cylinder, the height of the empty space, and the height of the flattened sphere.

B. What is the relationship between the volume of the sphere and the volume of the cylinder?

Remove the modeling dough from the cylinder, and save the cylinder for the next problem.

FIGURE 2.6.6 Students learn that the volume of a sphere of radius *r* is two-thirds that of a corresponding cylinder of radius *r* and height 2*r*. Reproduced from page 48 of *Filling and Wrapping* in the Connected Mathematics Project grade 7 materials.

EXAMPLE Let a right circular cone have a base with radius $r = 6$ cm, a height of $h = 8$ cm. Find (1) the volume of this cone, (2) the lateral surface area of this cone, and (3) the total surface area of this cone.

Solution   The volume is given by

$$V = \left(\frac{1}{3}\right)\pi r^2 h = \left(\frac{1}{3}\right)\pi(6\text{ cm})^2(8\text{ cm}) = 96\pi\text{ cm}^3 \approx 302\text{ cm}^3.$$

Problem 5.2

- Roll a piece of stiff paper into a cone shape so that the tip touches the bottom of your cylinder.

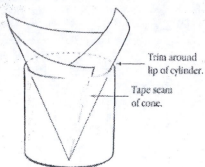

Trim around lip of cylinder.

Tape seam of cone.

- Tape the cone shape along the seam and trim it to form a cone with the same height as the cylinder.

- Fill the cone to the top with sand or rice, and empty the contents into the cylinder. Repeat this as many times as needed to completely fill the cylinder.

What is the relationship between the volume of the cone and the volume of the cylinder?

FIGURE 2.6.7 Students learn that the volume of a cone of radius r and height $2r$ is one-third that of a corresponding cylinder of radius r and height $2r$. Reproduced from page 49 of *Filling and Wrapping* in the Connected Mathematics Project grade 7 materials.

The lateral surface area is given by

$$A_{\text{lateral}} = \pi r \sqrt{r^2 + h^2} = \pi(6 \text{ cm})\sqrt{(6 \text{ cm})^2 + (8 \text{ cm})^2} = \pi(6 \text{ cm})(10 \text{ cm})$$

$$= 60\pi \text{ cm}^2 \approx 188 \text{ cm}^2.$$

The total surface area is given by

$$A_{\text{total}} = A_{\text{lateral}} + \pi r^2 = 60\pi \text{ cm}^2 + \pi(6 \text{ cm})^2 = 96\pi \text{ cm}^2 \approx 302 \text{ cm}^2. \quad \blacksquare$$

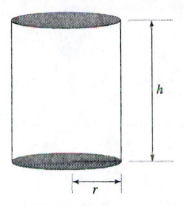

FIGURE 2.6.8 A right circular cylinder has height h and radius of base r. The volume is the area of the base, πr^2, times the height $h \cdot V = \pi r^2 h$. The lateral or side surface area is the circumference of the base, $2\pi r$, times the height $h \cdot A_{lateral} = 2\pi rh$. The total surface area is the lateral surface area plus the area of the two circles, one at the top and one at the base, $A_{total} = 2\pi rh + 2\pi r^2$.

FIGURE 2.6.9 Take a rectangular sheet of paper measuring b units long and h units high, where $b < h$. Rolling it into a cylinder one way yields a cylinder that is h units high, rolling it into a cylinder the other way yields a cylinder that is b units high. These two cylinders have equal lateral surface areas, but they enclose different volumes. The tall cylinder encloses a volume $V_{tall} = \frac{b^2 \cdot h}{4\pi}$, and the shorter cylinder encloses a volume of $V_{short} = \frac{h^2 \cdot b}{4\pi}$. Since $b < h$, one has $V_{tall} < V_{short}$.

In the previous example, the volume V was found to be 96π cm^3, and the total surface area A_{total} was found to be 96π cm^2. These two quantities are very different even though both have the number 96π. Note that the volume is measured in cubic centimeters and the total surface area is measured in square centimeters.

Spheres.   The formulas for both the **volume V of a sphere** and **the surface area S of a sphere** may be obtained using calculus. For a sphere of radius r, these formulas are $V = (4/3)\pi r^3$ and $S = 4\pi r^2$, respectively. Figure 2.6.6 illustrates a classroom activity that can be used to demonstrate the ratio of volume of a right circular cylinder of

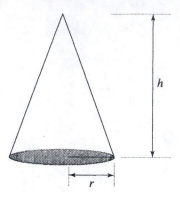

FIGURE 2.6.10 A right circular cone has height h and radius of base r. The volume is one-third the area of the base times the height h. Thus, $V = \frac{1}{3} \cdot \pi \cdot r^2 \cdot h$. The slant height l is the length of a line segment on the cone's surface from the vertex to the circle on the base. Using the Pythagorean theorem, one finds that $l = \sqrt{r^2 + h^2}$. The lateral surface area is given by $A_{\text{lateral}} = \pi r l = \pi r \sqrt{r^2 + h^2}$. The total surface area is the lateral surface area A_{lateral} plus the area of the base. Thus, $A_{\text{total}} = \pi r l + \pi r^2$.

height $h = 2r$ to the volume of a sphere of radius r is $3 : 2$.

$$V = \frac{4\pi r^3}{3} \quad \text{Volume of a sphere.}$$

$$S = 4\pi r^2 \quad \text{Surface area of a sphere.}$$

EXAMPLE Assuming the Earth is a sphere of radius 4,000 miles, find the volume V of the Earth in cubic miles and the surface area S of the Earth in square miles.

Solution   The volume V is given by

$$V = \frac{4 \cdot \pi \cdot r^3}{3} = \frac{4 \cdot \pi \cdot (4{,}000 \text{ miles})^3}{3} \approx 268{,}000{,}000{,}000 \text{ mi}^3.$$

The surface area S is given by

$$S = 4 \cdot \pi \cdot r^2 = 4 \cdot \pi \cdot (4{,}000 \text{ miles})^2 \approx 201{,}000{,}000 \text{ mi}^2. \qquad \blacksquare$$

EXAMPLE Let a first solid with volume V_{cl} be a right circular cylinder with radius of base r and height $h = 2r$. Let a second solid with volume V_{sp} be a sphere of radius r. Let a third solid with volume V_{cn} be a right circular cone with radius of base r and height $h = 2r$. Verify that the ratio $V_{\text{cl}} : V_{\text{sp}} : V_{\text{cn}}$ is given by $3 : 2 : 1$.

Solution   Use the formulas for the volumes of these solids to obtain $V_{\text{cl}} = \pi r^2 h = \pi r^2 (2r) = 2\pi r^3$ for the cylinder, $V_{\text{sp}} = \left(\frac{4}{3}\right) \pi r^3$ for the sphere, and $V_{\text{cn}} = \left(\frac{1}{3}\right) \pi r^2 h = \left(\frac{1}{3}\right) \pi r^2 (2r) = \left(\frac{2}{3}\right) \pi r^3$ for the cone. Then, $V_{\text{cl}} : V_{\text{sp}} : V_{\text{cn}}$ is given by $2\pi r^3 : \left(\frac{4}{3}\right) \pi r^3 : \left(\frac{2}{3}\right) \pi r^3$. Multiplying through by $\frac{3}{2\pi r^3}$ yields $3 : 2 : 1$, as desired. $\qquad \blacksquare$

EXERCISES 2.6

1. Answer questions 18–22 shown in Figure 2.6.2.
2. Find the volumes and surface areas of the prisms shown in Figure 2.6.5.
3. A given cube has each edge of 3-foot length.
   a. Find the volume of this cube.
   b. Find the surface area of this cube.
4. A can is in the shape of a right circular cylinder with a height of 5 inches and a radius of base equal to 3 inches. Find the volume enclosed by this can in cubic inches.
5. Make at least two different nets for a regular tetrahedron where each face is an equilateral triangle with sides of 1-inch length.
6. A right circular cylinder is to have a height of 4 inches and a base with a radius of .5 inches.
   a. Make a net for this surface.
   b. Find the lateral surface area of this cylinder.
   c. Find the total surface area of this cylinder.
   d. Find the volume of this cylinder.
7. A right circular cone has slant height of 5 inches and base of radius 1 inch.
   a. Find the lateral surface area of this cone.
   b. Find the total surface area of this cone.
8. A right rectangular prism (box) has dimensions given by 3 ft $\times$ 2 ft $\times$ 2 ft.
   a. Find the volume of the box in ft^3.
   b. Find the surface area of the box in ft^2.
9. A right rectangular prism (box) has dimensions given by 60 cm $\times$ 30 cm $\times$ 20 cm.
   a. Find the volume of the box in cm^3, and find the surface area in cm^2.
   b. Using 1 in = 2.54 cm, find the volume of the box using in^3, and find the surface area using in^2.
10. A storage tank in the shape of a right circular cylinder has a height of 10 feet and a base with diameter of 6 feet.
   a. Find the volume of this storage tank.
   b. Find the lateral surface area of this storage tank.
   c. Find the total surface area of this storage tank.
11. Find the volume of a right circular cone with a base of radius 6 feet and a height of 8 feet.
12. A triangular prism has an isosceles right triangle for bases where the legs of the triangles are 2 feet long. Assume the height of the prism is 3 feet.
   a. Assuming the prism is a right prism, find the volume of the prism.
   b. Assuming the prism is a right prism, find the surface area of the prism.
   c. If the prism is oblique, would you be able to calculate the volume without any additional information? Explain your reasoning.
13. A right square prism has vertices in xyz-space at the following eight points:

$$A_0 = (0,0,0) \qquad A_1 = (0,0,4)$$
$$B_0 = (1,0,0) \qquad B_1 = (1,0,4)$$
$$C_0 = (1,1,0) \qquad C_1 = (1,1,4)$$
$$D_0 = (0,1,0) \qquad D_1 = (0,1,4)$$

    a. Find the volume of the box (in cubic units).

    b. Find the surface area of the box (in square units).

14. An oblique square prism has vertices in xyz-space at the following eight points:

$$A_0 = (0,0,0) \qquad A_2 = (3,0,4)$$
$$B_0 = (1,0,0) \qquad B_2 = (4,0,4)$$
$$C_0 = (1,1,0) \qquad C_2 = (4,1,4)$$
$$D_0 = (0,1,0) \qquad D_2 = (3,1,4)$$

The points A_0, B_0, C_0, and D_0 yield a square in the plane $z = 0$, just as in the previous exercise. The other four vertices, A_2, B_2, C_2, and D_2, yield a square in the plane $z = 4$, which has been "shifted" 3 units in the positive x direction.

    a. Find the volume of this oblique prism (in cubic units).

    b. Find the surface area of this oblique prism (in square units).

    c. Compare your answer to part **a** of this exercise to part **a** of the previous exercise.

Classroom Discussion 2.6.2

Students on all levels are often confused about the formula for the lateral surface area of a right circular cone. In particular, they often don't like the slant height l in the formula $A_{\text{lateral}} = \pi r l = \pi r \sqrt{r^2 + h^2}$. One activity is to have one or more students construct a cone and then cut along a generator from vertex to base and lay the cone out flat. They end up with a part of a circle of radius l and can then measure the central angle corresponding to the portion of the circle covered by the flattened cone. The central angle should come out as $2\pi \cdot \left(\dfrac{r}{l}\right)$, measured in radians. Using this central angle in the formula for the sector of a circle, one obtains

$$A_{\text{lateral}} = \theta \cdot \left(\frac{l^2}{2}\right) = \pi r l. \quad \blacklozenge$$

CHAPTER 2 REVIEW

This chapter covered congruence, parallelism, quadrilaterals, areas, circles, volumes, and surface area. Two triangles are defined to be congruent if all corresponding sides are congruent and if all corresponding angles are congruent. Thus, to prove that two triangles are congruent directly from the definition requires showing that six different pairs of things are congruent. One can shorten the process of showing that two triangles are congruent by using one of the following criteria that establishes congruence for triangles: SAS, SSS, ASA, and SAA (=AAS). Congruence is an equivalence relation and thus has much in common with equality.

    The parallel postulate is fundamental to Euclidean geometry. Among other things, it allows one to show that all triangles have an angle sum of $180°$ (π radians). It is also fundamental to the definition and investigation of important quadrilaterals such as parallelograms, rectangles, squares, and trapezoids.

    Increasingly, the topics of area, volume, and surface area are being introduced to students before high school. These are all topics that virtually everyone must learn and deal with. This chapter covered area for many common figures such as rectangles, parallelograms, triangles, trapezoids, and circles. Volumes of boxes, prisms, cones, and spheres have been covered, as have surface areas of several common solids.

Selected formulas :

Area of rectangle: $\qquad A = b \cdot h$

Area of parallelogram: $\qquad A = b \cdot h$

Area of triangle: $\qquad A = \dfrac{b \cdot h}{2}$

Heron's formula $\qquad A = \sqrt{s(s - a)(s - b)(s - c)}$

Area of trapezoid: $\qquad A = \dfrac{(b_1 + b_2) \cdot h}{2}$

Pick's theorem: $\qquad A = I + \dfrac{B}{2} - 1$

Inscribed angle theorem: $\qquad \theta = \dfrac{m(\text{intercepted arc})}{2}$

Circumference of a circle: $\qquad C = \pi \cdot d = 2 \cdot \pi \cdot r$

Area of a circle: $\qquad A = \pi \cdot r^2$

Area of circular sector: $\qquad A = \dfrac{\theta \cdot r^2}{2}$

Volume of a box: $\qquad V = \ell \cdot w \cdot h$

Volume of a prism: $\qquad V = B \cdot h = (\text{area of base}) \cdot h$

Volume of a right circular cylinder: $\qquad V = \pi \cdot r^2 \cdot h$

Lateral surface area of a right circular cylinder: $\qquad A_{\text{lateral}} = 2 \cdot \pi \cdot r \cdot h$

Volume of a cone or pyramid: $\qquad V = \dfrac{1}{3} \cdot B \cdot h = \dfrac{1}{3}$
$$\cdot (\text{area of base}) \cdot (\text{height})$$

Volume of a right circular cone: $\qquad V = \dfrac{\pi \cdot r^2 \cdot h}{3}$

Lateral surface area of a right circular cone: $\qquad A_{\text{lateral}} = \pi \cdot r \cdot \ell$

Volume of a sphere: $\qquad V = \dfrac{4 \cdot \pi \cdot r^3}{3}$

Surface area of a sphere: $\qquad S = 4 \cdot \pi \cdot r^2$

CHAPTER 2 REVIEW EXERCISES

1. The triangle $\triangle ABC$ has sides of lengths $AB = 5$ cm, $BC = 7$ cm, and $CA = 3$ cm. List the angles of this triangle from smallest to largest.

2. The triangle $\triangle DEF$ has $m(\angle D) = 26°$ and $m(\angle E) = 34°$. List the sides of this triangle from smallest to largest.

3. Given that the point P lies on the perpendicular bisector of the segment $\overline{AB}$ and that $PA = 7$ inches, find PB.

4. The math team at Bright Middle School has eight members. On a national test, half of them scored 100 percent and half of them scored 99 percent. Let S be the set of members of this math team, and let $a \sim b$ mean that a and b scored the same on the math test. Here, a and b represent members (not necessarily distinct) of the math team.
   a. Is this relation reflexive?
   b. Is this relation symmetric?
   c. Is this relation transitive?
   d. Is this an equivalence relation?

5. The sum of the measures of the interior angles of a polygon is $360°$. How many sides does this polygon have?

6. The sum of the measures of the interior angles of a polygon is $720°$. How many sides does this polygon have?

7. A pentagon has four angles that measure $100°$ each. Find the measure of the fifth angle.

8. Assume that the two lines ℓ_1 and ℓ_2 in the following diagram are parallel with transversal ℓ_3. If the measure of angle x is $115°$, find the measures of the angles y, z, w, e, f, g, and h.

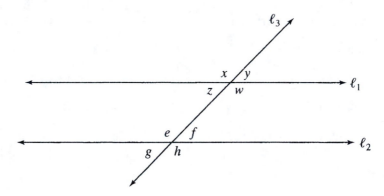

9. Assume that $ABCD$ in the following diagram is a parallelogram and that the points A, B, and F are collinear. If $m(\angle D) = 120°$ and $m(\angle F) = 25°$, find the measures of the angles $m(\angle A), m(\angle ABC), m(\angle BCD), m(\angle CBF)$, and $m(\angle BCF)$.

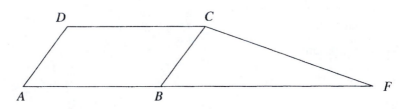

10. Find the area in square units of this trapezoid.

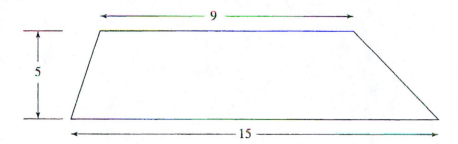

11. Find the area of the triangle $\triangle ABC$ in the xy-plane with vertices at $A = (7, -1), B = (-1, 2),$ and $C = (7, 7).$

12. The parallelogram $DEFG$ in the xy-plane has vertices given by $D = (1, 1), E = (7, 1), F = (8, 4),$ and $G = (2, 4).$ Find the area of this parallelogram.

13. Find the area enclosed by the quadrilateral $ABCD$ where $A = (0, 0), B = (2, 2), C = (2, 8),$ and $D = (0, 10).$

14. A rubber band on a geoboard encloses a region with nine boundary pegs and one interior point. What is the area of this region?

15. A circle with a radius of 2 feet has a given central angle of $60°$. Find the length of the arc that this angle intercepts.

16. A circle with a radius of 18 inches has a given central angle of $10°$. Find the area of the sector that this central angle cuts from the circle.

17. A given box has edges of lengths 2 feet, 20 inches, and 25 inches.
   a. Find the volume of this box in cubic inches.
   b. Find the surface area of this box in square inches.

18. A can is in the shape of a right circular cylinder with a height of 12 cm and a radius of base equal to 6 cm. Find the volume enclosed by this can in cubic centimeters.

19. Prove that a triangle in the Euclidean plane can have at most one obtuse angle.

20. Prove that a triangle is equilateral iff it is equiangular.

21. Given a triangle $\triangle ABC$, prove that the perpendicular bisectors of the three sides all meet at a point. (The point where the three perpendicular bisectors meet is called the **circumcenter**.)

22. In the diagram at the top of page 90, assume $\triangle ABC$ is isosceles with $AB = AC$ and assume that $BD = CE$. Here $D \in \overleftrightarrow{AB}$ and $E \in \overleftrightarrow{AC}$.
   Do only one of a or b. (Note: One of a and b is true and one is false.)
   a. Prove that $\overleftrightarrow{BC}$ is parallel to $\overleftrightarrow{DE}$.
   b. Prove that $\overleftrightarrow{BC}$ is not parallel to $\overleftrightarrow{DE}$.

23. Prove that the diagonals of a parallelogram bisect each other (i.e., prove Theorem 2.3.2).

24. Prove that a quadrilateral is a parallelogram iff each pair of consecutive angles are supplementary (i.e., prove Theorem 2.3.3).

25. Prove that the perpendicular bisector of a chord passes through the center of the circle.

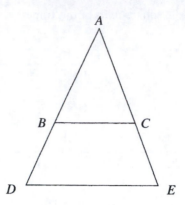

26. An **isosceles trapezoid** is a proper trapezoid with the two opposite sides that are not parallel being of equal length. Let $ABCD$ be an isosceles trapezoid with $\overline{AB}$ parallel to $\overline{DC}, AB > DC$, and $AD = BC$ as shown here.
   a. Prove that the base angles of an isosceles trapezoid are congruent (i.e., prove that $\angle A \cong \angle B$).
   b. Prove that the summit angles of an isosceles trapezoid are congruent (i.e., prove that $\angle C \cong \angle D$).

RELATED READING FOR CHAPTER 2

Beckmann, S. *Mathematics for Elementary Teachers, Vol. II: Geometry and Other Topics.* Preliminary Edition. New York: Addison Wesley, 2003.

Billstein, R. and J. Williamson. *Math Thematics: Book 1.* Evanston, IL: McDougal Littell, 1999.

Blue, M. N., et al. *Navigating through Geometry in Grades 6–8.* Reston, VA: National Council of Teachers of Mathematics, 2002.

Collier, C. P. *Geometry for Teachers.* 3rd ed. Prospect Heights, IL: Waveland Press Inc., 1994.

Geometer's Sketchpad [Software]. Designed by Nicholas Jackiw. Berkeley, CA: Key Curriculum Press, 1991.

Kleiman, G. "Designing Spaces." In *MathScape: Seeing and Thinking Mathematically.* Grade 6 materials. New York: Glencoe/McGraw-Hill, 1999.

Kleiman, G. "Getting in Shape." In *MathScape: Seeing and Thinking Mathematically.* Grade 7 materials. New York: Glencoe/McGraw-Hill, 1999.

Kleiman, G. "From The Ground Up." In *MathScape: Seeing and Thinking Mathematically.* Grade 7 materials. New York: Glencoe/McGraw-Hill, 1999.

Kleiman, G. "Shapes and Space." In *MathScape: Seeing and Thinking Mathematically.* Grade 8 materials. New York: Glencoe/McGraw-Hill, 1999.

Lappan, G., et al. *Filling and Wrapping.* Connected Mathematics Project. Needham, MA: Pearson/Prentice-Hall, 2004.

O'Daffer, P. G., and Stanley R. C. *Geometry: An Investigative Approach*. Reading, MA: Addison Wesley Longman, 1997.

Romberg, T. A., et al. *Reallotment*. Mathematics in Context. Chicago, IL: Encyclopedia Britannica Educational Corporation, 1998.

Romberg, T. A., et al. *Triangles and Beyond*. Mathematics in Context. Chicago, IL: Encyclopedia Britannica Educational Corporation, 1998.

Rubenstein, R. N., et al. *Teaching and Learning Middle Grades Mathematics*. Emeryville, CA: Key College Publishing, 2004.

Wenninger, M. J. *Polyhedron Models*. New York: Cambridge University Press, 1971.

CHAPTER 3

Similarity

3.1 **SIMILAR POLYGONS**
3.2 **APPLICATIONS OF SIMILAR TRIANGLES**
3.3 **PYTHAGOREAN THEOREM**
3.4 **AREA AND PERIMETER OF SIMILAR FIGURES**
3.5 **SIMILARITY FOR MORE GENERAL FIGURES**

We introduced congruence in Chapter 2 as a fundamental tool in the study of geometry. Congruent figures can be rigidly moved to coincide with each other. In this chapter, we introduce similarity as another important tool. Similar figures are the same shape but do not necessarily have the same size. Usually the second figure will be either a larger or smaller version of the first. However, similar figures may be the same size. When this happens, the figures are both similar and congruent. We define similarity for polygons in Section 3.1, and in Section 3.5 we provide a general definition of similarity for figures that are not necessarily polygons.

3.1 SIMILAR POLYGONS

Rep-tiles. There are many activities that one can do with similarity. One of these is to investigate rep-tiles. A **rep-tile** is a figure whose congruent copies can be put together to form a larger figure that is similar to the original figure. Figures 3.1.1 and 3.1.2 are from a seventh-grade Connected Mathematics Project unit. In particular, Figure 3.1.1 illustrates two different ways one may put together four copies of a triangle to get a larger triangle that is similar.

Classroom Discussion 3.1.1

The two methods illustrated in Figure 3.1.1 show that the original triangle is a rep-tile. The first method shown (middle triangle) works for 30°-60°-90° right triangles but does not generalize to the collection of all triangles. The second

92

Building with Rep-tiles

A rep-tile is a shape whose copies can be put together to make a larger, similar shape. The small triangle below is a rep-tile. The two large triangles are formed from copies of this rep-tile. Can you explain why each large triangle is similar to the small triangle?

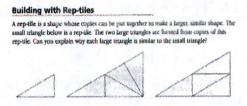

FIGURE 3.1.1 A given smaller 30°-60°-90° right triangle is used two different ways to get larger 30°-60°-90° right triangles similar to the original triangle. Reproduced from page 29 of *Stretching and Shrinking* in the Connected Mathematics Project grade 7 materials.

In this problem, your challenge is to figure out which of the shapes below are rep-tiles.

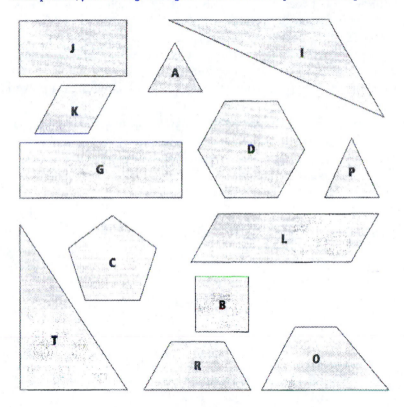

FIGURE 3.1.2 All but three of these polygons are rep-tiles. Reproduced from page 30 of *Stretching and Shrinking* in the Connected Mathematics Project grade 7 materials.

method shown (last triangle) does generalize. One possible exploration for college students is to have teams of two or three students work together to see how one may justify using the second method for triangles that are not assumed to be special. On the middle or high school levels, a mathematics teacher could have students work in teams with right triangles that are not 30°-60°-90° to discover that the first method does not work in general. For example, one can have students work with either $40°$-$50°$-$90°$ or $20°$-$70°$-$90°$ right triangles to discover why the first method fails for such triangles. ◆

Classroom Connection 3.1.1

One activity from the Connected Mathematics Project curriculum is indicated in Figure 3.1.2. Thus, one can have middle school students make several copies of each of the polygons in Figure 3.1.2 and then have them put some copies of each figure together to make a larger similar figure. Students should be able to convince themselves that all but C, D, and O are rep-tiles. ◆

Similar Polygons.   Let G_1 and G_2 be polygons and assume that there is a correspondence between their vertices, such that each pair of vertices joined by a side of G_1 corresponds to a pair of vertices joined by a side of G_2. The correspondence yields congruent polygons if each pair of corresponding angles is congruent and if each pair of corresponding sides is congruent. For similarity, one requires that the corresponding angles be congruent, but the corresponding sides are only required to be in the same proportion. The definition of similarity for polygons is given here.

> **Definition 3.1.1** Polygons G_1 and G_2 are similar if there is a correspondence between them with their corresponding angles congruent and with the ratio S of corresponding sides (of polygon G_2 to polygon G_1) the same for all pairs of corresponding sides. The ratio S is called the **scale factor**.

The notation for similarity is $\sim$. If $G_1 = P_1 P_2 \ldots P_n$, and $G_2 = Q_1 Q_2 \ldots Q_n$ are similar, one may write either $G_1 \sim G_2$ or $P_1 P_2 \ldots P_n \sim Q_1 Q_2 \ldots Q_n$.

$$P_1 P_2 \ldots P_n \sim Q_1 Q_2 \ldots Q_n$$

$$\angle P_i \cong \angle Q_i \text{ and } S = \frac{Q_i Q_{i+1}}{P_i P_{i+1}} \text{ for all } i = 1, 2, \ldots, n$$

$$S = \text{scale factor} = \frac{\text{length of side in second polygon}}{\text{length of corresponding side in first polygon}}$$

Thus, $\triangle ABC \sim \triangle A'B'C'$ with scale factor S means

$$\angle A \cong \angle A',$$

$$\angle B \cong \angle B',$$

$$\angle C \cong \angle C',$$

$$S = A'B'/AB = A'C'/AC = B'C'/BC.$$

Just as with congruence, the notation $\triangle ABC \sim \triangle A'B'C'$ implies that the vertex A corresponds to vertex A', B to B', and C to C' (see Figure 3.1.3).

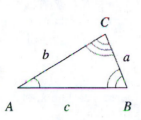
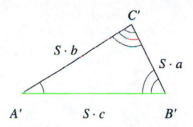

FIGURE 3.1.3 In order for the triangles $\triangle ABC$ and $\triangle A'B'C'$ to be similar, corresponding angles must be congruent, and corresponding sides must have the same ratio S. In this illustration, $c = AB$, $b = AC$, and $a = BC$.

EXAMPLE Let $\triangle ABC \sim \triangle A'B'C'$ with scale factor $S = 7$. If $AB = 3$, $BC = \sqrt{5}$, and $AC = 4$, find the lengths of the sides of $\triangle A'B'C'$.

Solution Using $S = \dfrac{A'B'}{AB} = \dfrac{A'C'}{AC} = \dfrac{B'C'}{BC}$, one obtains $A'B' = S \cdot AB$, $B'C' = S \cdot BC$ and $A'C' = S \cdot CA$. Hence, $A'B' = 7 \cdot 3 = 21$, $B'C' = 7 \cdot \sqrt{5}$, and $A'C' = 7 \cdot 4 = 28$. ∎

EXAMPLE Let the quadrilateral $ABCD$ have vertices at the points $A = (0,0)$, $B = (3,0)$, $C = (3,2)$, and $D = (0,2)$. Let the quadrilateral $EFGH$ have vertices at $E = (0,-12)$, $F = (0,-6)$, $G = (-4,-6)$, and $H = (-4,-12)$. Prove that quadrilaterals satisfy the similarity relation $ABCD \sim EFGH$.

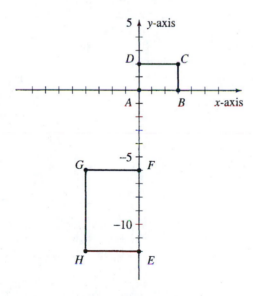

Proof. Show that corresponding angles are congruent and that corresponding sides have a common ratio S. First note that each of these quadrilaterals has right angles at all four vertices, since each quadrilateral has each side parallel to either the x-axis or the y-axis. Thus, $\angle A \cong \angle E$, $\angle B \cong \angle F$, $\angle C \cong \angle G$, and $\angle D \cong \angle H$. Using the distance formula for the xy-plane, one finds $AB = CD = 3$, $AD = BC = 2$, $EF = GH = 6$, and $FG = HE = 4$. Thus,

$$\frac{EF}{AB} = \frac{FG}{BC} = \frac{GH}{CD} = \frac{HE}{DA} = 2.$$

One concludes that $ABCD \sim EFGH$, since corresponding angles are congruent and corresponding sides have the same ratio $S = 2$. ∎

It follows easily from the definition of similarity that if one has two polygons G_1 and G_2, such that $G_1 \sim G_2$ with a scale factor of S, then one also has $G_2 \sim G_1$ with a scale factor of $\frac{1}{S} = S^{-1}$. Furthermore, $S = 1$ exactly for congruent polygons. In fact, one could define polygons to be congruent if they are similar and have a scale factor of $S = 1$.

EXAMPLE Assume the quadrilaterals $WXYZ$ and $W'X'Y'Z'$ satisfy $WXYZ \sim W'X'Y'Z'$ with a scale factor of $S = 3$. Find the lengths of the sides of the quadrilateral $WXYZ$ given that $W'X' = 6$, $X'Y' = 12$, $Y'Z' = 18$, and $Z'W' = 30$.

Solution   Since $S = 3$, the first quadrilateral will have sides that are $\frac{1}{S} = \frac{1}{3}$ as long as the corresponding sides of the second quadrilateral. Thus, using $WX = \left(\frac{1}{S}\right)W'X'$, $XY = \left(\frac{1}{S}\right)X'Y'$, $YZ = \left(\frac{1}{S}\right)Y'Z'$, and $ZW = \left(\frac{1}{S}\right)Z'W'$ yields $WX = 2$, $XY = 4$, $YZ = 6$, and $ZW = 10$. ∎

EXERCISES 3.1

1. Assume that $\triangle ABC \sim \triangle A'B'C'$ with $AB = 4$, $BC = 3$, $CA = 4$, and $A'B' = 12$.
   a. Find the scale factor S and explain your reasoning.
   b. Find $B'C'$ and explain your reasoning.
   c. Find $C'A'$ and explain your reasoning.

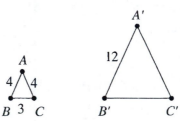

2. Assume that $\triangle EFG \sim \triangle E'F'G'$ with a scale factor of $S = 2$. If $EF = 5$, $FG = 3$, and $GE = 4$, find all of the lengths of the sides of $\triangle E'F'G'$. Also, explain the reasoning you used in finding each of the lengths of the sides of $\triangle E'F'G'$.

 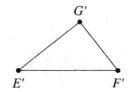

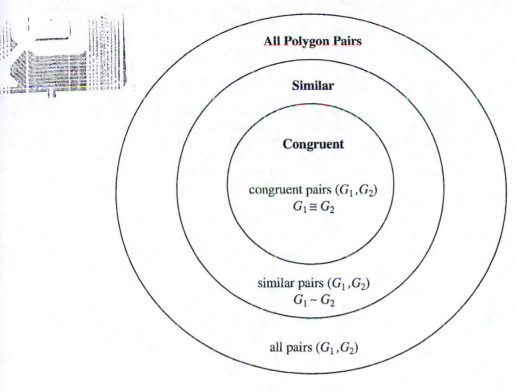

All Polygon Pairs

Similar

Congruent

congruent pairs (G_1, G_2)
$G_1 \cong G_2$

similar pairs (G_1, G_2)
$G_1 \sim G_2$

all pairs (G_1, G_2)

FIGURE 3.1.4 In this illustration, each point represents an ordered pair (G_1, G_2) of polygons, and each circle represents a collection of pairs (G_1, G_2). The circles are nested to indicate subset relations. Usually, two polygons will not be similar under any correspondence (outer ring). If they are similar, they need not be congruent (middle ring). However, if they are congruent, then they are also similar (innermost circle).

3. Let $\triangle XYZ \sim \triangle X'Y'Z'$ with a scale factor of $S_1 = 6$ and $\triangle X'Y'Z' \sim \triangle X''Y''Z''$ with a scale factor of $S_2 = 5$. What is the scale factor for $\triangle XYZ \sim \triangle X''Y''Z''$?

$$\triangle XYZ \sim \triangle X'Y'Z' \quad \triangle X'Y'Z' \sim \triangle X''Y''Z''$$
$$S_1 = 6 \qquad\qquad S_2 = 5$$

4. Let $\triangle RST \sim \triangle R'S'T'$ with a scale factor of $S = 3$. If $R'S' = 21$, $S'T' = 18$, and $T'R' = 24$, find all of the lengths of the sides of $\triangle RST$. Also, explain the reasoning you used in finding each of the lengths of the sides of $\triangle RST$.

5. Use a diagram to illustrate that all parallelograms are rep-tiles.

6. Given three n-gons, $P_1P_2 \ldots P_n$, $Q_1Q_2 \ldots Q_n$, and $R_1R_2 \ldots R_n$, assume that $P_1P_2 \ldots P_n \sim Q_1Q_2 \ldots Q_n$ with a scale factor of S_1 and assume that $Q_1Q_2 \ldots Q_n \sim R_1R_2 \ldots R_n$ with a scale factor of S_2. Find the scale factor S_3 for $P_1P_2 \ldots P_n \sim R_1R_2 \ldots R_n$ and justify your answer.

7. Assume the two quadrilaterals $KLMN$ and $K'L'M'N'$ satisfy $KLMN \sim K'L'M'N'$. Also, let $K = (0, 3)$, $L = (0, 0)$, $M = (3, 0)$, $N = (1, 1)$, $K' = (-9, 9)$, $L' = (0, 0)$, and $M' = (9, 9)$.
   a. Find the scale factor S.
   b. Find the point N'.

8. Let G_1 and G_2 be polygons with $G_1 \sim G_2$. If G_1 is a rep-tile, does it follow that G_2 is also a rep-tile? Explain your reasoning.

9. Let G_1 and G_2 be polygons. In each of the following cases, indicate where in Figure 3.1.4 you would place a dot to represent (G_1, G_2), assuming the following conditions. Place each dot in one of the following: innermost circle, middle ring, and outer ring.
   a. G_1 and G_2 are both isosceles right triangles but have different areas.
   b. G_1 and G_2 are polygons with different numbers of sides.
   c. G_1 and G_2 are both squares, and they both have the same area.
   d. G_1 is a rectangle with two sides of length 3 and two sides of length 5. G_2 is a rectangle with two sides of length 9 and two sides of length 25.

10. Give an example of two polygons with a correspondence of their sides such that all pairs of corresponding sides have the same ratio of $1/2$, but such that the polygons are *not* similar. (Hint: Use polygons with more than three sides.)

11. Give an example of two polygons with a correspondence of their sides such that all angles at corresponding vertices are congruent, but such that the polygons are *not* similar. (Hint: Use polygons with more than three sides.)

12. Prove that any two equilateral triangles are similar.

13. Prove that any two squares are similar.

14. Let G_1 and G_2 be rectangles with the length of the diagonals of G_2 equal to twice the length of the diagonals of G_1. Can one deduce that these are similar polygons? Why or why not?

Classroom Connection 3.1.2

Using the Geometer's Sketchpad (GSP), have middle or high school students construct two similar polygons G_1 and G_2, with $G_1 \sim G_2$ having a scale factor of 2. It is then easy for them to check that G_2 has four times the area of G_1. ◆

Classroom Connection 3.1.3

Have two middle or high school students of different heights stand next to each other and take a picture of them. Given the height of one student, use similarity and the picture to calculate the height of the second student. The ratio of the heights of the students will be the same as the ratio of the heights of the two figures in the picture. ◆

3.2 APPLICATIONS OF SIMILAR TRIANGLES

Results on similar triangles can be used to measure heights of buildings and other objects. Figure 3.2.1 illustrates the shadow method, and Figure 3.2.2 illustrates the mirror method. Both of these methods are well suited for student activities, and both are applications of the AA similarity theorem that we cover in Theorem 3.2.1.

EXAMPLE Using the shadow method illustrated in Figure 3.2.1, a student measures the shadow of the meterstick and finds the length to be 20 cm. The length of the shadow of the tall object is then immediately measured and found to be 140 cm. Find the height of the tall object.

Using Shadows to Find Heights

If an object is outdoors, you can use shadows to help estimate its height. The diagram below illustrates how the method works.

On a sunny day, an object casts a shadow. If you hold a meterstick perpendicular to the ground, it will also cast a shadow. The diagram below shows two triangles. One is formed by an object, its shadow, and an imaginary line. The other is formed by a meterstick, its shadow, and an imaginary line. These two triangles are similar.

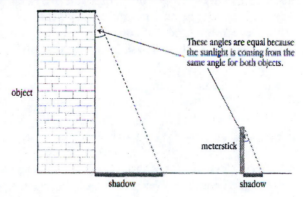

To find the height of the object, you can measure the lengths of the two shadows and apply what you know about similar triangles.

FIGURE 3.2.1 This illustrates how one can use shadows to find heights. It is thought that Thales (c. 624 BC–546 BC) used this method to estimate the height of an Egyptian pyramid. Reproduced from page 59 of *Stretching and Shrinking* in the Connected Mathematics Project grade 7 materials.

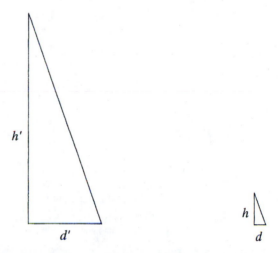

Solution   Let $h = 1$ m be the height of the meterstick, and let h' be the height of the tall object in meters. Let $d = 20$ cm be the length of the shadow of the meterstick, and let $d' = 140$ cm be the length of the shadow of the tall object. Then using similar triangles, one obtains

$$\frac{h'}{h} = \frac{d'}{d},$$

which becomes

$$\frac{h'}{1\text{ m}} = \frac{140\text{ cm}}{20\text{ cm}}.$$

Hence, the answer is $h' = 7$ m. Notice that one may measure the two heights in one set of units (i.e., both in meters) and the two shadow lengths in another set of units (i.e., both in centimeters). Of course, you will obtain the same answer if you measure all four lengths in the same set of units (e.g., all in meters).   ∎

EXAMPLE   A student is measuring the height of a brick wall using the mirror method illustrated in Figure 3.2.2. Assume the eyes of the student are 5 feet from the ground,

Using Mirrors to Find Heights

The shadow method is useful for estimating heights, but it only works outdoors and on a sunny day. In this problem, you will use a mirror to help estimate heights. The mirror method works both indoors and outdoors. All you need is a level spot near the object whose height you want to estimate.

The mirror method is illustrated below. Place a mirror on a level spot at a convenient distance from the object. Back up from the mirror until you can see the top of the object in the center of the mirror. The two triangles shown in the diagram are similar.

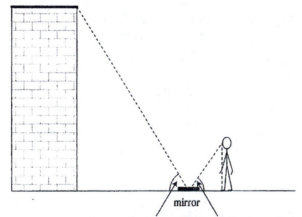

mirror

These angles are equal because light reflects off of a mirror at the same angle at which it hits the mirror.

To find the object's height, you need to measure three distances and then apply what you know about similar triangles.

FIGURE 3.2.2 This illustrates how one can use a mirror to find heights. Reproduced from page 61 of *Stretching and Shrinking* in the Connected Mathematics Project grade 7 materials.

the image is in the center of the mirror and located 6 feet from the wall's base, and the point on the ground right below the student's eyes is located 2 feet from the mirror's center. Find the height of the wall.

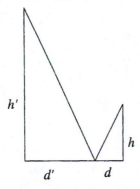

Solution   Let $h = 5$ ft be the height of the eyes above the ground, and let h' be the height of the wall. Let $d = 2$ ft be the distance from the point right below the student's eyes to the middle of the mirror, and let $d' = 6$ ft be the distance from the wall's base to the middle of the mirror. Then using similar triangles, one obtains

$$\frac{h'}{h} = \frac{d'}{d},$$

which becomes

$$\frac{h'}{5\text{ ft}} = \frac{6\text{ ft}}{2\text{ ft}}.$$

Hence, $h' = \dfrac{(6\text{ ft})\cdot(5\text{ ft})}{2\text{ ft}} = 15$ ft.   ■

Classroom Connection 3.2.1

Use the shadow method illustrated in Figure 3.2.1 on a sunny day to have middle or high school students estimate the height of a building. Apply this method by measuring the length of the building's shadow and at the same time measuring the length of the shadow of a yardstick held straight up. Students usually find this activity quite interesting. Use this method of estimating the height a building, and also the mirror method illustrated in Figure 3.2.2, to obtain two different estimates that can be compared.   ◆

Classroom Connection 3.2.2

The mirror method illustrated in Figure 3.2.2 for finding heights is a good activity for middle or high school students. It only requires rulers and small mirrors. The two key physical principles involved are that, in geometric optics, light rays travel in straight lines and that for mirrors, the angle of incidence is equal to the angle of reflection. How would you design classroom activities to illustrate these two physical principles?   ◆

The most well-known similarity theorem is the AA similarity theorem which is illustrated in Figure 3.2.3.

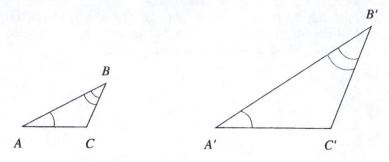

FIGURE 3.2.3 For triangles, AA implies similarity. Thus, $\angle A \cong \angle A'$ and $\angle B \cong \angle B'$ imply $\triangle ABC \sim \triangle A'B'C'$.

Theorem 3.2.1 (AA Similarity Theorem). *If two triangles have two pairs of corresponding angles that are congruent, then the triangles are similar. In particular, if* $\angle A \cong \angle A'$ *and* $\angle B \cong \angle B'$*, then* $\triangle ABC \sim \triangle A'B'C'$.

EXAMPLE   A first triangle has one angle of measure that is $75°$ and a second angle of measure that is $25°$. A second triangle has one angle of measure that is $80°$ and a second angle of measure that is $25°$. Are these triangles similar under some correspondence of their vertices?

Solution   The answer is yes. To see why the answer is yes, label the vertices of the first triangle A, B, and C, where $m(\angle A) = 75°$ and $m(\angle B) = 25°$. Label the vertices of the second A', B', and C', where $m(\angle B') = 25°$ and $m(\angle C') = 80°$. Then using the fact that the angle sum of a triangle is $180°$, one finds that $m(\angle A') = 180° - m(\angle B') - m(\angle C') = 180° - 25° - 80° = 75°$. Thus, $\angle A \cong \angle A'$ and $\angle B \cong \angle B'$, which shows that $\triangle ABC \sim \triangle A'B'C'$ using the AA similarity theorem. ∎

If two triangles have two sets of pairs of corresponding angles that are congruent, then the third pair of corresponding angles must also be congruent because the angle sum for triangles is $180°$. Thus, in Euclidean geometry, AA is equivalent to AAA for triangles.

The SAS similarity theorem is like the SAS postulate, except the two S's for the similarity result mean only that the ratio of corresponding sides is the same. This theorem is illustrated in Figure 3.2.4.

Theorem 3.2.2 (SAS Similarity Theorem). *If two triangles have two pairs of corresponding sides in the same ratio, and if the included angles determined by the pairs of sides are congruent, then the triangles are similar. In particular, if* $A'B'/AB = A'C'/AC$ *and* $\angle A \cong \angle A'$*, then* $\triangle ABC \sim \triangle A'B'C'$.

The SSS similarity theorem, illustrated in Figure 3.2.5, is like the SSS theorem for congruence except the three S's for the similarity result only mean the ratio of corresponding sides is the same.

Theorem 3.2.3 (SSS Similarity Theorem). *If two triangles have all three pairs of corresponding sides in the same ratio, then the triangles are similar. In particular, if* $A'B'/AB = A'C'/AC = B'C'/BC$*, then* $\triangle ABC \sim \triangle A'B'C'$.

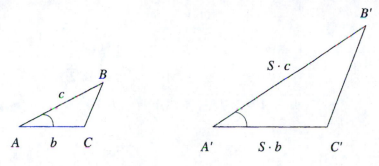

FIGURE 3.2.4 Illustration of the SAS similarity theorem. If $\angle A \cong \angle A'$, $A'B' = S \cdot AB$ and $A'C' = S \cdot AC$, then $\triangle ABC \sim \triangle A'B'C'$ with scale factor S. In this illustration, $c = AB$ and $b = AC$.

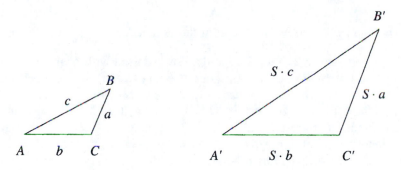

FIGURE 3.2.5 Illustration of the SSS similarity theorem. If $A'B' = S \cdot AB$, $A'C' = S \cdot AC$, and $B'C' = S \cdot BC$, then $\triangle ABC \sim \triangle A'B'C'$ with scale factor S. In this illustration, $a = BC$, $b = AC$, and $c = AB$.

EXAMPLE Let $\triangle ABC$ have vertices at $A = (-2, 0)$, $B = (2, 5)$, and $C = (3, 5)$. And assume that $\triangle XYZ$ has vertices at $X = (0, 4)$, $Y = (10, -4)$, and $Z = (10, -6)$. Do these triangles satisfy $\triangle ABC \sim \triangle XYZ$?

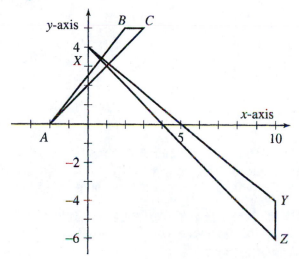

Solution   Using the distance formula yields

$$AB = \sqrt{(-2-2)^2 + (0-5)^2} = \sqrt{41}, \; BC = 1,$$

$$CA = \sqrt{(-2-3)^2 + (0-5)^2} = \sqrt{50} = 5\sqrt{2},$$

$$XY = \sqrt{(0-10) + (4-(-4))^2} = \sqrt{100+64} = \sqrt{164} = 2\sqrt{41},$$

$$YZ = 2, \text{ and}$$

$$ZX = \sqrt{(10-0)^2 + (-6-4)^2} = \sqrt{200} = 10\sqrt{2}. \text{ Thus,}$$

$$2 = \frac{XY}{AB} = \frac{YZ}{BC} = \frac{ZX}{CA}.$$

Hence, the SSS similarity theorem shows that $\triangle ABC \sim \triangle XYZ$ with $S = 2$. ▮

In Chapter 2, both SAS and SSS were used with triangles, and each S corresponded to a pair of congruent sides. It is easy to confuse these previous criteria with the SAS similarity theorem and the SSS similarity theorem, respectively. To avoid confusion, the convention is that SSS by itself will mean the congruence theorem and SSS similarity theorem will refer to Theorem 3.2.3. Of course, the same convention will hold when either using SAS by itself or when using the SAS similarity theorem. Unless the word *similarity* is used, the understanding is that one is referring to the congruence result.

An **altitude** of a triangle is a line segment that extends from a vertex to the line of the opposite side and is perpendicular to the opposite side. A given altitude may be contained in its triangle or may go outside its triangle. These two cases are illustrated here.

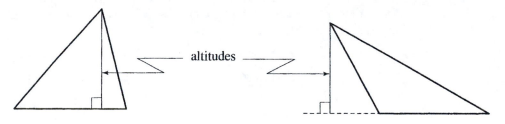

An altitude that is
contained in its triangle.

An altitude that is not
contained in its triangle.

If two triangles are similar with a scale factor S, then their altitudes are in the same ratio S. Thus, one has (length of corresponding altitude in second triangle) = $S \cdot$ (length of altitude in first triangle).

EXAMPLE   Assume that $\triangle ABC \sim \triangle A'B'C'$ with a scale factor of 7. If the altitude from C to $\overleftrightarrow{AB}$ in $\triangle ABC$ has a length of 4 units, what is the length of the altitude from C' to $\overleftrightarrow{A'B'}$ in $\triangle A'B'C'$?

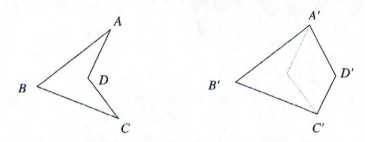

FIGURE 3.2.6 These quadrilaterals have corresponding pairs of sides that are congruent, but they are neither similar nor congruent. This shows that SSSS implies neither similarity nor congruence for quadrilaterals.

FIGURE 3.2.7 The two quadrilaterals shown here have corresponding pairs of angles that are congruent, but they are neither congruent nor similar. This shows that AAAA implies neither similarity nor congruence for quadrilaterals.

Solution   The length of the corresponding altitude in $\triangle A'B'C'$ is given by $S \cdot$ (4 units) = 7 · (4 units) = 28 units.   ■

It is natural to ask if there are results such as Theorems 3.2.1, 3.2.2, and 3.2.3 for quadrilaterals. Extensions of these theorems do exist but are somewhat more complicated than one might first guess. In particular, Figures 3.2.6 and 3.2.7 illustrate that for quadrilaterals, SSSS and AAAA imply neither congruence nor similarity.

EXERCISES 3.2

1. Given a first triangle $\triangle ABC$ and a second triangle $\triangle A'B'C'$ with $AB = 5$, $AC = 7$, $A'B' = 15$, and $\angle A \cong \angle A'$, then what must $A'C'$ be equal to for $\triangle ABC \sim \triangle A'B'C'$ to be valid?

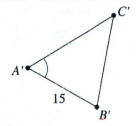

2. Let $\triangle HJK$ and $\triangle XYZ$ be given with $HJ = 3, JK = 2, KH = 4, XY = 4, YZ = 8$, and $ZX = 6$. Indicate which one of the following is correct and justify your answer.
   a. $\triangle HJK \sim \triangle XYZ$?
   b. $\triangle HJK \sim \triangle YZX$?
   c. $\triangle HJK \sim \triangle ZXY$?

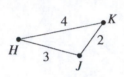

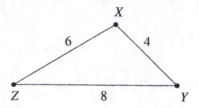

3. Assume a flagpole with a height of 20 feet casts a shadow of 35 feet, at the same time, assume a building casts a shadow of 140 feet. How tall is the building?

4. Given $\triangle LMN \sim \triangle PQR$ with $m(\angle L) + m(\angle M) = 150°$ and $m(\angle Q) + m(\angle R) = 120°$, find (a) $m(\angle L)$, (b) $m(\angle M)$, and (c) $m(\angle N)$.

5. Assume $\triangle DEF \sim \triangle D'E'F'$ with $DE = 6$ ft, $EF = 8$ ft, and $D'E' = 9$ ft. Find $E'F'$.

6. Prove that any two isosceles right triangles are similar.

7. Assume that $\triangle ABC \sim \triangle A'B'C'$ with a scale factor of 3. If the altitude from A' to $\overleftrightarrow{B'C'}$ in $\triangle A'B'C'$ has a length of 12 feet, what is the length of the altitude from A to $\overleftrightarrow{BC}$ in $\triangle ABC$?

8. A **regular polygon** is one with all sides that are congruent and all angles that are congruent. Thus, the regular triangles are equilateral triangles, and the regular quadrilaterals are squares. All equilateral triangles are similar to one another and all squares are similar to one another (see Exercises 3.1).

   a. Prove all regular pentagons are similar.

   b. Find a generalization of part a.

3.3 PYTHAGOREAN THEOREM

Pythagoras was a Greek mathematician and philosopher. He was born on Samos, an island in the Aegean Sea located near modern day Turkey. The exact dates of his life are not known with certainty. It is believed that he was born before 500 BC and that he died after 500 BC. He founded a society, now known as the Pythagoreans, that had a number of unusual customs. Among other things, this somewhat mystical society forbade the eating of beans. A fundamental belief of the Pythagoreans was that all numbers were, in effect, ratios of whole numbers. It is claimed that when some Pythagorean discovered that the square root of two was not such a ratio, then the Pythagoreans tried to keep this result secret. According to a story, which may or may not be true, after a certain society member divulged the secret, the society executed this particular individual.

Pythagoras and his followers were very interested in music and in astronomy and made some real advances in these fields as well as in mathematics. It seems that Pythagoras was the first to point out that the moon's orbit is not in the plane of the Earth's equator. The moon's orbit is inclined at an angle to that plane. Of course, Pythagoras is best remembered for the theorem named after him. Although the result of the Pythagorean theorem was evidently known before Pythagoras was born, this very important theorem, illustrated in Figure 3.3.1, goes by his name. Over the centuries there have been many different proofs of this theorem. In fact, one proof of this theorem was given by President Garfield.

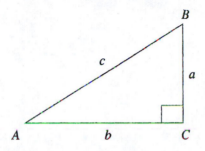

FIGURE 3.3.1 The Pythagorean theorem states that if $\triangle ABC$ has a right angle at vertex C, then $a^2 + b^2 = c^2$. The converse of the Pythagorean theorem is also valid. Thus, if one has a triangle $\triangle ABC$ satisfying $a^2 + b^2 = c^2$, then $\angle C$ is a right angle.

The Pythagorean theorem may be illustrated using area. Shown here is a right triangle and a copy of this triangle, where a square has been constructed on each side. This yields three squares that have respective areas a^2, b^2, and c^2.

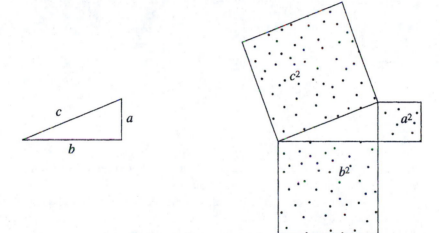

The Pythagorean theorem is the statement that for right triangles, the area c^2 of the square constructed on the longer (hypotenuse) side is equal to the sum of the areas a^2 and b^2 of the squares constructed on the legs.

Theorem 3.3.1 (Pythagorean Theorem). *If $\triangle ABC$ is a triangle with a right angle at C, then $a^2 + b^2 = c^2$.*

> **Proof.** Let D be the point on hypotenuse $\overline{AB}$ with $\overline{CD} \perp \overline{AB}$. Let $x = DB$ and let $y = AD$. Then $c = x + y$. Using the AA similarity theorem, one finds both that $\triangle ABC \sim \triangle ACD$ and that $\triangle ABC \sim \triangle CBD$.
>
> Using $\triangle ABC \sim \triangle CBD$ one finds that $BC/AB = BD/CB$.

Hence,
$$a/c = x/a,$$

which yields
$$a^2 = cx.$$

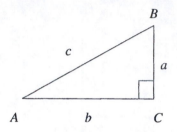 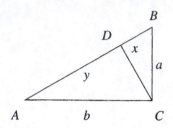

Using $\triangle ABC \sim \triangle ACD$ one finds that $AC/AB = AD/AC$.

Hence, $b/c = y/b,$

which yields $b^2 = cy.$

Using $a^2 = cx$ and $b^2 = cy$ one finds

$$a^2 + b^2 = cx + cy$$
$$a^2 + b^2 = c(x + y).$$

Since $x + y = c$, this becomes

$$a^2 + b^2 = c^2$$

as desired.

One of many alternative proofs of the Pythagorean theorem can be illustrated by starting with a square with each side of length $a + b$, then subdivide this square into a smaller inner square and into four right triangles as shown on the next page.

The inner square has each side of length c, and the four right triangles each have legs of lengths a and b and a hypotenuse of length c. The large square's area must be the sum of the areas of the triangles and of the inner square.

Area of the large square: $(a + b) \cdot (a + b) = a^2 + 2a \cdot b + b^2$

Area of inner square: c^2

Area of the four triangles: $4 \cdot \left(\frac{1}{2} \cdot a \cdot b \right) = 2a \cdot b$

Hence, using area of large square $=$ area of inner square $+$ area of triangles,

one finds $a^2 + 2a \cdot b + b^2 = c^2 + 2a \cdot b.$

Hence, $a^2 + b^2 = c^2$

as desired.

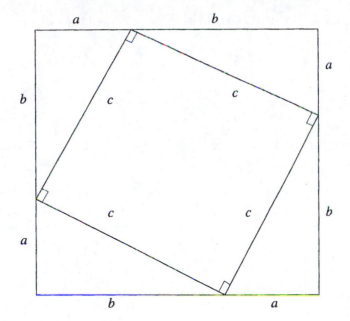

Much less familiar to students than the Pythagorean theorem is its converse. The converse follows using the SSS (congruence theorem) and the Pythagorean theorem.

Theorem 3.3.2 (Converse of the Pythagorean Theorem). *If triangle $\triangle ABC$ satisfies $a^2 + b^2 = c^2$, then $\angle C$ is a right angle.*

Proof. Given the triangle $\triangle ABC$ satisfying $a^2 + b^2 = c^2$, a new triangle $\triangle A'B'C'$ will be constructed. First construct a right angle at a point C' and then lay off lengths a and b to get points A' and B' with $a = B'C'$ and $b = A'C'$ as shown.

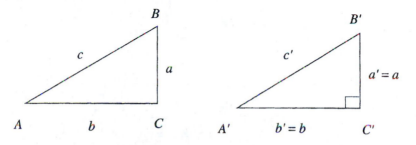

The second triangle, $\triangle A'B'C'$, has a right angle at C', and hence one may apply the Pythagorean theorem to $\triangle A'B'C'$ and obtain

$$(a')^2 + (b')^2 = (c')^2.$$

Thus, using $a' = a$, $b' = b$, and $a^2 + b^2 = c^2$, and because c and c' are positive, one finds

$$(a)^2 + (b)^2 = (c')^2$$

$$(c)^2 = (c')^2$$

$$c = c'.$$

Thus, all three corresponding pairs of sides of $\triangle ABC$ and $\triangle A'B'C'$ are congruent, and the two triangles must be congruent by SSS. Since $\angle C'$ and $\angle C$ are congruent and $\angle C'$ is a right angle, then $\angle C$ must also be a right angle, as desired.

Let the triangle $\triangle ABC$ have a right angle at C. Using this triangle, one may define the usual trigonometry functions. This method of defining these functions works well with acute angles but needs modification for more general angles.

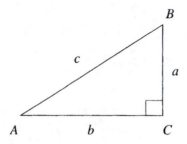

Definition 3.3.1 Let $\triangle ABC$ have a right angle at C as shown in the figure. Then:

$$\sin(A) = \frac{a}{c} = \frac{\text{opposite side}}{\text{hypotenuse}} \qquad \csc(A) = \frac{1}{\sin(A)} = \frac{c}{a}$$

$$\cos(A) = \frac{b}{c} = \frac{\text{adjacent side}}{\text{hypotenuse}} \qquad \sec(A) = \frac{1}{\cos(A)} = \frac{c}{b}$$

$$\tan(A) = \frac{a}{b} = \frac{\text{opposite side}}{\text{adjacent side}} \qquad \cot(A) = \frac{1}{\tan(A)} = \frac{b}{a}$$

EXAMPLE Let the right triangle $\triangle ABC$ have sides of lengths 5, 12, and 13 as shown here. Find $\sin(A)$, $\cos(A)$, and $\tan(A)$.

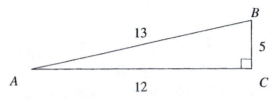

Solution   Using Definition 3.3.1, one finds $\sin(A) = \dfrac{a}{c} = \dfrac{5}{13}$, $\cos(A) = \dfrac{b}{c} = \dfrac{12}{13}$, and $\tan(A) = \dfrac{a}{b} = \dfrac{5}{12}$. ∎

From Glide Ratio to Tangent

The relationship between the glide ratio and the glide angle is very important in hang gliding as well as in other applications, such as the placement of a ladder. For this reason, there are several ways to express this ratio and angle.

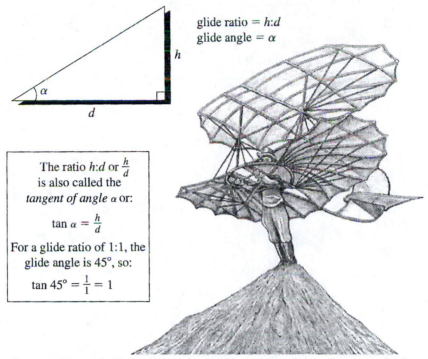

glide ratio = $h{:}d$
glide angle = α

The ratio $h{:}d$ or $\frac{h}{d}$ is also called the *tangent of angle α* or:

$$\tan \alpha = \frac{h}{d}$$

For a glide ratio of 1:1, the glide angle is 45°, so:

$$\tan 45° = \frac{1}{1} = 1$$

Suppose that another one of Otto's hang gliders, shown, has a glide ratio of 1:7. This means that the tangent of the glide angle is 1 to 7 (or $\frac{1}{7}$).

FIGURE 3.3.2 Glide ratios are related to angles using the tangent function. As this figure illustrates, middle school mathematics teachers must know and be able to explain at least some trigonometry. Reproduced from page 41 of *Looking at an Angle* in the Mathematics in Context grade 7/8 materials.

The tangent function is introduced in some middle school mathematics curricula. In particular, the Mathematics in Context (MiC) seventh/eighth grade materials introduce middle school students to the tangent function through the use of glide ratios. The tangent of the glide angle is the ratio h/d, where h is the vertical height at which the glider starts, and d is the horizontal distance traversed (compare Figure 3.3.2).

When dealing with angles that are not acute, one uses xy coordinates and measures the angle A going counterclockwise from the positive x-axis. In Figure 3.3.3, $\angle A$ is obtuse, and the point (x, y) is shown in the second quadrant. Of course, the value of x is negative in this particular quadrant. Let $r = \sqrt{x^2 + y^2}$ denote the distance of (x, y) from the origin.

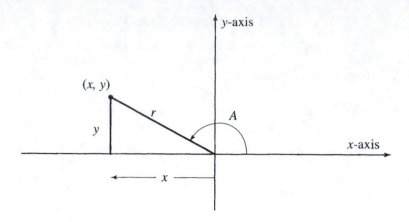

FIGURE 3.3.3 For angles that are not acute, one uses the *xy*-plane to define trigonometric functions. The angle *A* is measured counterclockwise from the positive *x*-axis.

For general angles in any quadrant of the *xy*-plane, one has the following definitions (these functions are defined whenever the given denominator is not zero):

$$\sin(A) = y/r \qquad\qquad \csc(A) = r/y$$

$$\cos(A) = x/r \qquad\qquad \sec(A) = r/x$$

$$\tan(A) = y/x \qquad\qquad \cot(A) = x/y$$

For any triangle $\triangle ABC$ (not just right triangles), the following law of sines and law of cosines are valid.

Law of Sines

$$\frac{\sin(A)}{a} = \frac{\sin(B)}{b} = \frac{\sin(C)}{c}$$

Law of Cosines

$$c^2 = a^2 + b^2 - 2ab\cos(C)$$

EXAMPLE A triangle has sides of lengths 5, 6, and 7. Find the (approximate) measure of the angle across from the side of length 6.

Solution Label the triangle with vertices *A*, *B*, and *C*, where $CB = 7$, $CA = 5$ and $AB = 6$. Then $a = 7$, $b = 5$, and $c = 6$.

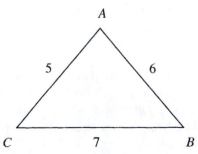

Using $c^2 = a^2 + b^2 - 2ab\cos(C)$, one obtains

$$36 = 49 + 25 - 2 \cdot 7 \cdot 5\cos(C) = 74 - 70\cos(C),$$

hence,

$$\cos(C) = \frac{36 - 74}{-70} = \frac{-38}{-70} = \frac{38}{70}.$$

Thus,

$$m(\angle C) = \cos^{-1}\left(\frac{38}{70}\right) \approx 57.1°. \qquad \blacksquare$$

The law of cosines allows one to calculate one side of a triangle given two sides and the included angle. For example, given a, b, and $\angle C$, one may use the law of cosines to find the side c. Alternatively, if one is given all three sides, then the law of cosines may be used to calculate the cosine of any of the angles. In particular, given a, b, and c, then $\cos(C) = [a^2 + b^2 - c^2]/[2ab]$ allows one to find $\cos(C)$. Then use the $\cos^{-1}$ function to find the measure of angle C. The law of cosines is clearly a generalization of the Pythagorean theorem since if $\angle C$ is a right angle, then $\cos(C) = 0$ and the formula $c^2 = a^2 + b^2 - 2ab\cos(C)$ becomes the familiar Pythagorean relation $c^2 = a^2 + b^2$.

The following argument proves that the law of cosines is valid when $\angle C$ is acute. Let $\triangle ABC$ be an arbitrary triangle. At least one of the two angles $\angle A$ or $\angle B$ is acute. One may assume without loss of generality that $\angle A$ is acute by interchanging A and B, if necessary. In this case, the triangle $\triangle ABC$ must look essentially like the triangle shown in Figure 3.3.4. Drop a perpendicular from B to the segment $\overline{AC}$. Assume this perpendicular intersects $\overline{AC}$ at D. Let $h = BD$ and $x = CD$. Then $b - x = DA$ as shown here.

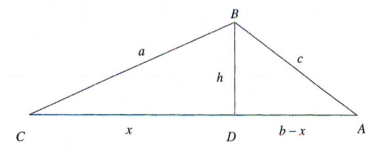

Applying the definition of cosine gives

$$\cos(C) = x/a.$$

Hence, $x = a\cos(C).$

Applying the Pythagorean theorem to $\triangle CDB$ yields

$$a^2 = x^2 + h^2.$$

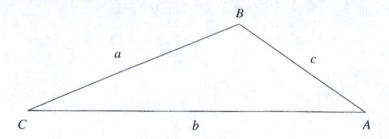

FIGURE 3.3.4 The law of sines, $\sin(A)/a = \sin(B)/b = \sin(C)/c$, and the law of cosines, $c^2 = a^2 + b^2 - 2ab\cos(C)$ hold for all triangles.

Applying the Pythagorean theorem to $\triangle ADB$ yields $c^2 = (b - x)^2 + h^2$,

hence, $$c^2 = b^2 - 2bx + x^2 + h^2.$$

Using $x^2 + h^2 = a^2$ and $x = a\cos(C)$ in this last equation yields

$$c^2 = b^2 - 2b(a\cos(C)) + a^2.$$

Hence, $$c^2 = a^2 + b^2 - 2ab\cos(C),$$

as desired.

One of the following exercises asks the reader to prove the law of cosines for the cases where $\angle C$ is not acute.

EXERCISES 3.3

1. Let $\triangle ABC$ be a right triangle with sides of lengths 3, 4, and 5. Let the right angle be at C and let A be the smallest angle. Find the following quantities.
   a. $\sin(A)$
   b. $\cos(A)$
   c. $\tan(A)$
   d. $\sin(B)$
   e. $\cos(B)$
   f. $\tan(B)$

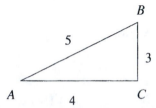

2. Let $\triangle ABC$ be a right triangle with right angle at C. How are $\sin(A)$ and $\cos(B)$ related to each other?
3. A given right triangle has legs of lengths 20 feet and 48 feet. Find the length of the hypotenuse.

4. Let $\triangle PQR$ have vertices at $P = (0,0)$, $Q = (7,0)$, and $R = (3,4)$. Use the law of cosines to find $\cos(R)$.

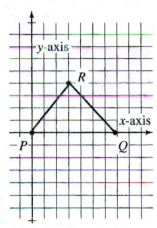

5. Let the triangle $\triangle EFG$ satisfy $EF = 8$ in, $EG = 10$ in, and $\cos(E) = \frac{1}{8}$. Find FG.

6. Using a calculator, find the measure (correct to within .1 degree) of each of the angles of $\triangle ABC$ given $BC = AC = 10$ and $AB = 14$.

7. A triangle has an angle of $60°$ and the sides of this angle have lengths 4 feet and 6 feet Find the length of the third side of this triangle.

8. Let an equilateral triangle be given with each side of length a. Find the length of the altitudes.

9. Is $\triangle LMN$ with vertices $L = (0,1)$, $M = (1,2)$, and $N = (-1,4)$ a right triangle? Why or why not?

10. Prove the law of cosines for the cases where
   a. $\angle C$ is a right angle, and
   b. $\angle C$ is an obtuse angle.

11. Let $ABCD$ be a rectangle with sides $AD = BC = 8$ and $AB = CD = 9$. Find the length of the diagonal from A to C and from B to D.

12. Prove that if two right triangles have one pair of legs congruent and their hypotenuses are congruent, then the right triangles are congruent. (This is the **HL theorem**).

13. Prove that the diagonals of a rectangle are always of equal length.

Classroom Connection 3.3.1

One can easily construct a discovery lesson using the GSP that leads middle or high school students to guess that the diagonals of a parallelogram are congruent iff the parallelogram is a rectangle. See the following Classroom Connection for how to do such a lesson if one does not have access to the GSP. ◆

Classroom Connection 3.3.2

Take two pairs of sticks (or geo-strips), such that each pair has the same length. Make holes near the end of each stick and fit them loosely together to get a quadrilateral that can be moved. Since opposite sides are stiff and have the same length, the figure always remains a parallelogram. Use this to construct a lesson to give physical demonstrations of the following:

1. SSSS does not imply congruence and also does not imply similarity.
2. This movable parallelogram is a rectangle exactly when the diagonals are of equal length.   ◆

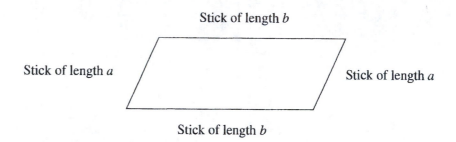

Stick of length b

Stick of length a Stick of length a

Stick of length b

3.4 AREA AND PERIMETER OF SIMILAR FIGURES

There is a famous passage in the **Meno** by Plato in which Socrates leads a previously untutored servant to construct a square with twice the area of a given square S. When Socrates first asks the servant what the length of the sides of the new square should be, the servant responds, "Obviously it should be twice that of the original square." Then Socrates draws a square and by doubling two of its sides, he constructs three more squares T, U, and V adjacent to S so that the total figure is a square with four times the area of the original square S. He leads the servant to realize that his answer would produce a square four times the size of the original square. Then Socrates draws the diagonals $\overline{AB}$, $\overline{BC}$, $\overline{CD}$, and $\overline{DA}$. He then asks the servant how many triangles are half the size of the original square and how many are inside the newly formed central square $ABCD$. The servant correctly answers four, and he quickly agrees that the square $ABCD$ has twice the area of the original square. Then Socrates asks the servant what line he should have used as a base to construct a square twice the size of a given square. The servant points out at the correct line $\overline{AB}$. Then Socrates says, "This line is known as the diagonal. So it is your view that the square whose side has length equal to the diagonal of the original square has twice the area?" The servant then promptly responds, "Yes." According to Socrates, the servant discovered the truth without being taught. Socrates himself gave no instructions or explanations; he simply directed the servant with his questions and did the constructions illustrated in Figure 3.4.1.

As the passage from the *Meno* shows, doubling the lengths of the sides of a square yields a square that has four times the area. Using similar arguments, if one takes a square and triples the lengths of the sides, then one obtains a square with nine times the area of the original square. In general, if two polygons G_1 and G_2 are similar with a scale factor S, then the area of the second polygon G_2 is S^2 times the area of the first. On the other hand, the perimeter of the second is only S times the perimeter of the second. These same relationships hold for familiar figures in the

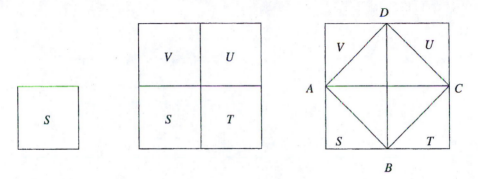

FIGURE 3.4.1 Socrates first draws a square S, then adds congruent squares T, U, and V to get a square with four times the area of S. The square $ABCD$ inside the union of squares S, T, U, V has one-half the area of the larger square as can be seen by calculating the areas of the triangles that make up the squares. Thus, the square $ABCD$ has twice the area of the original square S.

Euclidean plane, such as circles and ellipses.

> Similar polygons $G_1 \sim G_2$ with scale factor S
> $$S = \frac{\text{length of corresponding side in } G_2}{\text{length of corresponding side in } G_1}$$
> $$\text{perimeter of } G_2 = S \cdot (\text{perimeter of } G_1)$$
> $$\text{area of } G_2 = S^2 \cdot (\text{area of } G_1)$$

EXAMPLE Assume that pentagon $G_1 = P_1P_2P_3P_4P_5$ and pentagon $G_2 = Q_1Q_2Q_3 Q_4Q_5$ satisfy $G_1 \sim G_2$. Let the area of G_1 be 30 ft^2 and the area of G_2 be 12,000 ft^2. If $P_1P_2 = 2$ ft, find Q_1Q_2.

Solution Using (area G_2) $= S^2 \cdot$ (area G_1), one obtains $12{,}000 = S^2 \cdot (30)$. Hence, $S^2 = 400$ and $S = 20$. Consequently, $Q_1Q_2 = S \cdot P_1P_2$ yields $Q_1Q_2 = (20) \cdot (2$ ft$) = 40$ ft. ∎

Similarity is used throughout everyday life. When a person observes an object, such as a car, at a fairly large distance, they see a smaller version of the same car close up. The ratio of the distances to the two cars is the same as the inverse of the ratio of apparent heights of the cars. For example, if one sees a car at 100 feet and then at 25 feet, the ratio of the farther distance to the closer distance is $100/25 = 4$. On the other hand, the apparent height of the car at 100 feet appears as $25/100 = 1/4$ the height of the image at 25 feet. The area appears at the farther distance to be the square of this second ratio times the apparent area at the closer distance. Thus, at 100 feet the area of the car appears to be $(1/4)^2 = 1/16$ the area seen at 25 feet. This "scaling with distance" allows people to judge the distance to familiar objects. The accuracy of distance estimates goes down for objects that are far away. If one knows the apparent height of an object at some fixed distance, such at 25 feet, then when one sees this same object at very large distances, the ratio of 25 feet to the farther

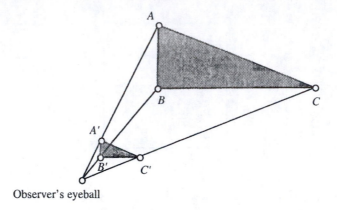

FIGURE 3.4.2 At four times the distance and four times the height, the larger triangle appears to have the same height and area as the smaller triangle.

distance (in feet) approaches zero as the farther distance increases toward infinity. Since it is hard to accurately judge the size of small things, it is fundamentally harder to make good judgments of distances when the objects are far away.

If two objects are at different distances and appear to have the same height, then the farther object must be taller. When two images appear to have the same height, the ratio of distances is the same as the ratio of heights. In Figure 3.4.2, triangle $\triangle ABC$ is shown at 4 units from the observer and appears the same as a similar triangle $\triangle A'B'C'$ located 1 unit away. The ratio of the height of the small triangle to the larger one is $A'B'/AB = 1/4$, which is the ratio of the distances away from the observer.

Dilations. If a number S and fixed point O are given, then a transformation known as a **dilation** is determined by mapping each point P to a point P' on the ray $\overrightarrow{OP}$, such that $OP' = S \cdot OP$. The point O is called the **center**, and the number S is known as either the **dilation factor** or scale factor.

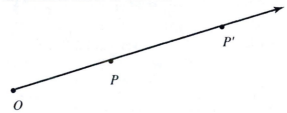

In Figure 3.4.2, two dilations are illustrated with the center at the observer's eyeball. One dilation is the transformation that takes the unprimed triangle $\triangle ABC$ to the primed triangle $\triangle A'B'C'$. The other dilation is the inverse transformation that takes the primed triangle to the unprimed triangle. The dilation (or scale) factor S is one-quarter if one is mapping $\triangle ABC$ to $\triangle A'B'C'$. The dilation factor is four if one is mapping the points $\triangle A'B'C'$ to $\triangle ABC$. Notice that with dilations, the center is not translated and there is no rotation of directions about the center. All of the points move either directly away from the center or directly toward the center.

Similarity Transformations.   Dilations are special cases of similarity transformations. A similarity transformation with scale factor S is a transformation that takes
each pair of points P, Q to a pair P', Q', such that $P'Q' = S \cdot PQ$. Similarity
transformations may involve compositions of translations, reflections, rotations, and
dilations.

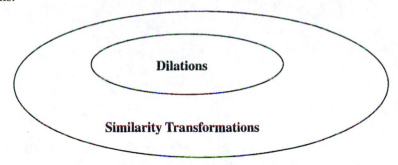

The use of coordinates is a very productive approach to studying similarity
transformations and also to investigating more general transformations. An especially important class of transformations consists of those that take the xy-plane
onto itself and map straight lines onto straight lines. Transformations of this type are
of the form $F: R^2 \to R^2$ where $F(x,y) = (ax + by + h, cx + dy + k)$ and where
a, b, c, d, h, and k are constants with $ad - bc \neq 0$. Since these transformations
take straight lines to straight lines, they take polygons to polygons. Figure 3.4.3
is a copy of a page from the Connected Mathematics Project unit *Stretching and
Shrinking*. Seventh-grade students are asked to fill in the table and answer questions
on some simple transformations of the form $F(x,y) = (ax + h, by + k)$. Students
get practice using these transformations of R^2 at the same time they are learning
about similarity.

EXAMPLE   Note that going from hat 1 to hat 6 in Figure 3.4.3 one multiplies both the
x-coordinate and y-coordinate by .5. This corresponds to doing the transformation
$F(x,y) = (.5x, .5y)$ on hat 1. Show that these two hats are similar by verifying that
$F(x,y) = (.5x, .5y)$ is a similarity transformation of the xy-plane with $S = .5$.

Solution   Show that given any two points, $P = (x_1, y_1)$ and $Q = (x_2, y_2)$ one has
$P'Q' = S \cdot PQ$ with $S = .5$. Here $P' = F(P) = F(x_1, y_1) = (.5x_1, .5y_1)$ and $Q' =$
$F(Q) = F(x_2, y_2) = (.5x_2, .5y_2)$.
Using the distance formula, one has

$$P'Q' = \sqrt{(.5x_1 - .5x_2)^2 + (.5y_1 - .5y_2)^2}$$
$$= \sqrt{(.5)^2[(x_1 - x_2)^2 + (y_1 - y_2)^2]}$$
$$= (.5) \cdot \sqrt{(x_1 - x_2)^2 + (y_1 - y_2)^2}$$
$$= (.5) \cdot PQ,$$

as desired.   ■

Point	Hat 1 (x, y)	Hat 2 $(x + 2, y + 2)$	Hat 3 $(x + 3, y - 1)$	Hat 4 $(2x, y + 2)$	Hat 5 $(2x, 3y)$	Hat 6 $(0.5x, 0.5y)$
A	$(0, 4)$	$(2, 6)$	$(3, 3)$	$(0, 6)$	$(0, 12)$	$(0, 2)$
B	$(0, 1)$					
C	$(6, 1)$					
D	$(4, 2)$					
E	$(4, 4)$					
F	$(3, 5)$					
G	$(1, 5)$					
H	$(0, 4)$					

Problem 2.3

Use the table and dot paper grids on Labsheets 2.3A and 2.3B.

- To make Mug's hat, plot points *A–H* from the Hat 1 column on the grid labeled Hat 1, connecting the points as you go.

- For Hats 2–6, use the rules in the table to fill in the coordinates for each column. Then, plot each hat on the appropriate grid, connecting the points as you go.

FIGURE 3.4.3 Note that students are being given gentle introductions to coordinate geometry and transformations. For example, going from hat 1 to hat 2 corresponds to the transformation given by $F(x, y) = (x + 2, y + 2)$. Transformations of the form $F(x, y) = (ax + h, by + k)$ take each axis to itself or a line parallel to the axis. They take polygons to similar polygons when $a = b$. Reproduced from page 21 of *Stretching and Shrinking* in the Connected Mathematics Project grade 7 materials.

One very simple way to obtain similar polygons with scale factor $S = |k|$ is to take transformations of the very special form $F(x, y) = (kx, ky)$ for some $k \neq 0$. It can be shown using reasoning, as in the previous example, that these transformations take each polygon G_1 in the xy-plane to a similar polygon G_2, and the scale factor is $|k|$. Also, these transformations take the origin $(0, 0)$ to itself. They are dilations with their centers at the origin. When the number k is greater than 1, they take each point away from the origin in a radial manner, with each point moved farther away by the factor k. For example, if $k = 2$, then each point is moved directly away from the origin to the point twice as far from the origin as it was originally located. If $0 < k < 1$, then each point is moved in toward the origin along the segment joining the point to the origin. For example, if $k = 1/2$, then each point moves half the distance toward the origin. When k is negative, the transformation $F(x, y) = (kx, ky)$ may be thought of as taking (x, y) to $(|k|x, |k|y)$ followed by a reflection across the origin.

The Geometer's Sketchpad is an excellent tool to illustrate dilations and, more generally, similarity transformations that need not be dilations. Use the Transform menu of the GSP to construct dilations and many other transformations. Use the Mark Center command under the Transform menu to define the center of the dilation. Use the Dilation command to define the dilation factor. Also use the Transform menu to illustrate that many, in fact most, similarity transformations are not dilations. In particular, one may construct a figure, such as a triangle, and then mark a center point. Using this center point, do a dilation with a dilation factor other than one. Then follow the dilation with some rotation about the same center point. Unless the rotation amount is chosen as something very special, such as 360°, the dilation followed by the rotation will yield a similarity transformation that is not a dilation.

EXERCISES 3.4

1. Consider the transformation $F(x, y) = (x + 1, y + 3)$. Let the parallelogram $PQRS$ have vertices $P = (0, 0)$, $Q = (4, 0)$, $R = (5, 2)$, and $S = (1, 2)$.
   a. Find the images P', Q', R', and S' of the points P, Q, R, and S, respectively.
   b. Is the image quadrilateral $P'Q'R'S'$ a parallelogram?

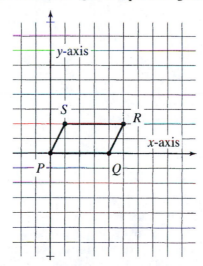

2. Consider the transformation $T(x, y) = (-2x, y)$. Let $KLMN$ be the square with vertices $K = (1, 0)$, $L = (3, 0)$, $M = (3, 2)$, and $N = (1, 2)$.
   a. Find the images K', L', M', and N' of the points K, L, M, and N, respectively.
   b. Is the image quadrilateral $K'L'M'N'$ a square? Why or why not?

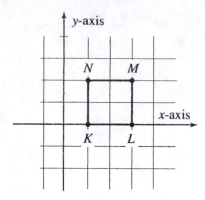

3. Consider the transformation $F(x,y) = (2x, 2y)$. Let $\triangle ABC$ have vertices $A = (1,1)$, $B = (3,4)$, and $C = (1,6)$. Let A', B', and C' be the respective images under F of A, B, and C.
   a. Find the coordinates of A', B', and C'.
   b. Find the distances AB, AC, BC, $A'B'$, $A'C'$, and $B'C'$.
   c. Are the two triangles $\triangle ABC$ and $\triangle A'B'C'$ similar? Why or why not?

4. Let $F(x,y) = (kx, ky)$ for some $k \neq 0$. Prove that for all pairs of points (x_1, y_1) and (x_2, y_2) the Euclidean distance between the pair of image points $F(x_1, y_1)$ and $F(x_2, y_2)$ is $|k|$ times the Euclidean distance between the original (x_1, y_1) and (x_2, y_2).

5. An equilateral triangle has an altitude of height 12 cm.
   a. Find the length of a side of this triangle.
   b. Find the perimeter of this triangle.
   c. Find the area of this triangle.

6. Assume you are given two regular hexagons. One has each side of length 2 units, and the other has each side of length 10 units.
   a. Find the perimeter of each of these hexagons.
   b. Find the area of each of these hexagons.

7. Let G_1 and G_2 be polygons, such that $G_1 \sim G_2$ with scale factor S. Assume that G_1 has a perimeter of 6,000 inches and an area of 4,000 square inches. Assume that G_2 has an area of 10 square inches.
   a. Find the scale factor S.
   b. Find the perimeter of G_2.

Classroom Connection 3.4.1

Divide a class of middle school students into five teams and have each team assigned one of the hats 2 to 6 in Figure 3.4.3. Each team should construct the figure for their hat and should find the area of the original hat 1 and the area for their hat. Each team should also find the ratio of the area for their hat to the area of the original hat. The class should be able to relate the ratio of areas for a given hat to the coefficients of x and y in the rule for constructing that hat. Classroom Connection 3.4.2 is somewhat more abstract and may be used as an additional exploration or as an alternative exploration. ◆

Classroom Connection 3.4.2

One activity with transformations of the form $F(x, y) = (ax, by)$ with positive numbers a and b is to ask middle school students to provide explanations of what effect the parameters a and b have on changing the shape of figures. If students have done the previous activity, they can relate the product $a \cdot b$ to the effect that F has on areas of figures. They can also guess how a and b should be related in order for the new figures to always be similar to the original figures. Of course, with middle school students, avoid using the function notation $F(x, y)$. Instead use a two-column table headed with (x, y) in the first column and with (ax, by) in the second column. One could start with teams of students using different values of a and b, such as $(a, b) = (2, 3)$ for Team 1, and $(a, b) = (2, 4)$ for Team 2, etc.   ◆

3.5   SIMILARITY FOR MORE GENERAL FIGURES

In Section 3.1, similarity was defined for polygons. Clearly the idea of similarity goes beyond polygons to more general pairs of figures that have the same shape but need not have the same size. This raises the question of how to define similarity for figures that are not polygons and may even be three dimensional.

Classroom Discussion 3.5.1

The page in Figure 3.5.1 from Math Thematics illustrates the importance of using similarity for figures that are not polygons. One possible activity is to have teams of three or four college students discuss the ways in which similarity for more general figures might be defined. Also ask them to state what criteria the more general definition should satisfy. In particular, at the end of this activity they should understand that the more general definition is equivalent to the usual definition for figures that are polygons.   ◆

The SSS similarity theorem says that if the corresponding pairs of sides of two triangles are in the same ratio, then the triangles are similar and hence the corresponding pairs of angles must be congruent. This means that for triangles, the original definition of two similar triangles could have been simplified to just require that corresponding pairs of sides be in the same ratio. The new definition of similarity will be given for two sets (i.e., figures) T_1 and T_2 that are assumed to be in some Euclidean space. These two sets may lie in the Euclidean plane, or they may be sets in Euclidean three-dimensional space. Recall that for two points P and Q, the Euclidean distance from P to Q is denoted by PQ. If P and Q lie in Euclidean three-dimensional space, i.e., in $R^3 = \{(x, y, z) \mid x, y, z \text{ are real numbers}\}$, then for $P = (x_1, y_1, z_1)$ and for $Q = (x_2, y_2, z_2)$ the Euclidean distance PQ from P to Q is given by the formula $PQ = \sqrt{(x_1 - x_2)^2 + (y_1 - y_2)^2 + (z_1 - z_2)^2}$.

Definition 3.5.1 Two sets T_1 and T_2 are said to be **similar** with **scale factor** S if there is a 1–1 and onto correspondence between them such that whenever A and B in T_1 correspond respectively to A' and B' in T_2, then the ratio $A'B'/AB$ is equal to the fixed number S.

In South Dakota, a model of the Sioux leader Chief Crazy Horse is being used to construct what may be the world's largest sculpture. When completed, the sculpture will measure 563 feet high by 641 feet long!

➤ **When creating a large piece of art, an artist often makes a model similar to what the completed artwork will be.**

10   **a.** The Crazy Horse model is similar to the sculpture. What do you know about their corresponding measurements?

b. The height of the model is 16.56 ft. Write and solve a proportion to find the length of the model. Round your answer to the nearest hundredth.

FIGURE 3.5.1 Here students are introduced to proportion and scaling via sculpture. Reproduced from page 427 of *Book 1* in the Math Thematics grade 6 materials.

When the two sets T_1 and T_2 are polygons, one cannot just think of the pair of points A and B in the first polygon as being vertices located at the ends of a single side of this polygon.

Classroom Discussion 3.5.2

Have a class discussion on why the new definition of similarity given in this section can be used to define two sets to be congruent by adding the requirement that the scale factor be equal to one to Definition 3.5.1.   ◆

Scaling of Volume and Surface Area.   For sets in Euclidean three-dimensional space with volume and surface area, one may ask how the volumes and surface areas are related given the scale factor S. The answer is that surface areas are related by a factor of S^2, and volumes are related by a factor of S^3.

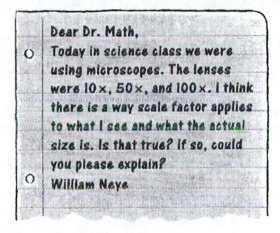

Dear Dr. Math,

Today in science class we were using microscopes. The lenses were 10×, 50×, and 100×. I think there is a way scale factor applies to what I see and what the actual size is. Is that true? If so, could you please explain?

William Neye.

FIGURE 3.5.2 Microscopes allow small things to have images that have been "scaled up" in size. In effect, the image one would see using only one's eyeball has been enlarged by a similarity transformation. Reproduced from page 34 of *Gulliver's Worlds* in the MathScape grade 6 materials.

T_1 is similar to T_2 with scale factor S

$$S = \frac{A'B'}{AB} = \frac{\text{distance between points in set } T_2}{\text{distance between corresponding points in set } T_1}$$

$$\text{surface area of } T_2 = S^2 \cdot (\text{surface area of } T_1)$$

$$\text{volume of } T_2 = S^3 \cdot (\text{volume of } T_1)$$

A simple example of two similar three-dimensional sets would be two cubes. If the first cube C_1 has edges with a length of 5 inches and the second cube C_2 has edges with a length of 10 inches, then $C_1 \sim C_2$ with scale factor $S = 2$. Of course, the first cube has volume $V_1 = (5)^3$ cubic inches = 125 cubic inches, and the second cube has volume $V_2 = (10)^3$ cubic inches = 1,000 cubic inches. Note that $V_2 = (2)^3 \cdot V_1 = (S)^3 \cdot V_1$, which, of course, is in agreement with the general formula for how volumes of similar figures are related. Since the surface area of a cube is always made up of six squares, it is easily checked that the surface area of the first cube is $A_1 = 6 \cdot (5)^2 = 150$ square inches and that the surface area of the second cube is $A_2 = 6 \cdot (10)^2 = 600$ square inches $= (2)^2 \cdot A_1 = (S)^2 \cdot A_1$, as expected.

With cameras and microscopes, the scale factor is often called a *multiplier* and may be denoted with a times sign. For example, $10x$ corresponds to enlarging a figure by a factor of 10 (i.e., $S = 10$). Of course, $.5x$ corresponds to $S = .5$ and hence to a reduction such that lengths are reduced to one-half their original size.

EXERCISES 3.5

1. You have two regular tetrahedrons. The first has edges that are 3 inches long, and the second has edges that are 6 inches long.
   a. What is the ratio of the surface area of the second tetrahedron to the surface area of the first?
   b. What is the ratio of the volume of the second tetrahedron to the volume of the first?

2. Answer question #10 shown in Figure 3.5.1. Explain your reasoning.

3. All spheres are similar. Consider two spheres such that the second has 1,000 times the volume of the first. What is the radius of the second sphere if the first sphere has a radius of 5 feet?

4. You have two similar right circular cones. The second cone has a total surface area that is 25 times the total surface area of the first cone. If the second cone has a height of 20 cm, what is the height of the first cone?

5. An engineer makes a scale model of a proposed new storage tank. The model is made with a scale factor S such that 1 cm in the model corresponds to 50 cm for the actual tank to be built. If the scale model holds 8 liters of liquid, how many liters of liquid will the actual tank hold after it is built?

6. A right circular cone C_1 has a lateral surface area of 15π square inches and a total surface area of 24π square inches.
   a. Find the radius of its base.
   b. Find the height of this cone.
   c. Find the volume of this cone.
   d. If C_2 is a second right circular cone with $C_1 \sim C_2$ and with the volume of C_2 equal to eight times the volume of C_1, then what is the scale factor for $C_1 \sim C_2$?

7. Assume that a 6-foot-tall iron statue of a horse has a weight of 3,456 pounds. If a similar copy of the statue is also made of iron and is only 6 inches tall, what is the weight of the smaller iron horse? Explain your reasoning.

8. A certain microscope with a camera attached is rated as 1,000x (i.e., the magnification factor is 1,000). A picture is taken of a very small spherically shaped object. If the object appears in the photograph as a circle with a diameter of approximately .1 cm, find the approximate diameter of the original object.

9. *Staphylococcus aureus* is a spherically shaped bacteria. A picture is taken of a single specimen of *Staphylococcus aureus* using a microscope with a camera attached having a rating of 2,000x (i.e., the magnification factor is 2,000). The picture appears as a circle with an approximate diameter of .16 cm.
   a. Find the approximate diameter of this specimen.
   b. Find the approximate volume of this specimen.

10. The function $F(x, y) = (-2x, -2y)$ takes a first polygon Q_1 lying in the xy-plane to a polygon Q_2. What is the ratio $\dfrac{\text{area}(Q_2)}{\text{area}(Q_1)}$?

11. The function $F(x, y, z) = (-x, y, z)$ takes a figure T_1 lying in xyz-space to a figure T_2. Let T_1 be a figure of a lady with her right hand touching her head and with her left hand at her left side. How would you describe the image T_2 of T_1? Give as detailed a description as possible. You don't have to "prove" that your description is accurate, but explain why you think it is correct.

Classroom Connection 3.5.1

Take a picture of a middle school student holding up his or her right hand. Then have the student look in the mirror, holding up his or her right hand (in the same way it was held up before), compare the picture to the image in the mirror. The picture appears to be of a person holding up his or her right hand, but the image in the mirror appears to be a person holding up his or her left hand. The picture and the image in the mirror are similar to each other but have the opposite orientation. Looking at a mirror, one sees an image corresponding to reflection across the mirror's plane. ◆

Classroom Connection 3.5.2

Mirror images can form the basis of a good class discussion on reflections across planes. If a person parts their hair on the right, then when they look at themselves in a mirror, the image they see has their hair parted on the left. Thus, it is no surprise that students often think that the image of themselves that they see in a mirror has the right and left sides of their bodies interchanged. They then may ask why their image does not have their heads and feet interchanged. A good question to ask when discussing this issue is: In the mirror, does the image of your right hand appear to be closer to your real left eye or to your real right eye? ◆

CHAPTER 3 REVIEW

Similarity is important in everyday life. The reason that a 3 inch × 5 inch picture of a friend "looks right," even though it is much smaller than the friend, is that all of the features are in the same ratio. Maps, photos, and scale models of things such as cars and planes are all examples of similarity in the real world. Similarity is particularly important in engineering where scale models are used for testing purposes and where similarity is used for measuring distances. Similarity has significant application within the field of mathematics. In this chapter we used it to give a proof of the Pythagorean theorem. Also, similarity is fundamental to trigonometry. The trigonometric functions are defined as ratios of sides, and these ratios are dependent on a triangle's shape, not its size.

The definition of similarity for triangles, and polygons in general, requires that all corresponding angles be congruent and that the length of each side in the second object, divided by the corresponding side of the first object, be a constant S. The ratio S is called the *scale factor* and may be any positive constant. Just as congruent figures may have different orientation, also similar figures may have the same or opposite orientation. In particular, the image you see in the mirror is similar to the image of yourself in a photo, but the two images have opposite orientations.

The definition of similarity for objects that are not polygons only requires that the distance between each pair of points in the second object, divided by the distance between the corresponding pair in the first object, be a constant S. As with polygons, S is called the *scale factor*. When two objects are similar with a scale factor S, then lengths are related by a factor of S, areas by a factor of S^2, and volumes by a factor of S^3.

Selected formulas:

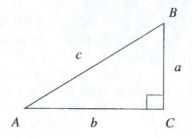

Pythagorean theorem: $a^2 + b^2 = c^2$

Sine: $\sin(A) = \dfrac{a}{c}$

Cosine: $\cos(A) = \dfrac{b}{c}$

Tangent: $\tan(A) = \dfrac{a}{b}$

Law of sines: $\dfrac{\sin(A)}{a} = \dfrac{\sin(B)}{b} = \dfrac{\sin(C)}{c}$ (for any triangle)

Law of cosines: $c^2 = a^2 + b^2 - 2ab\cos(C)$ (for any triangle)

CHAPTER 3 REVIEW EXERCISES

1. Assume that $\triangle ABC \sim \triangle A'B'C'$ with a scale factor of 3. If $BC = 3$ cm, find $B'C'$.
2. Assume that $\triangle HJK \sim \triangle H'J'K'$ with a scale factor of $S = 5$. If $H'J' = 25, J'K' = 15$, and $K'H' = 20$, find all of the lengths of the sides of $\triangle HJK$.
3. Assume that $\triangle LMN \sim \triangle L'M'N'$ with a scale factor of 2 and that the area of $\triangle L'M'N'$ is 5 cm^2. Find the area of $\triangle LMN$.
4. Assume that $\triangle ABC \sim \triangle A'B'C'$ with $AB = 5$ and $A'B' = 2$. If the area of $\triangle ABC$ is 100 in^2, find the area of $\triangle A'B'C'$.
5. Let $\triangle DEF \sim \triangle LMN$ with $DE = 7$ in, $EF = 6$ in, $FD = 8$ in, and $MN = 9$ in. Find NL and LM.
6. Let $\triangle XYZ$ and $\triangle PQR$ be given with $XY = 4, YZ = 3, ZX = 3, PQ = 9, QR = 9$, and $RP = 12$. Indicate which of the following are true and which are false.
    a. $\triangle XYZ \sim \triangle PQR$
    b. $\triangle YZX \sim \triangle PQR$
    c. $\triangle YXZ \sim \triangle PQR$
    d. $\triangle ZYX \sim \triangle PQR$
    e. $\triangle XZY \sim \triangle PQR$
7. Let $\triangle ABC$ and $\triangle DEF$ be given with $m(\angle A) = m(\angle F), AB = 4, BC = 6, CA = 8$, $DF = 8$, and $EF = 16$. Indicate which of the following are true and which are false.
    a. $\triangle ABC \sim \triangle DEF$
    b. $\triangle ABC \sim \triangle EFD$
    c. $\triangle ABC \sim \triangle FDE$
    d. $\triangle ABC \sim \triangle FED$

8. Assume a box-shaped building has a shadow of 100 feet. At the same time a person of 5.5 feet has a shadow of 4 feet. How tall is the building?

9. Let $\triangle XYZ \sim \triangle X'Y'Z'$. Let the area of $\triangle XYZ$ be $4\ \text{cm}^2$, and let the area of $\triangle X'Y'Z'$ be $2\ \text{cm}^2$. If $YZ = 6$ cm, find $Y'Z'$.

10. Let $\triangle ABC \sim \triangle A'B'C'$. Assume that the perimeter of $\triangle ABC$ is 10 meters and that the perimeter of $\triangle A'B'C'$ is 20 feet. If the altitude from A to side $\overleftrightarrow{BC}$ is 2 meters, find the length of the altitude from A' to $\overleftrightarrow{B'C'}$ measured in feet.

11. Let $PQRST$ and $P'Q'R'S'T'$ be pentagons with $PQRST \sim P'Q'R'S'T'$. Assuming that $P'Q'R'S'T'$ has an area of $10\ \text{ft}^2$ and that $PQRST$ has area $2\ \text{ft}^2$, find the ratio $\dfrac{P'Q'}{PQ}$.

12. Let $\triangle XYZ$ be a right triangle with sides of lengths 27, 36, and 45. Let the right angle be at Z and let X be the smallest angle. Find the following quantities.
   a. $\sin(X)$
   b. $\cos(X)$
   c. $\tan(X)$

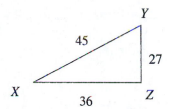

13. A given right triangle has legs with lengths of 10 feet and 24 feet.
   a. Find the length of the hypotenuse.
   b. If the smallest angle of this triangle is at vertex A, then find $\sin(A)$.

14. Let $\triangle ABC$ be a triangle in R^2 with vertices at $A = (1,1)$, $B = (-1,1)$, and $C = (0,0)$.
   a. Is this a right triangle? Explain your answer.
   b. Find $\cos(A)$.

15. Let $\triangle HJK$ be a triangle in R^2 with vertices at $H = (2,3)$, $J = (-2,-3)$, and $K = (0,3)$.
   a. Is this a right triangle? Explain your answer.
   b. Find $\cos(H)$.

16. Let $\triangle DEF$ be a triangle in R^3 and have vertices at $D = (0,0,5)$, $E = (5,0,5)$, and $F = (0,5,0)$.
   a. Is this a right triangle? Explain your answer.
   b. Find $\cos(F)$.

17. Let $\triangle ABC$ be a triangle in R^3 and have vertices at $A = (4,3,3)$, $B = (1,1,3)$, and $C = (1,3,3)$.
   a. Is this a right triangle? Explain your answer.
   b. Find $\cos(A)$.

18. A rectangle has two sides of 10-inch length and two sides of 24-inch length. Find the length of the diagonals.

19. A rectangle has two sides of 2-foot length and two sides of 3-foot length. Find the length of the diagonals.

20. A rhombus has all four sides of 8-inch length. It also has two 60° angles.

    a. Find the measure of the other two angles.

    b. Find the area of this rhombus.

21. A rhombus has all four sides of 6-cm length. It also has two $30°$ angles.

    a. Find the measure of the other two angles.

    b. Find the area of this rhombus.

22. Find the length of the shortest diagonal of a rhombus with sides of 8-inch length and two $60°$ angles.

23. Find the length of the longest diagonal of a rhombus with sides of 6-cm length and two $30°$ angles.

24. Which of the following are similarity transformations?

    a. A dilation with center O and a dilation factor of 3.

    b. A dilation with center O and a dilation factor of 3 followed by a translation moving all points a distance of 5 units.

    c. A rotation about the point O.

25. Which of the following are similarity transformations?

    a. A rotation about the point O followed by a dilation with center W and a dilation factor of 5.

    b. A transformation of the xy-plane holding the origin fixed, which stretches the x-axis by a factor of 2 and which stretches the y-axis by a factor of 3 [i.e., $F(x,y) = (2x, 3y)$].

    c. A transformation of the xy-plane holding the origin fixed, which stretches the x-axis by a factor of 2 and which stretches the y-axis by a factor of 2 [i.e., $F(x,y) = (2x, 2y)$].

26. Which of the following are dilations of the xy-plane with center at the origin?

    a. $F(x,y) = (7x + 1, 7y + 3)$.

    b. $F(x,y) = (7x, 7y)$.

    c. $F(x,y) = \left(\dfrac{x}{2}, \dfrac{y}{2}\right)$.

    d. $F(x,y) = (3y, 3x)$.

27. Which of the following are similarity transformations of the xy-plane?

    a. $F(x,y) = (3x + 1, 5y + 3)$.

    b. $F(x,y) = (2x + 1, 2y)$.

    c. $F(x,y) = \left(\dfrac{x}{3} + 5, \dfrac{y}{3} + 4\right)$.

    d. $F(x,y) = (3y, 3x)$.

28. Consider the transformation $F(x,y) = (x + 2, y + 3)$. Let the square $ABCD$ have vertices $A = (0,0)$, $B = (1,0)$, $C = (1,1)$, and $D = (0,1)$.

    a. Find the images A', B', C', and D' of the points A, B, C, and D, respectively.

    b. Graph both the square $ABCD$ and its image $A'B'C'D'$.

    c. Is the image quadrilateral $A'B'C'D'$ a square?

29. Consider the transformation $F(x,y) = (x + 2, 2y + 3)$. Let the square $EFGH$ have vertices $E = (0,0)$, $F = (1,0)$, $G = (1,1)$, and $H = (0,1)$.

    a. Find the images E', F', G', and H' of the points E, F, G, and H, respectively.

    b. Graph both the square $EFGH$ and its image $E'F'G'H'$.

    c. Is the image quadrilateral $E'F'G'H'$ a square?

30. Consider the transformation $F(x,y) = (3x + 2, 3y + 2)$. Let $\triangle XYZ$ have vertices at $X = (0,0)$, $Y = (2,0)$, and $Z = (1,1)$.

    a. Find the images X', Y', Z' of the points X, Y, Z, respectively.

    b. Graph both the triangle $\triangle XYZ$ and its image $\triangle X'Y'Z'$.

    c. Find the lengths of all sides of both $\triangle XYZ$ and $\triangle X'Y'Z'$.

    d. Are the two triangles similar? Explain why or why not.

31. You have two cubes C_1 and C_2. The edges of C_1 have a 5-cm length, and the volume of C_2 is 1,000,000 times the volume of C_1. Find the length of the edges of C_2.

32. You have two cubes D_1 and D_2. The edges of D_1 have a 5-foot length, and the surface area of D_2 is 1,000,000 times the surface area of D_1. Find the length of the edges of D_2.

33. You have two similar right circular cylinders K_1 and K_2. The height of the cylinder K_1 is 7 inches, and the lateral surface area of K_2 is 100 times the lateral surface area of K_1. Find the height of K_2.

34. An industrial artist wants to make a scale model of a truck that stands 12 feet tall and has a cargo volume of 1,000 ft^3. What will the model's height be if it is to have a cargo volume of 1 ft^3?

RELATED READING FOR CHAPTER 3

Billstein, R., and J. Williamson. *Math Thematics: Book 1*. Evanston, IL: McDougal Littell, 1999.

Blue, M. N., et al. *Navigating through Geometry in Grades 6–8*. Reston, VA: National Council of Teachers of Mathematics, 2002.

Collier, C. P. *Geometry for Teachers*. 3rd ed. Prospect Heights, IL: Waveland Press Inc., 1994.

Kleiman, G. "Gulliver's Worlds," In *MathScape: Seeing and Thinking Mathematically*. Grade 6 materials. New York: Glencoe/McGraw-Hill, 1999.

Lappan, G., et al. *Stretching and Shrinking*. Connected Mathematics Project. Needham, MA: Pearson/Prentice-Hall, 2004.

O'Daffer, P. G., and S. R. Clements. *Geometry: An Investigative Approach*. Reading, MA: Addison Wesley Longman, 1997.

Romberg, T. A., et al. *Looking at an Angle*. Mathematics in Context. Chicago, IL: Encyclopedia Britannica Educational Corporation, 1998.

Rubenstein, R. N., et al. *Teaching and Learning Middle Grades Mathematics*. Emeryville, CA: Key Curriculum Press, 2004.

Rigid Motions and Symmetry

Transformations that preserve distance are called **isometries**. They are also known as **rigid motions** and are defined to be transformations that take each pair of points to a pair that is exactly the same distance apart. Thus, for each pair of points P and Q with images P' and Q', the distance PQ between the original points is the same as the distance $P'Q'$ between the image points. Reflections across lines, reflections across points, rotations, translations, and glide reflections are examples of isometries of the Euclidean plane. Using rigid motions, one can define symmetries of objects. An object is said to have line symmetry with respect to a given line m if the reflection across m takes the object onto itself. In like fashion, an object has point symmetry if reflection across a point takes the object onto itself. Rotational symmetry is defined using rotations.

The **symmetries** of an object are the rigid motions that take the object onto itself. In this book, the identity map that leaves each point fixed is considered to be a symmetry and is often known as the **trivial symmetry**. The advantage of considering the identity map to be a symmetry is that when this is included as a symmetry, then the collection of symmetries of an object becomes an example of what is known as a group. In particular, the collection of all isometries taking T onto itself is a group where the group operation is composition.

Section 4.1 introduces reflections across lines in the Euclidean plane. These reflections reverse orientation. On the other hand, reflection across a point in the plane does not reverse orientation. Section 4.2 considers translations, rotations, and glide reflections of the Euclidean plane. We study symmetries in Section 4.3 and explore isometries of Euclidean three-space in Section 4.4.

4.1   REFLECTIONS OVER LINES AND ORIENTATION

One way to understand isometries is to first understand reflections across lines in the Euclidean plane. It is easy to demonstrate these reflections with the use of paper. Draw a straight line m on a sheet of paper lying on a flat surface, then turn the sheet over and place it down with the points on the line being laid on top of where they were previously.

Classroom Connection 4.1.1

Using sheets of tracing paper or transparencies, have middle or high school students draw three or four geometric figures such as triangles, quadrilaterals, circles, etc. Then have them draw a straight line on a sheet of regular paper and have them turn the sheet over and place it down with the points on the line being laid on top of where they were previously. This gives a physical demonstration of a reflection of a plane across a line.

In this example it is assumed that one does not tear or stretch the page that is turned over. The students can check that the geometric figures they constructed go to congruent figures. If one marks two points as P and Q on the tracing paper or transparency before it is turned, then they will go to new points P' and Q' where the Euclidean distance $P'Q'$ will be equal to the original Euclidean distance PQ. Reflections across lines are said to preserve distances, angles, and areas, since in each case the original figure and the image figure have the same measure for corresponding items.   ◆

Reflections across Lines.   Given a line m in the Euclidean plane and a point P off m, draw a perpendicular from P to the point F on m. The point F is the closest point on m to P and is said to be the **foot** of P on m. Extend the segment $\overline{PF}$ to get P' on the other side of m from P, such that $PF = P'F$. The **reflection** across m takes P to P' and takes P' to P. It leaves each point of m fixed. The reflection across m is illustrated in Figure 4.1.1.

Transformations.   Reflection across a line is an example of a transformation. In geometry, a **transformation** is a $1 - 1$ function that is also an onto function. Thus, a transformation $F: R^2 \rightarrow R^2$ is a function that takes points of R^2, to points of R^2 such that (1) different points have different images and (2) each point of R^2 is the image under F of some point. If one uses P' to denote the image $F(P)$ of P under F, then (1) can be stated as $P \neq Q$ implies $P' \neq Q'$, and (2) can be stated as the requirement that each point in R^2 is the image P' of some P.

> **Definition 4.1.1** A transformation $F: R^2 \rightarrow R^2$ of the Euclidean plane onto itself is an isometry or rigid motion if for all points P and Q in R^2, the equality $PQ = F(P)F(Q)$ holds.

Orientation.   Reflections of the Euclidean plane across lines are isometries. However, they do not preserve orientation. In other words, a figure that is lopsided or distorted on the right side will have an image that is lopsided or distorted on the left. Demonstrate this by drawing a figure on tracing paper or a transparency. Turn the

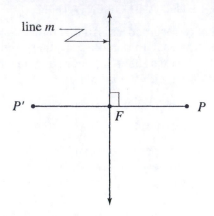

FIGURE 4.1.1 The foot of P on m is the point F, which is the closest point on m to P. The segment $\overline{PF}$ is perpendicular to m at F. Extending $\overline{PF}$ to the other side of m, one obtains a point P' with $PF = P'F$. Thus, F is the midpoint of the segment $\overline{PP'}$. The reflection across m takes P to P' and takes P' to P. The reflection across the line m is denoted by r_m.

sheet over and lay it down with the points on the line l going to themselves, you then find a figure that is the **mirror image** of the original figure. The image figure is congruent to the original figure, but it has the opposite orientation (compare Figure 4.1.2).

Each isometry of the Euclidean plane will either preserve or reverse orientation. Orientation-preserving transformations are said to be **direct**. An isometry that is not direct is said to be **orientation-reversing** or **opposite**. A reflection across a line of the Euclidean plane is an opposite isometry.

Given a transformation $F: R^2 \to R^2$ of the Euclidean plane onto itself, then one may use a triangle and its image to check to see if it is orientation-preserving or orientation-reversing. Let $\triangle ABC$ be a triangle and assume that going around the triangle in the order A, B, C is moving counterclockwise. Then find the image triangle $\triangle A'B'C'$ and traverse this triangle in the order A', B', C'. If the image triangle is traversed in the same direction (i.e., counterclockwise) as the original triangle, then the transformation F is orientation-preserving. If the image triangle is traversed in the opposite (i.e., clockwise) direction, then the transformation F is orientation-reversing. This method works quite well for two-dimensional Euclidean geometry, but in higher dimensions the situation is somewhat more complicated.

EXAMPLE Let $F: R^2 \to R^2$ be the transformation given by $F(x, y) = (x + 2, -y)$. Prove that F is an isometry and determine if it is orientation-preserving or orientation-reversing.

Solution Part 1 (prove that it is an isometry). To prove that F is an isometry, one must verify that Definition 4.1.1 is satisfied. In other words, one needs to take two points P, Q in general form and show that $PQ = F(P)F(Q)$. Thus, let $P = (x_1, y_1)$ and $Q = (x_2, y_2)$. Notice that to give this proof, one must have P and Q expressed in general form with coordinates having x's and y's as opposed

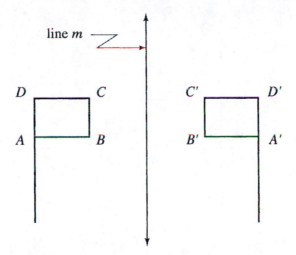

FIGURE 4.1.2 This figure illustrates the reflection r_m across the line m. Here the figure on the left is a form of the letter P that had been constructed using a vertical line segment and the rectangle $ABCD$. Reflecting the left figure over m yields an image on the right that is the **mirror image** (in the Euclidean plane) of the left figure. The points $A, B, C,$ and D are reflected to corresponding points $A', B', C',$ and D'. For the left figure, traversing the rectangle $ABCD$ in alphabetical order results in going counterclockwise. Traversing the image rectangle $A'B'C'D'$ in alphabetical order results in going clockwise. This change from going counterclockwise to going clockwise results because reflection across a line in the Euclidean plane reverses orientation.

to choosing actual fixed numbers for the coordinates of P and Q. Applying the definition of F to first P and then Q, one obtains $F(P) = F(x_1, y_1) = (x_1 + 2, -y_1)$, and $F(Q) = F(x_2, y_2) = (x_2 + 2, -y_2)$. Now, using the definition of distance, notice that $PQ = \sqrt{(x_1 - x_2)^2 + (y_1 - y_2)^2}$

and that

$$F(P)F(Q) = \sqrt{[(x_1 + 2) - (x_2 + 2)]^2 + [(-y_1) - (-y_2)]^2}$$
$$= \sqrt{(x_1 - x_2)^2 + (-y_1 + y_2)^2}$$
$$= \sqrt{(x_1 - x_2)^2 + (y_1 - y_2)^2} = PQ$$

Hence, $PQ = F(P)F(Q)$, which proves F is an isometry.

Part 2 (investigate the orientation-preserving or reversing character of this transformation). To find if F is orientation-preserving or reversing, choose any triangle and find if its image is traversed in the same direction or in the opposite direction. For example, one may choose $\triangle ABC$ where $A = (1, 1), B = (3, 1),$ and $C = (2, 2)$. Graphing these points, note that traversing the triangle in the order A, B, C is going counterclockwise.

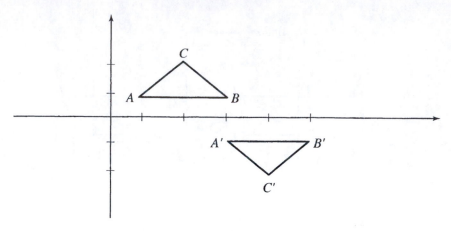

Applying the definition of F to the points A, B, and C, one finds that the images are given by $A' = F(A) = F(1,1) = (3, -1), B' = F(B) = F(3,1) = (5, -1)$, and $C' = F(C) = F(2,2) = (4, -2)$. Graphing these points, one finds that traversing the triangle $\triangle A'B'C'$ in the order A', B', C' is moving in the opposite (i.e., clockwise) direction. Thus, F is orientation-reversing.   ∎

Recall that following one function F_1 by another F_2 is called **composition** and is denoted by the symbol "$\circ$." Thus, $F_2 \circ F_1(P) = F_2(F_1(P))$. Clearly, the composition of isometries is an isometry since distance is preserved as each transformation is applied. The order in which a composition of isometries is done usually makes a difference. For example, in the xy-plane, let m be the line $x = 1$, and let k be the line $x = 2$. In order to show that the composition $r_k \circ r_m$ is different from the composition $r_m \circ r_k$, it is necessary to find only one point P, such that $r_k \circ r_m(P) \neq r_m \circ r_k(P)$. If one sets $P = (0,0)$, it is easy to check (using the definition of reflection across a line and the given lines m and k) that $r_k \circ r_m(0,0) = r_k(r_m(0,0)) = r_k(2,0) = (2,0)$ and $r_m \circ r_k(0,0) = r_m(4,0) = (-2,0)$. Since $(2,0) \neq (-2,0)$, one has $r_k \circ r_m \neq r_m \circ r_k$. Hence, the two isometries r_m and r_k do not commute.

An example of a direct (i.e., orientation-preserving) isometry is easily constructed by taking one reflection across a line followed by another reflection across a line. Thus, for any two lines m and k, the composition $r_k \circ r_m$ is direct. In fact, taking any even number of reflections across lines in the Euclidean plane always results in a direct isometry. An odd number of reflections across lines yields an orientation-reversing isometry.

Using xy-coordinates and letting v denote the vertical line that is the y-axis, reflections across the y-axis are easily seen to interchange the right-hand side of the plane with the left-hand side. More precisely, $r_v(x,y) = (-x,y)$. Similarly, if h represents the x-axis, then r_h interchanges the top half of the xy-plane and the bottom half of the xy-plane. More precisely, $r_h(x,y) = (x, -y)$.

Reflections across Points.   The reflection across the point C in the Euclidean plane takes C to itself. For each point P other than C, the reflection across the point C takes P to P', where C is the midpoint of the segment $\overline{PP'}$ connecting P to P'. Thus, to obtain the image point P', given P and C, construct the segment $\overline{PC}$ and then extend this segment a distance PC beyond C. In the Euclidean plane, reflection across a point C is the same as rotation about the point C by $180°$.

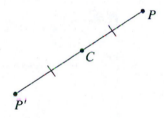

For example, if C is taken to be the origin $C = (0,0)$ of the xy-plane, then reflection about the origin takes the point $P = (x,y)$ to the point $P' = (-x,-y)$.

EXERCISES 4.1

1. Let the triangle $\triangle ABC$ have vertices given by $A = (0,0)$, $B = (2,0)$, and $C = (1,2)$. The transformation $r_m: R^2 \to R^2$ given by $r_m(x,y) = (-x + 4, y)$ is reflection across the line m given by the equation $x = 2$.
   a. Find the area of $\triangle ABC$.
   b. Find the images under r_m of A, B, and C (i.e., find A', B', and C').
   c. Find the area of the triangle $\triangle A'B'C'$.

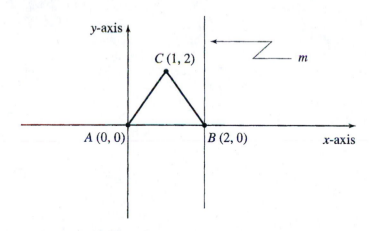

2. Let the parallelogram $EFGH$ have vertices given by $E = (0,0)$, $F = (3,0)$, $G = (4,1)$, and $H = (1,1)$. The transformation $T: R^2 \to R^2$ given by $T(x,y) = (-x, -y)$ is reflection across the origin. It can also be thought of as rotation about the origin by $180°$.
   a. Find the area of the parallelogram $EFGH$.
   b. Find the images of E, F, G, and H under the transformation F.
   c. Is the quadrilateral $E'F'G'H'$ a parallelogram?
   d. Find the area of $E'F'G'H'$.

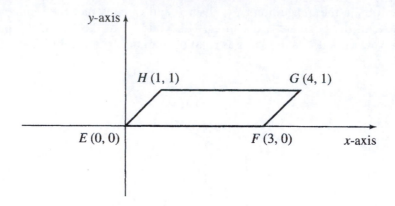

3. Let $ABCD$ be a rectangle that is not a square. Thus, $AB = CD$, $AD = BC$, and $AB \neq AD$. Find two lines of symmetry. In other words, find two lines m and k such that reflection over each of m and k takes the rectangle exactly onto itself.

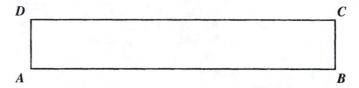

4. Explain why, for any line m, the composition of r_m with itself is the identity mapping. In other words, explain why $r_m \circ r_m(P) = P$ for all points P.

5. Let v represent the y-axis and let h represent the x-axis.
   a. Find the image of $(2,3)$ under the reflection r_h.
   b. Find the image of $(2,-3)$ under the reflection r_v.
   c. Find the image of $(2,3)$ under the composition mapping $r_v \circ r_h$.
   d. What is the image of an arbitrary point (x,y) under $r_v \circ r_h$?
   e. Give a simple geometric description of the composition $r_v \circ r_h$.

6. Let k be the line $y = x$ in the xy-plane. Find
   a. $r_k(1,0)$.
   b. $r_k(0,1)$.
   c. $r_k(3,4)$.
   d. $r_k(x,y)$.

7. Let k be the line $y = x$ in the xy-plane.
   a. Graph the curve C given by $y = 10^x$.
   b. Graph the image C' of the curve C under the reflection r_k.
   c. What is the equation of the curve C'?

8. Let $T_1 : R^2 \to R^2$ be reflection across the origin $(0,0)$, and let $T_2 : R^2 \to R^2$ be the reflection across the point $(1,1)$. Find
   a. $T_1(1,0)$.
   b. $T_2(1,0)$.
   c. $T_1 \circ T_2(1,0)$.
   d. $_{(d)} T_2 \circ T_1(1,0)$.

Classroom Connection 4.1.2

The GSP is especially good for teaching middle and high school students about transformations in geometry. Using the GSP, mark a line m and then reflect figures across this line. One may easily construct a GSP lesson that illustrates that lengths, angles, and areas are preserved under reflections across lines. Furthermore, easily include a part of the lesson to illustrate that orientation is not preserved under reflections of the Euclidean plane across a line.   ◆

4.2   TRANSLATIONS, ROTATIONS, AND GLIDE-REFLECTIONS

Rigid motions of the plane are transformations of the Euclidean plane that leave distances unchanged; all of them may be obtained by doing repeated reflections across lines. In this section, one finds how repeated reflections across lines yield translations, reflections, and glide reflections. To get an intuitive feel for rigid motions, use either sheets of tracing paper or transparencies. The following Classroom Connection demonstrates that translations of the xy-plane correspond to moving the point with coordinates (x, y) to the point with coordinates $(x + a, y + b)$. For purposes of illustration, the case $(a, b) = (4, 2)$ has been used in this Classroom Connection.

Classroom Connection 4.2.1

Give students a paper sheet with xy-coordinates and also a transparency that contains the same xy-coordinates. Have them place the transparency over the paper sheet with the origin of the transparency located at some point with coordinates different from $(0, 0)$ on the paper sheet. For example, place the origin of the transparency over the point with coordinates $(4, 2)$ on the paper sheet. Also have the coordinate axes of the transparency parallel to the coordinate axes of the paper sheet. The positive x-axis on both sheets should point in the same direction, and the positive y-axis on both sheets should point in the same direction. Now for different points have the students make a table that is both the coordinates for the paper sheet and for the transparency. They should soon be able to see that, in general, the point with coordinates (x, y) on the transparency lies over the point with coordinates $(x + 4, y + 2)$ on the paper sheet. This demonstrates the point with the original coordinates (x, y) on the transparency corresponds to the point with new coordinates $(x + 4, y + 2)$ on the paper sheet.   ◆

Translations.   A translation can be defined as the composition of two reflections across parallel lines. Let m and k be parallel lines and let c be the constant distance between them. Consider the reflection across m followed by the reflection across k (i.e., $r_k \circ r_m$). The net result is that each point of the plane is translated in the direction of m toward k by an amount $2c$, which is twice the separation of the two lines. This is illustrated in Figure 4.2.1.

EXAMPLE   Let the line m be given by $x = 6$ and the line k be given by $x = 10$. Find the image of the point (x, y) under the translation given by $r_k \circ r_m$. In particular, find the image of the point $P = (1, 1)$.

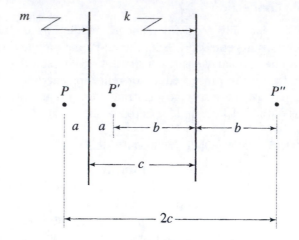

FIGURE 4.2.1 The parallel lines m and k are distance $c = a + b$ apart. The reflection r_m takes P to P' and the reflection r_k takes P' to P''. The distance from P to m and also from P' to m is shown as a. The distance from P' to k and also the distance from P'' to k is shown as b. Then $PP' = 2a$, $P'P'' = 2b$, and $PP'' = 2a + 2b = 2c$. Thus, $r_k \circ r_m (P) = P''$ is the image of P after it has been moved a distance $2c$ in a direction perpendicular to the parallel lines and in the direction m toward k.

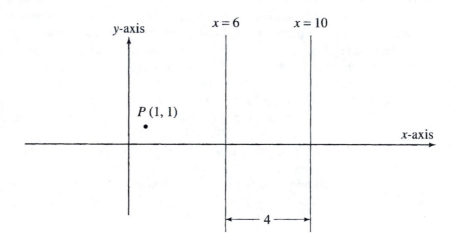

Solution The lines m and k are parallel to the y-axis, and the translation must move points in the perpendicular direction, which means parallel to the x-axis. In other words, the y-coordinate of each point remains the same under the translation $r_k \circ r_m$, but the x-coordinate will be changed by the translation. The distance between the two lines is $c = 10 - 6 = 4$. Notice that in going from the first line m toward the second line k, one moves in the positive x-direction. Thus, the translation moves each point the distance $2c = 8$ units in the positive x-direction. Thus, the general point (x, y) has image given by $(x + 8, y)$. In particular, $r_k \circ r_m(P) = r_k \circ r_m(1, 1) = (9, 1)$. ■

A translation τ_{AB} is determined when one point A and its image B are given. If L is the midpoint of the segment $\overline{AB}$, let m be the line perpendicular to $\overline{AB}$ at L, and let k be the line perpendicular to $\overline{AB}$ at B. The translation taking A to B is $\tau_{AB} = r_k \circ r_m$.

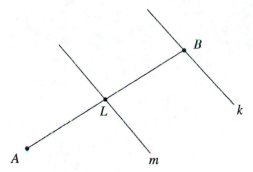

Using xy-coordinates, translations are of the form $F(x,y) = (x + h, y + k)$. Notice that this transformation takes the origin $(0,0)$ to the point (h,k).

EXAMPLE Given that the translation $F(x,y) = (x + h, y + k)$ takes the point $(7,9)$ to the point $(11,23)$, find the constants h and k.

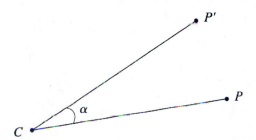

FIGURE 4.2.2 Illustration of rotation by an angle α about the point C. The point P goes to a point P' that is the same distance from C as P. Thus, CP' = CP. If α is positive, then the rotation is counterclockwise about the center C. If α is negative, the rotation is clockwise. Figures go to congruent figures and orientation is preserved. The rotation about C by angle α is denoted by $R_{C,\alpha}$.

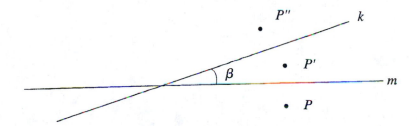

FIGURE 4.2.3 Illustration of two reflections across intersecting lines m and k with an angle β between them. The net result of $r_k \circ r_m$ is a rotation of size 2β about the point of intersection. In this figure, $P' = r_m(P)$ and $P'' = r_k(P') = r_k \circ r_m(P)$.

Solution   Using $F(x, y) = (x + h, y + k)$, one has that

$$F(7, 9) = (7 + h, 9 + k).$$

Also, one is given that

$$F(7, 9) = (11, 23).$$

These last two equations yield

$$(7 + h, 9 + k) = (11, 23).$$

Solving for h and k, one obtains $h = 11 - 7 = 4$ and $k = 23 - 9 = 14$.   ■

Rotations.   A **rotation** by angle α about the point C is denoted by $R_{C,\alpha}$, see Figure 4.2.2. By convention, angles are considered positive when measured counterclockwise and negative when measured clockwise.

On page 36, you made two reflections of an irregular shape over two perpendicular lines. The lines that you folded the paper along are lines of reflection. You saw that these two reflections have the same result as rotating the original shape.

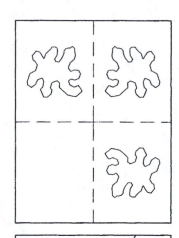

7. **a.** If the two lines of reflection are not perpendicular, is the resulting image still a rotation of the original shape? Use cutting and folding as you did on page 36 to discover what happens.

   b. Describe the resulting image if the two lines of reflection are parallel.

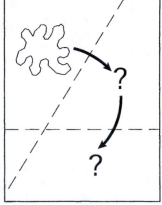

FIGURE 4.2.4 Two reflections across intersecting lines yield a rotation about the point of intersection. The angle of rotation is twice the angle between the two lines. Reproduced from page 38 of *Triangles and Beyond* in the Mathematics in Context grade 7/8 materials.

Whenever one has the composition of two reflections across lines that intersect in a point C, the net result is a rotation about C by an angle that is twice that of the angle between the two lines. Let the intersecting lines be m and k, and let the angle between them be β, then the rotation resulting from $r_k \circ r_m$ is in the direction of m to k and is of size 2β. Figures 4.2.3 and 4.2.4 illustrates that reflections across two intersecting lines yield a rotation.

EXAMPLE Let the line m be the positive x-axis, and let the line k be the line $y = x$ in the xy-plane. If $R_{C,\alpha} = r_k \circ r_m$ is the rotation resulting from reflection across the line m followed by reflection across the line k, then find the center C and angle α of this rotation.

▶ **A rotation** turns a figure about a fixed point—the **center of rotation**—a certain amount in one direction, either clockwise or counterclockwise. The new figure is the **image** of the original figure.

EXAMPLE

The rotation shown can be described in two ways.

Clockwise Rotation

The image is a **90° clockwise rotation** of the original figure about point P.

Counterclockwise Rotation

The image is a **270° counter-clockwise rotation** of the original figure about point P.

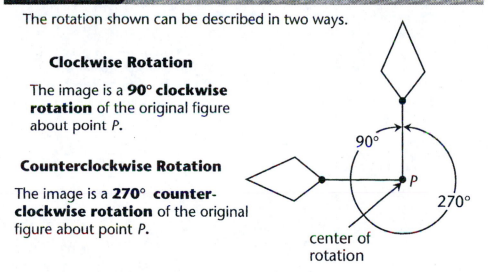

FIGURE 4.2.5 A rotation by α degrees in one direction about a point P is equivalent to a rotation about the point P in the opposite direction by an amount $360° - \alpha$. More generally, a rotation by α degrees in one direction is equivalent to a rotation by $n \cdot 360° + \alpha$ in that same direction for any integer n. It is also equivalent to a rotation by $n \cdot 360° - \alpha$ in the opposite direction for any integer n. Reproduced from page 264 of *Book 2* in the Math Thematics grade 7 materials.

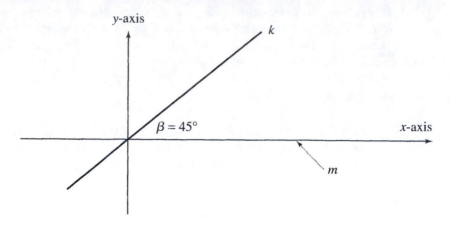

Solution   The intersection of the two lines m and k is clearly the origin, hence $C = (0,0)$. The angle between the lines m and k is 45° (counterclockwise from m to k) hence, the angle of rotation is $\alpha = 2\beta = 2 \cdot (45°) = 90°$. Consequently, $R_{C,\alpha} = r_k \circ r_m$ is the rotation about the origin by 90° in the counterclockwise direction.   ■

Of course, rotation about C by any integer multiple of 360° has the net effect of not moving any point. Points are just rotated a certain number of times about C and come to rest where they started. Consequently, rotation about C by an angle α (measured in degrees) is equivalent to rotation by $\alpha + n \cdot 360°$ for any integer n. Furthermore, as Figure 4.2.5 illustrates, rotation about the point C by an angle α in one direction is equivalent to a rotation in the opposite direction by $360° - \alpha$. Figure 4.2.6 has illustrations demonstrating rotations, rotational symmetry and reflections.

EXAMPLE   What is this equivalent clockwise rotation to a given rotation of 45° in the counterclockwise direction?

Solution   There are an infinite number of correct answers to this question. The usual answer given for this question is $360° - 45° = 315°$ in the clockwise direction. Adding or subtracting any (integer) multiple of 360° to 315° also yields a correct answer. In particular, some other correct answers are 675° clockwise and $-45°$ clockwise.   ■

Glide Reflections.   A **glide reflection** is a translation τ_{AB} followed by a reflection r_{AB} across the line $\overleftrightarrow{AB}$. The notation is γ_{AB}. Thus, $\gamma_{AB} = r_{AB} \circ \tau_{AB}$. Usually, two isometries do not commute. Thus, generally the order in which one follows the other makes a difference. However, translation in the direction of a given line and reflection across that line do commute. Thus, $r_{AB} \circ \tau_{AB} = \tau_{AB} \circ r_{AB}$.

$$\gamma_{AB} = r_{AB} \circ \tau_{AB} = \tau_{AB} \circ r_{AB} \quad \text{(glide reflection)}$$

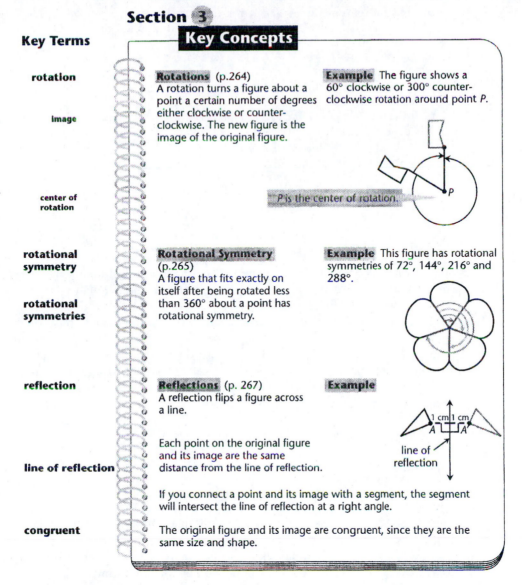

Section 3

Key Terms

Key Concepts

rotation

image

Rotations (p.264)
A rotation turns a figure about a point a certain number of degrees either clockwise or counter-clockwise. The new figure is the image of the original figure.

Example The figure shows a 60° clockwise or 300° counter-clockwise rotation around point *P*.

P is the center of rotation.

center of rotation

rotational symmetry

rotational symmetries

Rotational Symmetry (p.265)
A figure that fits exactly on itself after being rotated less than 360° about a point has rotational symmetry.

Example This figure has rotational symmetries of 72°, 144°, 216° and 288°.

reflection

line of reflection

Reflections (p. 267)
A reflection flips a figure across a line.

Each point on the original figure and its image are the same distance from the line of reflection.

If you connect a point and its image with a segment, the segment will intersect the line of reflection at a right angle.

Example

1 cm 1 cm
A   *A*

line of reflection

congruent

The original figure and its image are congruent, since they are the same size and shape.

FIGURE 4.2.6 Note that in Math Thematics, the authors do not consider the trivial symmetry (i.e., the identity map) to be a symmetry. On the other hand, the identity map is considered to be a symmetry in the present book. Reproduced from page 270 of *Book 2* in the Math Thematics grade 7 materials.

EXERCISES 4.2

1. Find the parameters h and k for the translation $F(x, y) = (x + h, y + k)$, which takes the origin $(0,0)$ to the point $(5,7)$.

2. Find the parameters h and k for the translation $F(x, y) = (x + h, y + k)$, which takes the point $(1, 2)$ to the point $(4, 4)$.

3. Find the parameters h and k for the translation $F(x,y) = (x + h, y + k)$ if $F = r_w \circ r_m$, where m is the line $x = 0$ and w is the line $x = 2$. Give a short description of this translation.

4. The map $F(x,y) = (-x + 6, y)$ is a reflection over some line m. Find the equation of the line m.

5. $R_{C,\alpha}(x,y) = (-y, x)$ is a rotation about the origin $C = (0,0)$ by some angle α. Find the angle α.

6. Is the map $F(x,y) = (2x, y)$ an isometry of the Euclidean plane? Why or why not?

7. Let $A = (0,0)$, $B = (2,0)$, and $P = (-3,2)$.
    a. Find the coordinates of $\tau_{AB}(P)$.
    b. Find the coordinates of $\gamma_{AB}(P)$.

8. Let $C = (1,1)$, $D = (3,3)$, and $Q = (3,4)$.
    a. Find the coordinates of $\tau_{CD}(Q)$.
    b. Find the coordinates of $\gamma_{CD}(Q)$.

Classroom Connection 4.2.2

The GSP is very good for teaching middle and high school students about rotations around points and reflections across points in the Euclidean plane. Have students construct two perpendicular lines m and k using the GSP and then reflect first across one line and then the other. Ask students how they think the composition of these reflections is related to some rotation. Assuming they had difficulty, how would you guide them to the conclusion that the composition is the same as a $180°$ rotation about the point of intersection of the lines (i.e., $O = m \cap k$)? How would you guide them to the conclusion that reflection across the point O is also the same as rotation by $180°$ about O? After students understand the above, then have them do reflections across two lines ℓ_1 and ℓ_2, which form an angle of $\beta (\neq 90°)$ at their point $C = \ell_1 \cap \ell_2$ of intersection. Compare that with the result of a rotation by 2β about C.   ◆

4.3 SYMMETRIES

Consider a figure consisting of two equal-size circles with an added line segment joining their closest points. Let a line v be the perpendicular bisector of the line segment and consider the reflection across this line. This reflection will take the total figure exactly onto itself (see Figure 4.3.1).

Symmetry occurs throughout the world we live in and plays an important role in art as well as the study of natural phenomena. A set T in the Euclidean plane may have line symmetry as illustrated in Figure 4.3.1, symmetry under a rotation or symmetry about a point. In addition, infinite sets such as lines may contain translational symmetries. In two dimensions, symmetry about a point is equivalent to symmetry under rotation by $180°$ about that point.

> **Definition 4.3.1** Let T be a set in the Euclidean plane. A symmetry of T is an isometry F such that F takes T onto itself. If $F = r_m$ is reflection across a line m, then T is said to be **symmetric with respect to the line** m. If F is reflection across a point C, then T is said to have **point symmetry with respect to** C. If $F = R_{C,\alpha}$ is a rotation about the point C, then T is said to have **rotational**

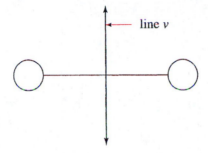

line *v*

FIGURE 4.3.1 A set is made up of two equal circles and a line segment joining their closest points. Reflection across the line that is the perpendicular bisector of the line segment just takes the set onto itself. The set is said to be **symmetric** with respect to the line *v*.

symmetry about the point C. If $F = \tau_{AB}$ is a translation, then T is said to have a **translational symmetry**.

If the set T is symmetric with respect to all rotations about a point C, then T is said to have **complete rotational symmetry** about the point C.

Let a polygon with vertices $P_1, P_2, \ldots, P_k$ be given. Of course, no three consecutive vertices are collinear and one takes $P_{k+1} = P_1$ in order to close the polygon. When considering symmetries of this polygon, notice that each side $\overline{P_i P_{i+1}}$ must be mapped to itself or to some other side that is of the same length. Also, each vertex P_i must either be mapped to itself or to another vertex.

Trivial Symmetry.   The identity map I takes each point to itself. In particular, the identity map for the Euclidean plane using xy-coordinates is $I(x, y) = (x, y)$. For any set T, the identity map is always a symmetry, according to the definition used in this book—it is the **trivial symmetry**. The reader is cautioned that not all authors consider the trivial symmetry to be a symmetry.

Classroom Connection 4.3.1

Let $T = \triangle ABC$ be a triangle with $AB = AC$, but $AB \neq BC$. Thus, $\triangle ABC$ is an isosceles triangle that is not equilateral. Middle or high school students who have studied reflections over lines should be able to discover that the trivial symmetry I and the reflection over the perpendicular bisector of the segment $\overline{BC}$ are the only two symmetries of the triangle T. A possible exploration for students is to have them work in teams of three or four to solve the general problem of classifying all possible symmetries of various types of triangles. Encourage them to think of a way to subdivide the collection of all triangles into separate types of triangles where the answer is the same for all triangles of a certain type (i.e., scalene, isosceles, and equilateral). The following are three key tools in studying symmetries of triangles (or more generally of all types of polygons) that are easy to justify. Each vertex must be mapped to a vertex, each vertex angle must be mapped to a vertex angle of equal measure, and each side must be mapped to a side of the same length. ◆

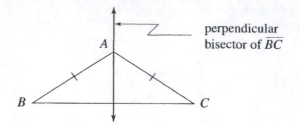

perpendicular
bisector of $\overline{BC}$

Classroom Connection 4.3.2

As Figure 4.3.2 illustrates, a good middle school activity is to have students find the symmetries of several different plane figures. In particular, pass out a sheet with several figures to the class and have them find all of the (a) rotational symmetries, (b) line symmetries, and (c) point symmetries. Have students use either transparencies or sheets of tracing paper to help them find the symmetries. ◆

A group is an important algebraic structure that is used in geometry and in many other areas of mathematics.

Group

A **group** $(G, *)$ consists of a set G of objects together with an operation $*$ taking pairs of objects to a new object $a * b$, such that the following four rules are satisfied.

1. Closed	$a * b \in G$	for all $a, b \in G$
2. Associative	$a * (b * c) = (a * b) * c$	for all $a, b, c \in G$
3. Identity	$e * a = a * e = a$	for some fixed $e \in G$ and all $a \in G$
4. Inverse	$a * a^{-1} = a^{-1} * a = e$	for all $a \in G$ there is such an $a^{-1} \in G$

---

11 **✓ CHECKPOINT**   Find all rotational symmetries of each figure.

a.

b.

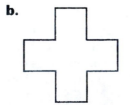

c.

FIGURE 4.3.2 Reproduced from page 265 of *Book 2* in the Math Thematics grade 7 series.

Of course, the element e in a group $(G, *)$ is called the *identity*, and the element a^{-1} is said to be the inverse of the element a.

A group $(G, *)$ is a **commutative group** if, in addition to the rules 1–4 above, it also satisfies the additional rule (5) known as the **commutative rule** (or **law**)

5. Commutative $a * b = b * a$ for all $a, b \in G$

If $(G, *)$ is a group, and if G is finite, then the elements of G may be listed in some order, say $G = \{a_1, a_2, \ldots, a_n\}$. In this case, one may construct a multiplication table by listing the elements of G in order vertically to label rows and then in order across the top to label columns. In this table, put the result of the multiplication $a_i * a_j$ in the ith row and j^{th} column.

$*$	a_1	a_2	$\ldots$	a_n
a_1	$a_1 * a_1$	$a_1 * a_2$	$\ldots$	$a_1 * a_n$
a_2	$a_2 * a_1$	$a_2 * a_2$	$\ldots$	$a_2 * a_n$
$\cdot$	$\cdot$	$\cdot$		$\cdot$
$\cdot$	$\cdot$	$\cdot$		$\cdot$
$\cdot$	$\cdot$	$\cdot$		$\cdot$
a_n	$a_n * a_1$	$a_n * a_2$		$a_n * a_n$

In the previous Classroom Connection, the set $T = \triangle ABC$ was an isosceles triangle with $AB = AC$, but $AB \neq BC$. The set G of isometries consisted of the trivial symmetry I and the reflection r over the perpendicular bisector of the segment $\overline{BC}$. Let multiplication $*$ be composition. Then for $G = \{I, r\}$, using a_1 and a_2 as I and r, respectively, one has the following multiplication table for this group.

$*$	I	r
I	I	r
r	r	I

The set of all symmetries of any set T is a group using composition for the group operation and regarding the trivial symmetry as the group's identity. Hence, if G is the set of all symmetries of T, then $(G, \circ)$ is a group. Thus, the following four rules are satisfied. It is worth noting that the associative law (i.e., 2) always holds when G consists of functions and when the operation is a composition of functions.

1. Closed: $F_2 \circ F_1 \in G$ for all $F_1, F_2 \in G$
2. Associative: $F_1 \circ (F_2 \circ F_3) = (F_1 \circ F_2) \circ F_3$ for all $F_1, F_2, F_3 \in G$
3. Identity: $F \circ I = I \circ F = F$ for all $F \in G$
4. Inverse: $F \circ F^{-1} = F^{-1} \circ F = I$ if $F \in G$ then $F^{-1} \in G$

EXAMPLE   Let T be the set of four points $\{(2, 0), (-2, 0), (0, 1), (0, -1)\}$ in the xy-plane as illustrated in the following diagram. Find the group of symmetries of T and give the multiplication table for this group.

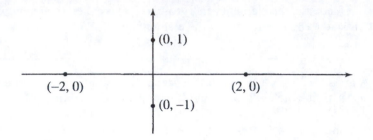

Solution   Note that $(2,0)$ and $(-2,0)$ are the only points in T at distance 4 from each other and that $(0,1)$ and $(0,-1)$ are the only points in T at distance 2 from each other. Hence, the set $\{(2,0),(-2,0)\}$ is mapped onto itself, and the set $\{(0,1),(0,-1)\}$ is mapped onto itself. It follows that $(2,0)$ must either go to itself or it must be interchanged with $(-2,0)$ under an isometry taking T to T. Likewise, $(0,1)$ must either go to itself or it must be interchanged with $(0,-1)$. Since one has two choices for the image of $(2,0)$ and two choices for the image of $(0,1)$, the set of symmetries has at most four elements. Furthermore, it is easy to find four different isometries taking T to T. Thus, the set of symmetries G of T has exactly four elements. In particular, $G = \{I, r_v, r_h, R_{(0,0),180}\}$ where I is the identity, r_v is reflection across the y-axis, r_h is reflection across the x-axis and $R_{(0,0),180}$ is rotation about the origin $(0,0)$ by $180°$. The multiplication table for $(G, \circ)$ is given here.

$\circ$	I	r_v	r_h	$R_{(o,o),180}$
I	I	r_v	r_h	$R_{(o,o),180}$
r_v	r_v	I	$R_{(o,o),180}$	r_h
r_h	r_h	$R_{(o,o),180}$	I	r_v
$R_{(o,o),180}$	$R_{(o,o),180}$	r_h	r_v	I

EXERCISES 4.3

1. Find the group of symmetries of an equilateral triangle. Explain your reasoning.
2. Consider the set shown in Figure 4.3.1 consisting of two circles and a line segment joining the points on each circle. Find the group of symmetries of this set. Explain your reasoning.
3. Give two isometries F and G of R^2 such that F and G do not commute (i.e., $F \circ G \neq G \circ F$). Explain your reasoning.
4. Find the group of symmetries of a square. Explain your reasoning.

Classroom Discussion 4.3.1

It is not hard to see that an equilateral triangle has exactly six symmetries, counting the identity, and a square has exactly eight symmetries. It is natural to ask how many symmetries a regular polygon of k sides has. This can form the basis of a worthwhile classroom discussion. An alternative approach is to form teams of four or five students and have each team come up the number of symmetries of a given

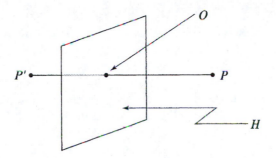

FIGURE 4.4.1 If the point P is reflected across the plane H to the image point P', then the segment $\overline{PP'}$ is perpendicular to the plane H at the point O, and O is the midpoint of the segment $\overline{PP'}$. Objects that are reflected across planes are taken to their mirror images. Thus, reflection across a plane is orientation-reversing.

regular polygon. The first team could consider a regular pentagon, the second team a regular hexagon, etc. The class should be able to identify a pattern and be able to justify the pattern. ◆

4.4 ISOMETRIES IN SPACE

Mirror Images.   In order to reflect across a plane H in Euclidean three-space, each point P is mapped to the point P'. Find this by going perpendicular from P to the plane H and then continue across H to the point P' at the same distance from H as P. This is illustrated in Figure 4.4.1. When you look at yourself in a mirror, you see an image that is the **mirror image** of yourself. For example, if you part your hair on the left, then your image in the mirror is of someone who parts his or her hair on the right. Mirror images also have a dramatic effect on writing, as the following Classroom Connection illustrates.

Classroom Connection 4.4.1

Have each student write some simple sentences on a transparency. The writing is then on the sheet's front. Now have each student turn their transparency over and look at the back of the sheet. The writing, as seen from the back, appears as the mirror image and is difficult to read. Now have each student hold their sheet up to a mirror with the front of the sheet facing the mirror. The writing shown in the mirror is the mirror image of the original sentences and is hard to read. Finally, have each student hold their transparency up with the back of the sheet facing the mirror. Now the writing shown in the mirror is the same as the original writing. ◆

EXAMPLE   If $F: R^3 \to R^3$ is a reflection across the yz-coordinate plane (i.e., across the plane $x = 0$), then the y and z axes are held fixed, and the x-axis is mapped onto itself with the positive x-axis and negative x-axis interchanged. In particular, this reflection is given in function form by $F(x, y, z) = (-x, y, z)$. Prove that this is an isometry.

> **Proof.** To prove that this is an isometry, take two points P, Q given in general form and prove that the distance from P to Q is the same as that from point $F(P)$ to $F(Q)$. In other words, prove that $PQ = F(P)F(Q)$. Thus,

let $P = (x_1, y_1, z_1)$ and let $Q = (x_2, y_2, z_2)$. Notice that to give this proof, one must have P and Q expressed in general form with coordinates having x's, y's, and z's, as opposed to choosing actual fixed numbers for the coordinates of P and Q. Using the distance formula for space one obtains $PQ = \sqrt{(x_1 - x_2)^2 + (y_1 - y_2)^2 + (z_1 - z_2)^2}$. Using $F(x, y, z) = (-x, y, z)$, one finds that $F(P) = F(x_1, y_1, z_1) = (-x_1, y_1, z_1)$ and that $F(Q) = F(x_2, y_2, z_2) = (-x_2, y_2, z_2)$. Hence, the distance from the point $F(P)$ to the point $F(Q)$ is given by the following:

$$F(P)F(Q) = \sqrt{[(-x_1) - (-x_2)]^2 + (y_1 - y_2)^2 + (z_1 - z_2)^2}$$

$$= \sqrt{(-x_1 + x_2)^2 + (y_1 - y_2)^2 + (z_1 - z_2)^2}$$

$$= \sqrt{(x_1 - x_2)^2 + (y_1 - y_2)^2 + (z_1 - z_2)^2},$$

which is equal to PQ, as desired.   ∎

Consider the set of coordinate axes shown here. This is a right-handed system, since when using a right hand, one may point the thumb in the positive x-axis direction and point the index finger in the positive y-axis direction and the middle finger in the positive z-axis direction.

Right-handed coordinate system

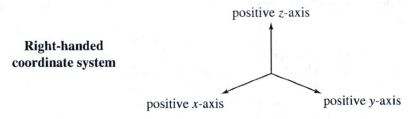

positive z-axis

positive x-axis

positive y-axis

A left-handed coordinate system can be obtained from the previous by reversing one of the coordinate axes (i.e., by pointing that axis in the opposite direction) while holding the other two axes fixed. For example, in the previous diagram, one could have the positive z-axis point in the opposite direction and have the other two axes point as they were originally.

positive x-axis   positive y-axis

Left-handed coordinate system

positive z-axis

Orientation-Reversing.   An **orientation reversing isometry** will convert a right-handed system into a left-handed system and vice versa. If you look in a mirror and hold up your right hand with thumb, index finger, and middle finger all perpendicular to one another, then the image in the mirror will be holding up its left hand with thumb, index finger, and middle finger all mutually perpendicular. In other words, your mirror image has the opposite "handedness" and is thus an orientation-reversed version of yourself. More precisely, looking at the image in a mirror corresponds to reflection across the mirror's plane. In three dimensions, reflection across a plane reverses orientation of three-dimensional objects just as reflection across a line in two dimensions reverses orientation of two-dimensional objects.

Recall that if the point O is given and one wishes to reflect across that point, then each point P goes to a point P' such that O is the midpoint of the segment $\overline{PP'}$ joining P to P'. In the two-dimensional Euclidean plane, a rotation about a point by $180°$ corresponds to reflection across the point. Thus, in the Euclidean plane reflection across a point is orientation-preserving. However, in three-dimensional Euclidean space, reflection across a point reverses orientation.

Translations in Three Dimensions.   A translation in the Euclidean xy-plane is an orientation-preserving isometry that takes the ray that is the positive x-axis to a ray pointing in the same direction and that takes the positive y-axis to a ray pointing in the same direction. In three dimensions, using xyz-coordinates, the situation is essentially the same except that the ray that is the positive z-axis must now be included. Thus, a **translation** in three dimensions is an orientation-preserving isometry that takes each individual positive coordinate axis to a ray pointing in the same direction. In function form, $F : R^3 \rightarrow R^3$ is a translation if there are constants a, b, c such that F is given by $F(x, y, z) = (x + a, y + b, z + c)$.

Rotations in Three Dimensions.   All proper (i.e., orientation-preserving) rotations in three-dimensional Euclidean space are known to be **axial rotations**. In other words, there is some line l of fixed points, and for any plane H perpendicular to l, the plane H is rotated about the point of intersection of H and l. The line l is the **axis** of the rotation. For example, consider a rotation about the z-axis where points on the positive x-axis are taken to points on the positive y-axis. The function form of this transformation is given by $F(x, y, z) = (-y, x, z)$. Clearly, this F takes each point on the z-axis to itself since points on the z-axis are of the form $(0, 0, z)$ and $F(0, 0, z) = (0, 0, z)$. The unit point $(1, 0, 0)$ on the x-axis is mapped to the unit point $(0, 1, 0)$ on the y-axis. However, the unit point $(0, 1, 0)$ on the y-axis is mapped to the point $(-1, 0, 0)$ on the negative x-axis. This particular transformation is often referred to as a rotation about the z-axis by $+90°$ since it induces a rotation in the xy-plane by $+90°$ about the origin. Here the xy-plane is thought of as being viewed by an observer located on the positive z-axis. This observer sees the rotation as a $+90°$ (i.e., counterclockwise) rotation of the xy-plane

A cube or ordinary die may be used to illustrate rotation about various coordinate axes. Consider a cube with dots as illustrated here.

top: 3 dots

left back: 5 dots   right back: 6 dots

left front: 1 dot   right front: 2 dots

bottom: 4 dots

Now think of a right-handed xyz-coordinate system with origin located at the cube's center. Let the positive x-axis go out the face with one dot, let the positive y-axis go out the face with two dots, and let the positive z-axis go up through the face with three dots. A rotation about the z-axis by $+90°$, as described in the previous example, will take the face with three dots to itself but will reposition the three dots. It will take the face with one dot to where the face with two dots was originally. The face with five dots will rotate to where the face with one dot was originally located.

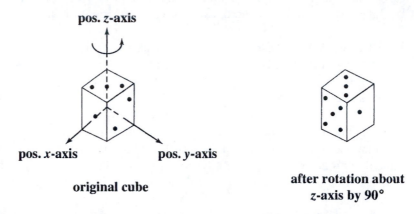

pos. z-axis

pos. x-axis   **pos. y-axis**

original cube

**after rotation about
z-axis by 90°**

A rotation about the x-axis that takes points on the positive y-axis to points on the positive z-axis is given by $F(x, y, z) = (x, -z, y)$. This rotation about the x-axis induces a rotation in the yz-plane by $+90°$. Here an observer looking down at the yz-plane is located on the positive x-axis. Of course, the y-axis is the first axis in the yz-plane, and the z-axis is the second axis. This rotation is illustrated here with a cube having faces with dots as previous. Notice that after rotation about the x-axis, the new top face has two dots. The face with four dots has been rotated to where the face with two dots was before.

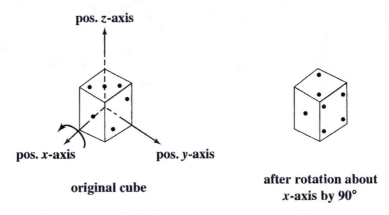

pos. z-axis

pos. x-axis   **pos. y-axis**

original cube

**after rotation about
x-axis by 90°**

A rotation by $+90°$ about the y-axis will take the positive z-axis onto the positive x-axis. Here an observer is located on the positive y-axis, and the z-axis is the first axis in the zx-plane. In function form, this rotation is given by $F(x, y, z) = (z, y, -x)$.

pos. z-axis

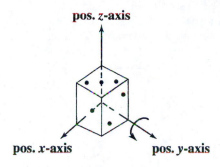

pos. x-axis   **pos. y-axis**

original cube

**after rotation about
y-axis by 90°**

Classroom Discussion 4.4.1

As previously discussed, the transformation $F_1(x, y, z) = (-y, x, z)$ is a rotation by $+90°$ about the z-axis, and $F_2(x, y, z) = (x, -z, y)$ is a rotation by $+90°$ about the x-axis. Show your students a cube (or die) with dots as illustrated previously. Ask them what they think the cube will look like, viewed from the same direction as in the original illustration, after doing a rotation by $+90°$ about the z-axis and then a rotation by $+90°$ about the x-axis. This corresponds to the composition $F_2 \circ F_1$. It might help to tell the students that when doing a rotation, the faces of the cube move but each coordinate axis stays fixed in its original position. After they understand the first part of this exploration, ask them what the cube would look like if first rotated by $+90°$ about the x-axis and then by $+90°$ about the z-axis. This second sequence of rotations correspond to the composition $F_1 \circ F_2$. These two sequences of rotations show that $F_2 \circ F_1 \neq F_1 \circ F_2$, which illustrates that rotations in Euclidean space do not necessarily commute. In Chapter 5, rotations about the origin of xyz-space are represented using 3×3 orthogonal matrices. Thus, F_1 and F_2 are represented by orthogonal matrices A and B, respectively. Of course, in this case, one finds that $AB \neq BA$. ◆

**Rotation by 90°
first around z-axis and
then around x-axis.**

**Rotation by 90°
first around x-axis and
then around z-axis.**

A point P is fixed by the transformation $F: R^3 \to R^3$ if it is mapped to itself. To find the set of points that are fixed by a given transformation, solve the equation $F(P) = P$ [i.e., solve for (x, y, z) using $F(x, y, z) = (x, y, z)$]. For example, if $F = F_2 \circ F_1$ where $F_1(x, y, z) = (-y, x, z)$ and $F_2(x, y, z) = (x, -z, y)$, then

$$F(x, y, z) = F_2[F_1(x, y, z)] = F_2(-y, x, z) = (-y, -z, x).$$

Look at the first structure in each box. Tell which of the other structures are rotations of the sample. Answer *yes* or *no* for each structure.

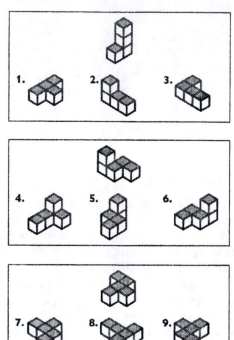

FIGURE 4.4.2 Reproduced from page 35 of *Designing Spaces* in the MathScape grade 6 materials.

Thus, $F(x, y, z) = (x, y, z)$ becomes $(-y, -z, x) = (x, y, z)$. Setting the first, second, and third components of $(-y, -z, x)$ equal to the respective first, second, and third components of (x, y, z) yields the following three equations: $-y = x, -z = y$, and $x = z$. If one thinks of the x' as a variable and sets it equal to t (i.e., let $x = t$), then the three equations $-y = x, -z = y$, and $x = z$ become equivalent to the following:

$$x = t$$

$$y = -t$$

$$z = t.$$

This represents a line in R^3. Letting t take on different values gives different points on the line. For example, if $t = 1$, then $x = 1, y = -1$, and $z = 1$. This means that the point $(1, -1, 1)$ is mapped to itself. Likewise, if one sets $t = 0$, then one finds $x = 0, y = 0$, and $z = 0$, which means that the origin $(0, 0, 0)$ is another fixed point of this transformation F.

Figure 4.4.2 demonstrates that three-dimensional rotations are now part of the middle school mathematics curriculum.

EXERCISES 4.4

1. Let $F(x, y, z) = (x + a, y + b, z + c)$ be the translation of xyz-space that translates the point $(3, 4, 5)$ to the point $(6, -2, 7)$. Find the constants a, b, c.

2. Let $F : R^3 \to R^3$ be given by $F(x, y, z) = (x, y, -z)$. This F represents reflection across some plane H. Find the plane H (you may either describe H in words or give the equation of H).

3. Assume that F_i is a reflection across the plane H_i, for $i = 1, 2, \ldots, n$.
   a. What condition on n will make the composition $F_n \circ F_{n-1} \circ \cdots \circ F_2 \circ F_1$ orientation-preserving?
   b. What condition on n will make the composition $F_n \circ F_{n-1} \circ \cdots \circ F_2 \circ F_1$ orientation-reversing?

4. The plane H given by $x = y$ in xyz-space contains the z-axis and also the line $x = y$ in the xy-coordinate plane (i.e., the line $x = y$ in the plane $z = 0$). Express the reflection across the plane H in the form $F : R^3 \to R^3$.

5. The following cube is rotated by $+90°$ three times, first about the x-axis, then the y-axis, and then the z-axis. Indicate in the illustration how many dots each face has after the sequence of three rotations has been completed. Here the origin is at the cube's center, the positive x-axis goes through the face with one dot, the positive y-axis goes through the face with two dots, and the positive z-axis goes through the face with three dots.

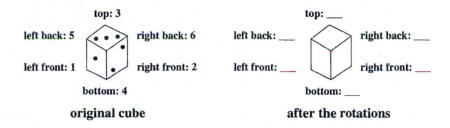

original cube — top: 3, left back: 5, right back: 6, left front: 1, right front: 2, bottom: 4

after the rotations — top: ___, left back: ___, right back: ___, left front: ___, right front: ___, bottom: ___

CHAPTER 4 REVIEW

Isometries are transformations that preserve distance, angles, and areas; however, they do not necessarily preserve orientation. Thus, an isometry will take a set W to an image W' that either looks like W or is a mirror image of W. This chapter's approach was to first investigate reflections over lines in the Euclidean plane. A single reflection over a line in the Euclidean plane reverses orientation. Thus, after such a reflection, a triangle $\triangle ABC$ that is traversed clockwise by going from vertex to vertex in alphabetic order will have an image $\triangle A'B'C'$ that is traversed counterclockwise in traversing the vertices in alphabetic order. On the other hand, the two triangles will have the same area, corresponding sides will have the same length, and corresponding angles will be congruent.

Using repeated reflections across lines in the Euclidean plane, one may build up all of the other isometries of the plane. An odd number of reflections across lines in the Euclidean plane ends up reversing orientation, and an even number of such reflections does not reverse orientation. Reflecting across two parallel lines yields a translation that is perpendicular to the lines such that each point is taken a distance that is twice the distance between the lines. The translation is in the direction of the first line toward the second. Reflecting across two lines that have just the point C in common results in a rotation about C. If the angle between the lines is β, then each point other than C is rotated by an angle of 2β. Since both translations and rotations come from an even number (i.e., two) of reflections across lines, they are orientation-preserving. Glide reflections come from doing a translation taking point A to point B and then reflecting across the line $\overleftrightarrow{AB}$. Glide reflections are orientation-reversing. The collection of all isometries of the Euclidean plane forms a noncommutative group using the operation of composition.

Isometries in three-dimensional Euclidean space include reflections across planes, translations, rotations about a point, and combinations of these types. A reflection across a plane reverses orientation and thus takes a figure W to one of its mirror images W'. Imagine an irregularly shaped solid object is picked up and put somewhere. This movement corresponds to an isometry, and it will be a nontrivial isometry if the object has either been moved to a new location or has been given a nontrivial rotation. Furthermore, since solid objects cannot have their orientation reversed, the isometry will be orientation-preserving.

Symmetries are increasingly studied in both middle and high school mathematics. An object's group of symmetries is the set of all isometries that map the object onto itself. In this book, any figure in the Euclidean plane or space always has the trivial isometry, which does not move any point, as one of its symmetries.

When investigating symmetries of a given bounded figure in the plane, one looks for lines of symmetry, points of symmetry, and rotational angles of symmetry. Unbounded figures may also have translations as symmetries. A figure in the plane has a line m as a line of symmetry if the reflection across m takes the figure onto itself. It has point symmetry with respect to point C, if the reflection across point C takes it onto itself. A figure that is mapped onto itself under a rotation about C by an angle α has $R_{C,\alpha}$ as one of its symmetries. The figure has complete rotational symmetry about C if every rotation about C is a symmetry. Solid objects in Euclidean space may have many types of symmetries, including planar, point, rotational, and, in the case of unbounded sets, translational.

Selected notational conventions and rules:

Reflection across line m:	r_m
Translation taking A to B:	τ_{AB}
Rotation by angle α about C:	$R_{C,\alpha}$
Glide reflection:	$\gamma_{AB}(= r_{AB} \circ \tau_{AB})$
Composition of functions:	$F_2 \circ F_1(P) = F_2[F_1(P)]$ (i.e., do F_1, then do F_2)
Associative rule:	$a * (b * c) = (a * b) * c$
Identity rule:	$a * e = e * a = a$
Inverse rule:	$a * a^{-1} = a^{-1} * a = e$
Commutative rule:	$a * b = b * a$

CHAPTER 4 REVIEW EXERCISES

1. The point $P = (5,1)$ in the xy-plane is reflected across the x-axis.
   a. Find the image P' of P.
   b. Graph both P' and P.

2. The point $Q = (6,2)$ in the xy-plane is reflected across the y-axis.
   a. Find the image Q' of Q.
   b. Graph both Q' and Q.

3. Let $S = (1,0)$ and $T = (2,1)$.
   a. Reflect S and T across the line $x = 3$ and find their images S' and T'.
   b. Reflect S' and T' across the line $x = 6$ and find their images S'' and T''.
   c. Graph all three of the segments $\overline{ST}, \overline{S'T'}$, and $\overline{S''T''}$.

4. Let $S = (1,0)$, $T = (2,1)$, and $U = (2,0)$.
   a. Reflect $\triangle STU$ across the line $x = 3$ to get $\triangle S'T'U'$.
   b. Reflect $\triangle S'T'U'$ across the line $x = 6$ to get $\triangle S''T''U''$.
   c. Graph all three of the triangles $\triangle STU, \triangle S'T'U'$, and $\triangle S''T''U''$.

5. Let $A = (2,3)$, $B = (6,3)$, and $C = (4,4)$ be the vertices of $\triangle ABC$.
   a. Find the image triangle $\triangle A'B'C'$ after reflection across the x-axis.
   b. Graph both $\triangle ABC$ and $\triangle A'B'C'$.
   c. Assuming that each triangle is traversed such that the vertices are in alphabetic order, which is traversed clockwise and which is traversed counterclockwise?

6. Let $A = (0,0)$, $B = (1,0)$, $C = (1,1)$, and $D = (0,1)$. Thus, $ABCD$ is the unit square in the first quadrant of the xy-plane.
   a. Find the image $A'B'C'D'$ of $ABCD$ after reflection across the line $x = 2$.
   b. Graph both $ABCD$ and $A'B'C'D'$.

7. Let $L = (1,1)$, $M = (2,1)$, and $N = (2,2)$ be the vertices of triangle $\triangle LMN$ in the xy-plane.
   a. Find the image triangle $\triangle L'M'N'$ after reflection across the line $y = -3$.
   b. Graph both $\triangle LMN$ and $\triangle L'M'N'$.

8. Let $E = (0,0), F = (3,0), G = (3,2)$, and $H = (0,2)$ be the vertices of a rectangle in the xy-plane.
   a. Find the vertices of the image $E'F'G'H'$ of $EFGH$ after rotation about the origin by $+90°$ (i.e., counterclockwise by $90°$).
   b. Graph both $EFGH$ and $E'F'G'H'$.

9. Let $W = (1,0), X = (5,0), Y = (5,3)$, and $Z = (1,3)$ be the vertices of a rectangle in the xy-plane.
   a. Find the vertices of the image $W'X'Y'Z'$ of $WXYZ$ after rotation about the origin by $+270°$.
   b. Graph both $WXYZ$ and $W'X'Y'Z'$.

10. Let $P = (2,2), Q = (4,2)$, and $R = (4,4)$ be the vertices of triangle $\triangle PQR$ in the xy-plane.
    a. Find the image triangle $\triangle P'Q'R'$ after reflection across the line $y = x$.
    b. Graph both $\triangle PQR$ and $\triangle P'Q'R'$.

11. Let $\overline{ST}$ be the segment with endpoint $S = (1,0)$ and $T = (0,4)$.
    a. Find the image $\overline{S'T'}$ of the segment $\overline{ST}$ after reflection across the line $y = -x$.
    b. Graph both $\overline{ST}$ and $\overline{S'T'}$.

12. Let $A = (0,0)$, $B = (1,0)$, $C = (1,1)$, and $D = (0,1)$ be the vertices of the unit square in the first quadrant of the xy-plane.
    a. Find the image $A'B'C'D'$ of $ABCD$ after reflection across the line $y = -x$.
    b. Graph both $ABCD$ and $A'B'C'D'$.

13. Let v represent the y-axis, and let h represent the x-axis.
   a. Find the image of $(5,7)$ under the reflection r_h.
   b. Find the image of $(5,-7)$ under the reflection r_v.
   c. Find the image of $(-5,-7)$ under the mapping r_h.
   d. Find the image of the point $(5,7)$ under the composition mapping $r_h \circ r_v \circ r_h$.
   e. Find the image of an arbitrary point (x,y) under the mapping $r_h \circ r_v \circ r_h$.
   f. Give a simple geometric description of the mapping $r_h \circ r_v \circ r_h$.

14. Find the parameters h and k for the translation $F(x,y) = (x+h, y+k)$, which takes the origin $(2,5)$ to the point $(10,20)$.

15. The map $F(x,y) = (x, -y+10)$ is a reflection over some line m. Find the equation of the line m.

16. Find the number of elements in the group of symmetries of a regular hexagon and explain your reasoning. Don't forget to count the identity (i.e., trivial) symmetry.

17. Two circles of equal radius have exactly one point in common, and together they form a figure eight with the top circle congruent to the bottom circle. Let P be the point of contact of these two circles. Let ℓ_1 be the line tangent to both circles at P and let ℓ_2 be the line perpendicular to ℓ_1 at the point P.
   a. Find all lines of symmetry of this figure eight.
   b. Find all (nontrivial) rotational symmetries of this figure.
   c. Find the number of elements of the group of symmetries of this figure.

18. Two circles of unequal radii are arranged as a figure eight. Thus, one circle (say, the top circle) is smaller than the other circle.
   a. Find all lines of symmetry of this figure eight made up of circles with different radii.
   b. Find all rotational symmetries of this figure.
   c. Find the number of elements of the group of symmetries of this figure.

19. Let $O = (0,0)$, $A = (1,0)$, $B = \left(-\frac{1}{2}, \frac{\sqrt{3}}{2}\right)$, and $C = \left(-\frac{1}{2}, -\frac{\sqrt{3}}{2}\right)$. Let T be the union of the three line segments $\overline{OA}$, $\overline{OB}$, and $\overline{OC}$.
   a. Graph the set T.
   b. Find the symmetries of this set T.

20. The following cube is rotated by $+180°$ about the z-axis and then by $+90°$ about the x-axis. Indicate in the illustration, how many dots each face has after these rotations have been completed. Here the origin is at the cube's center, the positive x-axis goes through the face with one dot, the positive y-axis goes through the face with two dots, and the positive z-axis goes through the face with three dots.

top: 3
left back: 5   right back: 6
left front: 1   right front: 2
bottom: 4
original cube

top: ___
left back: ___   right back: ___
left front: ___   right front: ___
bottom: ___
after the rotations

RELATED READING FOR CHAPTER 4

Billstein, R., and J. Williamson. *Math Thematics: Book 2*. Evanston, IL: McDougal Littell, 1999.

Coxeter, H. M. S. *Introduction to Geometry*. New York: John Wiley & Sons, 1969.

Day, R., et al. *Navigating through Geometry in Grades 9–12*. Reston, VA: National Council of Teachers of Mathematics, 2001.

Dayoub, I. M., and J. W. Lott. *Geometry: Constructions and Transformations*. Menlo Park, CA: Dale Seymour Publications, 1997.

Kay, D. C. *College Geometry: A Discovery Approach*. 2nd ed. New York: Addison Wesley Longman, 2001.

Kleiman, G. "Designing Spaces". In *MathScape: Seeing and Thinking Mathematically*. Grade 6 materials. New York: Glencoe/McGraw-Hill, 1999.

Romberg, T. A., et al. *Triangles and Beyond*. Mathematics in Context. Chicago, IL: Encyclopedia Britannica Educational Corporation, 1998.

Vectors and Transformations

We studied similarity transformations in Chapter 3. These take each pair of points P, Q to points to a pair P', Q' such that the distance $P'Q'$ is S times the distance PQ between the original points. Thus, $P'Q' = S \cdot PQ$. In Chapter 4, we investigated rigid motions, which are similarity transformations with scale factor $S = 1$. In this chapter, we study more general transformations taking the xy-plane onto itself and transformations taking xyz-space onto itself. Before considering these general transformations, however, we will examine vectors and matrices.

In Section 5.1, we introduce vectors as oriented line segments where one identifies all oriented line segments that point in the same direction and have the same length. Vector addition is introduced using the "head-to-tail" method. In Section 5.2, we consider the coordinate approach to vectors, and in Section 5.3 we introduce matrices as rectangular arrays of numbers and some basic fundamentals of matrix algebra. In Section 5.4, we consider transformations in Euclidean space that take lines to lines. Square matrices with nonzero determinants are of special importance in studying these transformations. In Section 5.5, we deal with isometries using matrices.

5.1 VECTORS AS ORIENTED LINE SEGMENTS

A **vector** is an oriented line segment. If a segment $\overline{PQ}$ is given, it has two possible orientations. One possibility is to have P as the starting point or "**tail**" and have

Q as the endpoint or "**head**," see Figure 5.1.1. The notation for this case is $\overrightarrow{PQ}$. In earlier chapters, $\overrightarrow{PQ}$ represented the ray with beginning point P, which contained Q. In this chapter, we don't consider rays, so $\overrightarrow{PQ}$ will always refer to the vector with tail P and head Q. Another possibility is to have Q as the tail and P as the head. This second case is denoted by $\overrightarrow{QP}$ and is taken as the negative of $\overrightarrow{PQ}$. Thus, one writes $\overrightarrow{QP} = -\overrightarrow{PQ}$.

$$\boxed{\overrightarrow{QP} = -\overrightarrow{PQ}}$$

When using vectors, one identifies oriented line segments that are parallel, of equal length, and that are oriented in the same direction. In particular, if the lines $\overleftrightarrow{PQ}$ and $\overleftrightarrow{RS}$ are parallel with $\overleftrightarrow{PR}$ parallel to $\overleftrightarrow{QS}$, then $\overrightarrow{PQ} = \overrightarrow{RS}$. In other words, if $PQSR$ is a parallelogram, then $\overrightarrow{PQ} = \overrightarrow{RS}$. This is illustrated in Figure 5.1.2.

When two vectors $\overrightarrow{PQ}$ and $\overrightarrow{RS}$ lie on the same line, then they are equal if there is some vector $\overrightarrow{TW}$ not on the line containing the four points P, Q, R, and S, such that both $\overrightarrow{PQ} = \overrightarrow{TW}$ and $\overrightarrow{RS} = \overrightarrow{TW}$. This is illustrated in Figure 5.1.3.

Addition of Vectors.   Using the movement property of vectors, one may define **addition of vectors**. To add the vector $\overrightarrow{AB}$ to the vector $\overrightarrow{CD}$, one moves the second vector $\overrightarrow{CD}$ to start at the endpoint of the first vector to get $\overrightarrow{CD} = \overrightarrow{BE}$. Then one defines the sum of $\overrightarrow{AB}$ and $\overrightarrow{CD}$ to be the vector starting at A and ending at E, see Figure 5.1.4.

FIGURE 5.1.1 The vector $\overrightarrow{PQ}$ from P to Q is the oriented line segment starting at P and ending at the point Q. Orienting the line segment in the opposite direction yields $\overrightarrow{QP} = -\overrightarrow{PQ}$.

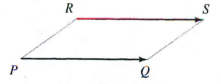

FIGURE 5.1.2 If $\overleftrightarrow{PQ}$ // $\overleftrightarrow{RS}$ and $\overleftrightarrow{PR}$ // $\overleftrightarrow{QS}$, then $PQSR$ is a parallelogram and $\overrightarrow{PQ} = \overrightarrow{RS}$. Vectors that lie on parallel lines, that are of the same length, and that are oriented in the same direction are taken to be equal. Opposite sides of a parallelogram are equal as vectors when oriented in the same direction. This, in effect, allows one to move any vector parallel to itself in order to get the vector to have any "tail" that one desires.

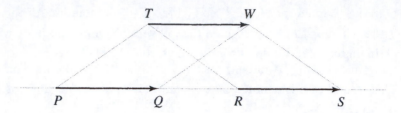

FIGURE 5.1.3 To see if two vectors on the same line are equal, use some other vector off the line that contains the two vectors. In particular, vectors $\overrightarrow{PQ}$ and $\overrightarrow{RS}$ lying on the same line are equal if there is some vector $\overrightarrow{TW}$ not on the same line with $\overrightarrow{PQ} = \overrightarrow{TW}$ and $\overrightarrow{RS} = \overrightarrow{TW}$.

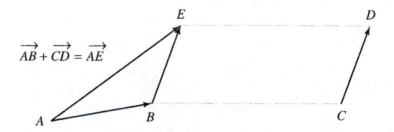

$$\overrightarrow{AB} + \overrightarrow{CD} = \overrightarrow{AE}$$

FIGURE 5.1.4 Addition of vectors is done by the head-to-tail method in which the second vector is moved to have its tail at the head of the first vector. Thus, if $\overrightarrow{CD} = \overrightarrow{BE}$, then $\overrightarrow{AB} + \overrightarrow{CD} = \overrightarrow{AB} + \overrightarrow{BE} = \overrightarrow{AE}$.

An important special case of vector addition is when the tail of the second vector is already the head of the first vector. In other words, $\overrightarrow{AB} + \overrightarrow{CD}$ where $C = B$. In this case, one has $\overrightarrow{AB} + \overrightarrow{BD} = \overrightarrow{AD}$.

In order to **subtract** vectors, use the equality $\overrightarrow{DC} = -\overrightarrow{CD}$ and the previous definition of addition of vectors.

$$\overrightarrow{AB} - \overrightarrow{CD} = \overrightarrow{AB} + \overrightarrow{DC}$$

The **zero vector** is considered equal to any vector whose head and tail are the same. In other words, the zero vector $\overrightarrow{0}$ is equal to any vector of the form $\overrightarrow{AA}$ where the head equals the tail. In effect, all single points are vectors of zero length and are identified as the zero vector.

$$\boxed{\overrightarrow{0} = \overrightarrow{AA} \text{ (zero vector)}}$$

EXAMPLE Show $\overrightarrow{0} + \overrightarrow{0} = \overrightarrow{0}$.

Solution Let A be any point and set $\overrightarrow{0} = \overrightarrow{AA}$. Using the definition of vector addition, one has $\overrightarrow{0} + \overrightarrow{0} = \overrightarrow{AA} + \overrightarrow{AA} = \overrightarrow{AA} = \overrightarrow{0}$, as desired. ∎

Norm of a Vector.   The **norm** or **length** of the vector $\overrightarrow{AB}$ is the length AB of the (unoriented) line segment with endpoints A and B. This norm is often denoted by $|\overrightarrow{AB}|$. Thus, $|\overrightarrow{AB}| = AB$.

$$\boxed{|\overrightarrow{AB}| = AB \text{ (norm or length of } \overrightarrow{AB})}$$

EXAMPLE   Find $|\overrightarrow{AB}|$, given points $A = (4,3)$ and $B = (7,7)$ in the xy-plane.

Solution $|\overrightarrow{AB}| = AB = \sqrt{(4-7)^2 + (3-7)^2} = \sqrt{9 + 16} = \sqrt{25} = 5.$ ■

Another operation to define is that of multiplying a real number a times a vector $\overrightarrow{AB}$. If a is positive, then $a \cdot \overrightarrow{AB}$ is a vector $\overrightarrow{AF}$ pointing in the same direction as $\overrightarrow{AB}$ but a times as long. If $a = 1.5$, then $a \cdot \overrightarrow{AB}$ is a vector $\overrightarrow{AF}$, which is 50 percent longer than the original $\overrightarrow{AB}$. Of course, if $a = 0$ then $0 \cdot \overrightarrow{AB}$ is the zero vector $\overrightarrow{0} = \overrightarrow{AA}$. When $a < 0$, then $a \cdot \overrightarrow{AB}$ points in the opposite direction of $\overrightarrow{AB}$ and is a times as long. In other words, $a \cdot \overrightarrow{AB} = |a| \cdot (-\overrightarrow{AB}) = |a| \cdot \overrightarrow{BA}$. Thus, if $a = -2$, one has that $(-2) \cdot \overrightarrow{AB}$ is twice as long as the original vector $\overrightarrow{AB}$ and points in the direction of $\overrightarrow{BA}$.

EXAMPLE   Find $2 \cdot \overrightarrow{OB}$ given $O = (0,0)$ and $B = (3,0)$.

Solution   The original vector $\overrightarrow{OB}$ has the tail at the origin and the head at the point, 3 units from the origin on the positive x-axis. In other words, the vector $\overrightarrow{OB}$ is 3 units long and points in the direction of the positive x-axis. Thus, the vector $2 \cdot \overrightarrow{OB}$ will be six units long and will also point in the direction of the positive x-axis. Hence, $2 \cdot \overrightarrow{OB} = \overrightarrow{OC}$ where $C = (6,0)$.   ■

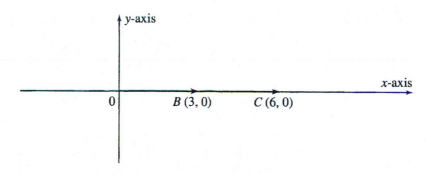

EXERCISES 5.1

1. Let $\overrightarrow{AB}$ be any given vector, and let $\overrightarrow{0}$ represent the zero vector. Using the definitions of the zero vector and of vector addition, explain why the following is valid:

$$\overrightarrow{AB} + \overrightarrow{0} = \overrightarrow{0} + \overrightarrow{AB} = \overrightarrow{AB}.$$

2. Let $\overrightarrow{EF}$ be the vector with the tail at $E = (0,0)$ and the head at $F = (1,2)$. If $\overrightarrow{EF} = \overrightarrow{HG}$ and $H = (1,0)$, find the coordinates of the point G.

3. Let $L = (0,0)$ and $M = (1,1)$. If $\overrightarrow{LN} = 3 \cdot \overrightarrow{LM}$, find the coordinates of the point N.

4. Let $K = (0,0)$ and $W = (2,1)$. If $\overrightarrow{KX} = (-2) \cdot \overrightarrow{KW}$, find the coordinates of the point X.

5. Given $Y = (1,0)$ and $Z = (2,1)$, find the norm $|\overrightarrow{YZ}|$.

6. Given any vector $\overrightarrow{AB}$, show there is always a vector $\overrightarrow{CD}$, such that the following is valid: $\overrightarrow{AB} + \overrightarrow{CD} = \overrightarrow{CD} + \overrightarrow{AB} = \overrightarrow{0}$. (Hint: Think of good candidates for C and D.)

7. Let $R = (0,0), S = (1,0)$, and $T = (1,1)$. If $\overrightarrow{RU} = \overrightarrow{RS} + \overrightarrow{RT}$, find the coordinates of the point U.

8. In this section, we considered vectors lying in Euclidean two-dimensional space. Of course, one can also study vectors in higher-dimensional Euclidean spaces in much the same fashion as in two dimensions. One defines the vector $\overrightarrow{PQ}$ to be the oriented line segment starting at P and ending at Q. As in the second dimension, one sets $\overrightarrow{PQ} = \overrightarrow{RS}$ if $\overleftrightarrow{PQ} // \overleftrightarrow{RS}$ and if $\overleftrightarrow{PR} // \overleftrightarrow{QS}$. This means that all four of the points P, Q, S, R lie in a plane and that $PQSR$ is a parallelogram in this plane. Again, one has that the movement property and vector addition may be defined using the head-to-tail method. In xyz-space, let $O = (0,0,0)$, $P = (1,0,0)$, and $Q = (1,0,1)$. What would be the coordinates of the point W if $\overrightarrow{OW} = \overrightarrow{OP} + \overrightarrow{OQ}$?

Classroom Connection 5.1.1

Illustrate vectors to middle or high school students using wooden sticks with the ends painted different colors to show the orientation. For example, paint one end red to represent the tail and the other end blue to represent the head. Have students take two sticks representing two vectors and ask them to illustrate vector addition. The directions the sticks are pointed is very important. If the sticks are parallel and pointing in the same direction, then the sum of the two vectors represented by the sticks is a vector of length equal to the sum of the two lengths of the sticks. If the sticks represent vectors that are perpendicular, then one may use the Pythagorean theorem to find the length of the vector that is the sum of the two perpendicular vectors. Use these sticks to illustrate the following inequality, $|\overrightarrow{AB}| - |\overrightarrow{CD}| \le |\overrightarrow{AB} + \overrightarrow{CD}| \le |\overrightarrow{AB}| + |\overrightarrow{CD}|$. Have the students hold one stick, representing $\overrightarrow{AB}$ fixed, and place the end C of the other stick as near as possible to the point B of the first stick. Then rotate the $\overrightarrow{CD}$ stick in a circle, holding the point C fixed as near as possible to point B. At various points during the rotation, one can check that the inequality $|\overrightarrow{AB}| - |\overrightarrow{CD}| \le |\overrightarrow{AB} + \overrightarrow{CD}| \le |\overrightarrow{AB}| + |\overrightarrow{CD}|$ holds. In fact, the equality $|\overrightarrow{AB} + \overrightarrow{CD}| = |\overrightarrow{AB}| + |\overrightarrow{CD}|$ holds exactly when $\overrightarrow{CD}$ points in the same direction as $\overrightarrow{AB}$. Under what conditions will $|\overrightarrow{AB}| - |\overrightarrow{CD}| = |\overrightarrow{AB} + \overrightarrow{CD}|$? Note that this last equality never holds when $|\overrightarrow{CD}| > |\overrightarrow{AB}|$. ◆

5.2   REPRESENTING VECTORS WITH COORDINATES

Using coordinates is especially useful in studying vectors and doing applications involving vectors. If one uses (x, y) coordinates, then one may use the movement property to place a given vector with its tail at the origin $O = (0, 0)$. After placing the tail at the origin, the head of the vector is located at some point $Q = (x, y)$. The vector $\overrightarrow{OQ}$ from O to Q is represented by the ordered pair x and y written in column form. Thus, one writes $\overrightarrow{OQ} = \begin{bmatrix} x \\ y \end{bmatrix}$ and says that $\begin{bmatrix} x \\ y \end{bmatrix}$ is the **coordinate representation** of the vector $\overrightarrow{OQ}$, see Figure 5.2.1.

Notice that when studying vectors, one uses several identifications. Sometimes the different things being identified are said to be different representations of the vector. In particular, when considering a given vector in the Euclidean plane, one is identifying all oriented line segments in the Euclidean plane that are of the same length and are pointing in the same direction. One is also identifying, or representing, a given oriented line segment with an ordered pair written as a column that is two high.

$$\overrightarrow{OQ} \quad \Leftrightarrow \quad \begin{cases} \text{oriented line segment from } O \text{ to } Q. \\[1em] \text{any oriented line segment parallel to the one from } \\ O \text{ to } Q \text{ with the same length and oriented in the } \\ \text{same direction as from } O \text{ to } Q. \\[1em] \begin{bmatrix} x \\ y \end{bmatrix} \text{ where } O = (0, 0) \text{ and } Q = (x, y). \end{cases}$$

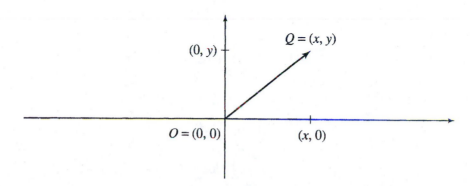

FIGURE 5.2.1 The vector from the origin $O = (0, 0)$ to the point $Q = (x, y)$ is represented by $\overrightarrow{OQ} = \begin{bmatrix} x \\ y \end{bmatrix}$. Thus, when the tail is at the origin, the coordinates of a vector are the coordinates of the head written as a column.

Addition of vectors using coordinates is done by adding each component separately. Thus, if $\overrightarrow{OA} = \begin{bmatrix} x_1 \\ y_1 \end{bmatrix}$ and $\overrightarrow{OB} = \begin{bmatrix} x_2 \\ y_2 \end{bmatrix}$, one has $\overrightarrow{OA} + \overrightarrow{OB} = \begin{bmatrix} x_1 + x_2 \\ y_1 + y_2 \end{bmatrix}$.

$$\begin{bmatrix} x_1 \\ y_1 \end{bmatrix} + \begin{bmatrix} x_2 \\ y_2 \end{bmatrix} = \begin{bmatrix} x_1 + x_2 \\ y_1 + y_2 \end{bmatrix} \text{ (vector addition using coordinates)}$$

EXAMPLE Let $A = (4, -3)$ and $B = (2, 5)$. Find the vector $\overrightarrow{OA} + \overrightarrow{OB}$.

Solution $\overrightarrow{OA} + \overrightarrow{OB} = \begin{bmatrix} 4 \\ -3 \end{bmatrix} + \begin{bmatrix} 2 \\ 5 \end{bmatrix} = \begin{bmatrix} 4 + 2 \\ -3 + 5 \end{bmatrix} = \begin{bmatrix} 6 \\ 2 \end{bmatrix}.$ ∎

Of course, the "addition-of-coordinates" method is equivalent to the head-to-tail method. A way to illustrate this equivalence is outlined in Classroom Connection 5.2.1 at the end of this section.

Similarly, for **subtraction of vectors** one has $\overrightarrow{OA} - \overrightarrow{OB} = \begin{bmatrix} x_1 - x_2 \\ y_1 - y_2 \end{bmatrix}$.

$$\begin{bmatrix} x_1 \\ y_1 \end{bmatrix} - \begin{bmatrix} x_2 \\ y_2 \end{bmatrix} = \begin{bmatrix} x_1 - x_2 \\ y_1 - y_2 \end{bmatrix} \text{ (vector subtraction using coordinates)}$$

EXAMPLE Let $C = (2, -4)$ and $D = (6, -5)$. Find the vector $\overrightarrow{OC} - \overrightarrow{OD}$.

Solution $\overrightarrow{OC} - \overrightarrow{OD} = \begin{bmatrix} 2 \\ -4 \end{bmatrix} - \begin{bmatrix} 6 \\ -5 \end{bmatrix} = \begin{bmatrix} 2 - 6 \\ -4 - (-5) \end{bmatrix} = \begin{bmatrix} -4 \\ 1 \end{bmatrix}.$ ∎

One use of vector subtraction is to find the coordinate representation of a vector $\overrightarrow{PQ}$ with tail P and head Q. If

$$P = (x_1, y_1) \text{ and } Q = (x_2, y_2),$$

then one writes

$$\overrightarrow{OP} = \begin{bmatrix} x_1 \\ y_1 \end{bmatrix} \text{ and } \overrightarrow{OQ} = \begin{bmatrix} x_2 \\ y_2 \end{bmatrix}.$$

From the definition of vector addition, one has

$$\overrightarrow{OP} + \overrightarrow{PQ} = \overrightarrow{OQ}.$$

Hence,

$$\overrightarrow{PQ} = \overrightarrow{OQ} - \overrightarrow{OP}.$$

Thus, the coordinate representation of the vector from P to Q is given by

$$\overrightarrow{PQ} = \begin{bmatrix} x_2 \\ y_2 \end{bmatrix} - \begin{bmatrix} x_1 \\ y_1 \end{bmatrix} = \begin{bmatrix} x_2 - x_1 \\ y_2 - y_1 \end{bmatrix}.$$

$$\overrightarrow{PQ} = \overrightarrow{OQ} - \overrightarrow{OP} = \text{(head)} - \text{(tail)}$$

EXAMPLE If $P = (3, 2)$ and $Q = (2, -1)$, find the column representation of the vector $\overrightarrow{PQ}$ from P to Q.

Solution $\overrightarrow{PQ} = \overrightarrow{OQ} - \overrightarrow{OP} = \begin{bmatrix} 2 \\ -1 \end{bmatrix} - \begin{bmatrix} 3 \\ 2 \end{bmatrix} = \begin{bmatrix} 2 - 3 \\ -1 - 2 \end{bmatrix} = \begin{bmatrix} -1 \\ -3 \end{bmatrix}.$ ∎

Multiplication of a real number a times the vector $\overrightarrow{OQ} = \begin{bmatrix} x \\ y \end{bmatrix}$ is done by multiplying each component by the number a. Thus, $a\overrightarrow{OQ} = a\begin{bmatrix} x \\ y \end{bmatrix} = \begin{bmatrix} ax \\ ay \end{bmatrix}.$

$$a\begin{bmatrix} x \\ y \end{bmatrix} = \begin{bmatrix} ax \\ ay \end{bmatrix} \quad \text{(real number } a \text{ times a vector)}$$

EXAMPLE Given $Q = (5, 7)$, find $3\overrightarrow{OQ}$.

Solution $3\overrightarrow{OQ} = 3\begin{bmatrix} 5 \\ 7 \end{bmatrix} = \begin{bmatrix} 3 \cdot 5 \\ 3 \cdot 7 \end{bmatrix} = \begin{bmatrix} 15 \\ 21 \end{bmatrix}.$ ∎

The coordinates of a point or a vector's components may be variables, such as x and y, as well as given constants.

EXAMPLE Given $A = (x, y)$ and $B = (3, 1)$, find $2\overrightarrow{OA} + 5\overrightarrow{OB}$.

Solution $2\overrightarrow{OA} + 5\overrightarrow{OB} = 2\begin{bmatrix} x \\ y \end{bmatrix} + 5\begin{bmatrix} 3 \\ 1 \end{bmatrix} = \begin{bmatrix} 2x \\ 2y \end{bmatrix} + \begin{bmatrix} 15 \\ 5 \end{bmatrix} = \begin{bmatrix} 2x + 15 \\ 2y + 5 \end{bmatrix}.$ ∎

The coordinate approach to vectors given in this section has been in terms of columns—$\begin{bmatrix} x \\ y \end{bmatrix}$—instead of in terms of rows—written as $[\, x \quad y \,]$ or (x, y). Any of these three notational conventions could have been chosen, and, in fact, at times mathematicians will use one of the two row conventions instead of the column convention. There are advantages and disadvantages with each of the conventions. In this book, we chose the column convention because it is used more often when dealing with matrices, which we discuss in Section 5.3.

EXERCISES 5.2

1. If $\overrightarrow{OA} = \begin{bmatrix} 3 \\ 2 \end{bmatrix}$ and $\overrightarrow{OB} = \begin{bmatrix} 0 \\ 2 \end{bmatrix}$, find $\overrightarrow{OA} + \overrightarrow{OB}$.

2. If $\overrightarrow{OC} = \begin{bmatrix} -3 \\ 2 \end{bmatrix}$ and $\overrightarrow{OD} = \begin{bmatrix} 0 \\ 5 \end{bmatrix}$, find $\overrightarrow{OC} + \overrightarrow{OD}$.

3. If $\overrightarrow{OE} = \begin{bmatrix} 7 \\ -2 \end{bmatrix}$ and $\overrightarrow{OF} = \begin{bmatrix} 2 \\ -2 \end{bmatrix}$, find $\overrightarrow{OE} - \overrightarrow{OF}$.

4. If $\overrightarrow{OG} = \begin{bmatrix} -5 \\ 2 + \pi \end{bmatrix}$ and $\overrightarrow{OH} = \begin{bmatrix} x \\ y \end{bmatrix}$, find $\overrightarrow{OG} - \overrightarrow{OH}$.

5. If $\overrightarrow{OJ} = \begin{bmatrix} 3 \\ -4 \end{bmatrix}$ and $\overrightarrow{OK} = \begin{bmatrix} 1 \\ 3 \end{bmatrix}$, find $3\overrightarrow{OJ} - 4\overrightarrow{OK}$.

6. If $L = (3,2)$ and $M = (0,2)$, find $\overrightarrow{LM}$.

7. If $P = (5, -2)$ and $Q = (5, 2)$, find both (a) $\overrightarrow{PQ}$ and (b) $\overrightarrow{QP}$. Write your answers in column form.

8. In Euclidean three-dimensional space (i.e., xyz-space), one identifies oriented line segments of the same length that are pointing in the same direction. Furthermore, when using coordinates, one has a corresponding representation with ordered triples of real numbers written in column form. For example, the vector with the tail at the origin $(0, 0, 0)$ and the head at (x, y, z) is represented by $\begin{bmatrix} x \\ y \\ z \end{bmatrix}$. Find the oriented triple of real numbers written in column form that represents the oriented line segment in xyz-space with the tail at the point $(0, 1, 0)$ and the head at the point $(0, 1, 1)$. Draw a diagram and explain your reasoning.

Classroom Connection 5.2.1

Students on all levels may wonder about the different ways of adding vectors that have been discussed. In particular, some students find it hard to understand that the head-to-tail method and the addition-of-coordinates method are equivalent. One can illustrate that, when using coordinates to do vector addition, the addition of corresponding components corresponds to the head-to-tail method studied in Section 5.1. Start with graph paper and two sticks (or toothpicks or thin straws) with the ends marked to indicate the orientation. Lay the first stick with tail at the origin, making some angle of $A°$ with the positive x-axis. Note the coordinates (x_1, y_1) of this stick's head. Then lay the second stick along the positive x-axis, with its tail at the origin. Note the coordinates of this stick's head. These coordinates will be of the form $(x_2, 0)$. Now move the second stick, keeping it parallel to the x-axis such that its new tail lies at the head point (x_1, y_1) of the first stick. Then the new head of the second stick should be at the point $(x_1 + x_2, y_1)$. This illustrates the vector equation $\begin{bmatrix} x_1 \\ y_1 \end{bmatrix} + \begin{bmatrix} x_2 \\ 0 \end{bmatrix} = \begin{bmatrix} x_1 + x_2 \\ y_1 \end{bmatrix}$. In general, one can easily modify this method to illustrate the more general formula $\begin{bmatrix} x_1 \\ y_1 \end{bmatrix} + \begin{bmatrix} x_2 \\ y_2 \end{bmatrix} = \begin{bmatrix} x_1 + x_2 \\ y_1 + y_2 \end{bmatrix}$. When dealing with students who are reasonably familiar with coordinates and vectors in three dimensions, this approach can be modified to show the three-dimensional version $\begin{bmatrix} x_1 \\ y_1 \\ z_1 \end{bmatrix} + \begin{bmatrix} x_2 \\ y_2 \\ z_2 \end{bmatrix} = \begin{bmatrix} x_1 + x_2 \\ y_1 + y_2 \\ z_1 + z_2 \end{bmatrix}$. ◆

5.3 MATRICES

A column of numbers is called a **column matrix** or sometimes a **column vector**. The phrase "column vector" is easy to understand in light of the previous section, where the column $\begin{bmatrix} x \\ y \end{bmatrix}$ was used to represent the vector with the tail at the origin and the head at the point with coordinates (x, y). If the column is k high, then the matrix is said to be $k \times 1$. Thus, in Figure 5.3.1, A is a 2×1 matrix, and B is a 3×1 matrix.

A row of numbers is called a **row matrix** or sometimes a **row vector**. If the row is r long, then the matrix is said to be $1 \times r$. Thus, in Figure 5.3.2, C is a 1×2 matrix, and D is a 1×3 matrix.

Matrices.   A **matrix** is an array of numbers arranged as a rectangle. If the height of the rectangle is k and the length is r, then the matrix is said to be $k \times r$. An example of a 3×2 matrix is shown in Figure 5.3.3.

A common notational convention is to write a_{ij} for the element of matrix A that lies in the ith row and j^{th} column. Hence, if $A = \begin{bmatrix} a_{11} & a_{12} \\ a_{21} & a_{22} \end{bmatrix} = \begin{bmatrix} 1 & 2 \\ 4 & 6 \end{bmatrix}$, then $a_{11} = 1$, $a_{12} = 2$, $a_{21} = 4$, and $a_{22} = 6$. Notice that the first subscript refers to the row and the second to the column.

$$A = \begin{bmatrix} 5 \\ 6 \end{bmatrix} \qquad B = \begin{bmatrix} 2 \\ 3 \\ -7 \end{bmatrix}$$

FIGURE 5.3.1 Here A and B may be called either *column vectors* or *column matrices*. The column vector A represents a vector in R^2 with the tail at the origin and the head at the point (5, 6). The column vector B represents a vector (i.e., oriented line segment) in R^3 with the tail at the origin (0, 0, 0) and the head at (2, 3, −7).

$$C = [2 \ \ 7] \qquad\qquad D = [4 \ -10 \ \ 0]$$

FIGURE 5.3.2 Here C and D may be called either *row matrices* or *row vectors*. The matrix C is a 1×2 matrix, and D is a 1×3 matrix.

$$E = \begin{bmatrix} -2 & \sqrt{3} \\ 2/3 & 0 \\ 4.5 & -\pi \end{bmatrix}$$

FIGURE 5.3.3 This matrix E is 3×2. Note that the elements of a matrix may be positive, negative, zero, rational, or irrational.

Equality of Matrices.   Two matrices that are of equal size $k \times r$ and are such that each element of the first is equal to the corresponding element of the other are said to be equal. In other words, $A = B$ iff they are both $k \times r$ and if each element a_{ij} of A located in the ith row and j^{th} column of A is equal respectively to the element b_{ij} of B located in the corresponding ij^{th} position of B. So, $A = B$ iff $a_{ij} = b_{ij}$ for all $1 \leq i \leq k$ and $1 \leq j \leq r$ where both A and B are $k \times r$.

Addition of Matrices.   Addition is only defined for matrices when the two matrices are of the same size. If A and B are both $k \times r$ matrices, then one may form the sum $C = A + B$, and the sum will again be a $k \times r$ matrix. Calculate the elements of the matrix C by adding corresponding elements of A and B. Let a_{ij} represent the element in the ith row and j^{th} column of A, and let b_{ij} represent the element in the ith row and j^{th} column of B. Then $c_{ij} = a_{ij} + b_{ij}$ represents the element in the ith row and j^{th} column of C.

$$\begin{bmatrix} a_{11} & a_{12} & \cdots \\ a_{21} & a_{22} & \cdots \\ \cdots & \cdots & \cdots \end{bmatrix} + \begin{bmatrix} b_{11} & b_{12} & \cdots \\ b_{21} & b_{22} & \cdots \\ \cdots & \cdots & \cdots \end{bmatrix} = \begin{bmatrix} a_{11} + b_{11} & a_{12} + b_{12} & \cdots \\ a_{21} + b_{21} & a_{22} + b_{22} & \cdots \\ \cdots & \cdots & \cdots \end{bmatrix}$$

(addition of matrices)

EXAMPLE   Given $A = \begin{bmatrix} 1 & 4 \\ 5 & -2 \end{bmatrix}$ and $B = \begin{bmatrix} 2 & -2 \\ 0 & \pi \end{bmatrix}$, find $A + B$.

Solution $A + B = \begin{bmatrix} 1 & 4 \\ 5 & -2 \end{bmatrix} + \begin{bmatrix} 2 & -2 \\ 0 & \pi \end{bmatrix} = \begin{bmatrix} 1 + 2 & 4 - 2 \\ 5 + 0 & -2 + \pi \end{bmatrix}$

$= \begin{bmatrix} 3 & 2 \\ 5 & -2 + \pi \end{bmatrix}$.   ∎

Subtraction is also only defined when the two matrices are both of the same size $k \times r$. The difference $A - B$ is defined by subtracting corresponding elements of B from elements of A.

$$\begin{bmatrix} a_{11} & a_{12} & \cdots \\ a_{21} & a_{22} & \cdots \\ \cdots & \cdots & \cdots \end{bmatrix} - \begin{bmatrix} b_{11} & b_{12} & \cdots \\ b_{21} & b_{22} & \cdots \\ \cdots & \cdots & \cdots \end{bmatrix} = \begin{bmatrix} a_{11} - b_{11} & a_{12} - b_{12} & \cdots \\ a_{21} - b_{21} & a_{22} - b_{22} & \cdots \\ \cdots & \cdots & \cdots \end{bmatrix}$$

(subtraction of matrices)

EXAMPLE   Given $A = \begin{bmatrix} 0 & 3 \\ 5 & 7 \\ 9 & 11 \end{bmatrix}$ and $B = \begin{bmatrix} 2 & 3 \\ 0 & 4 \\ 5 & 6 \end{bmatrix}$, find $A - B$.

Solution $A - B = \begin{bmatrix} 0 & 3 \\ 5 & 7 \\ 9 & 11 \end{bmatrix} - \begin{bmatrix} 2 & 3 \\ 0 & 4 \\ 5 & 6 \end{bmatrix} = \begin{bmatrix} 0 - 2 & 3 - 3 \\ 5 - 0 & 7 - 4 \\ 9 - 5 & 11 - 6 \end{bmatrix} = \begin{bmatrix} -2 & 0 \\ 5 & 3 \\ 4 & 5 \end{bmatrix}$.   ∎

Scalar multiplication is multiplication of a real number a times the $k \times r$ matrix B. This type of multiplication is defined as a times each entry of B.

$$a \begin{bmatrix} b_{11} & b_{12} & \cdots \\ b_{21} & b_{22} & \cdots \\ \cdots & \cdots & \cdots \end{bmatrix} = \begin{bmatrix} ab_{11} & ab_{12} & \cdots \\ ab_{21} & ab_{22} & \cdots \\ \cdots & \cdots & \cdots \end{bmatrix} \quad \text{(scalar multiplication)}$$

EXAMPLE Given $B = \begin{bmatrix} \sqrt{2} & -4 \\ 0 & 1 \\ -5 & .7 \end{bmatrix}$, find $3B$.

Solution $3B = 3 \begin{bmatrix} \sqrt{2} & -4 \\ 0 & 1 \\ -5 & .7 \end{bmatrix} = \begin{bmatrix} 3 \cdot \sqrt{2} & 3 \cdot (-4) \\ 3 \cdot 0 & 3 \cdot 1 \\ 3 \cdot (-5) & 3 \cdot (.7) \end{bmatrix} = \begin{bmatrix} 3\sqrt{2} & -12 \\ 0 & 3 \\ -15 & 2.1 \end{bmatrix}.$ ∎

Matrix Multiplication.   The product AB is only defined when the length of matrix A is the same as the height of matrix B. If A is $k \times r$ and B is $r \times s$, then the product $C = AB$ will be $k \times s$.

$$[k \times r]\,[r \times s] = [k \times s]$$

$$\text{size of } AB = (\text{height of } A) \times (\text{length of } B)$$

The ij^{th} element of the product AB is defined by multiplying each element of the ith row of A times the corresponding element of the jth column of B and then summing the products. In particular, if A has a_{ij} as its ij^{th} element and B has b_{ij} as its ij^{th} element, the product AB will be a matrix C with c_{ij} as its ij^{th} element where $c_{ij} = \sum\limits_{h=1}^{h=r} a_{ih}b_{hj}$. Using the notational convention $A = [a_{ij}]$, $B = [b_{ij}]$, and $C = [c_{ij}]$, one has $[a_{ij}]\,[b_{ij}] = \left[\sum\limits_{h=1}^{h=r} a_{ih}b_{hj} \right]$.

$$AB = [a_{ij}]\,[b_{ij}] = \left[\sum_{h=1}^{h=r} a_{ih}b_{hj} \right] \quad \text{(multiplication of matrices)}$$

EXAMPLE Given $A = \begin{bmatrix} 1 & 2 \\ -1 & 4 \end{bmatrix}$ and $B = \begin{bmatrix} 0 & 2 & 3 \\ 1 & -2 & 5 \end{bmatrix}$, determine if the product AB is defined, if it is defined, find this product.

Solution   The product AB is defined since the length of A is the same as the height of B. Note that the product will be a matrix C that is $2 \times 3 = (\text{height of } A) \times$

(length of B).

$$C = AB = \begin{bmatrix} 1 & 2 \\ -1 & 4 \end{bmatrix}\begin{bmatrix} 0 & 2 & 3 \\ 1 & -2 & 5 \end{bmatrix}$$

$$= \begin{bmatrix} 1 \cdot 0 + 2 \cdot 1 & 1 \cdot 2 + 2 \cdot (-2) & 1 \cdot 3 + 2 \cdot 5 \\ (-1) \cdot 0 + 4 \cdot 1 & (-1) \cdot 2 + 4 \cdot (-2) & (-1) \cdot 3 + 4 \cdot 5 \end{bmatrix}$$

$$= \begin{bmatrix} 2 & -2 & 13 \\ 4 & -10 & 17 \end{bmatrix}.$$

Notice that with the previous matrices, the product AB is defined, but the product BA is not defined since the length of B is not equal to the height of A. Clearly, in this case $AB \neq BA$ since BA is not even defined. On the other hand, when the length of A equals the height of B and the height of A equals the length of B, then one may multiply them in either order AB or BA. However, even when both AB and BA are defined, the two different products may or may not be equal.

EXAMPLE Given $A = \begin{bmatrix} 1 & 0 \\ 2 & 3 \end{bmatrix}$ and $B = \begin{bmatrix} 0 & \pi \\ 4 & 5 \end{bmatrix}$, find both AB and BA. Also, check to see if these products are equal or unequal.

Solution $AB = \begin{bmatrix} 1 & 0 \\ 2 & 3 \end{bmatrix}\begin{bmatrix} 0 & \pi \\ 4 & 5 \end{bmatrix} = \begin{bmatrix} 1 \cdot 0 + 0 \cdot 4 & 1 \cdot \pi + 0 \cdot 5 \\ 2 \cdot 0 + 3 \cdot 4 & 2 \cdot \pi + 3 \cdot 5 \end{bmatrix}$

$= \begin{bmatrix} 0 & \pi \\ 12 & 2\pi + 15 \end{bmatrix}$ and $BA = \begin{bmatrix} 0 & \pi \\ 4 & 5 \end{bmatrix}\begin{bmatrix} 1 & 0 \\ 2 & 3 \end{bmatrix} = \begin{bmatrix} 0 \cdot 1 + \pi \cdot 2 & 0 \cdot 0 + \pi \cdot 3 \\ 4 \cdot 1 + 5 \cdot 2 & 4 \cdot 0 + 5 \cdot 3 \end{bmatrix}$

$= \begin{bmatrix} 2\pi & 3\pi \\ 14 & 15 \end{bmatrix}$. Clearly $AB \neq BA$.

EXAMPLE Given $E = \begin{bmatrix} 1 & 0 & 2 \\ 3 & 4 & 5 \\ -2 & 7 & 9 \end{bmatrix}$ and $F = \begin{bmatrix} 11 \\ 3 \\ 4 \end{bmatrix}$ determine which of FE and EF is defined. Also, find the product that is defined.

Solution   Notice that the length of F is 1, which fails to equal 3, the height of E. Consequently, the product FE is not defined. On the other hand, since the length of E is 3, which is also the height of F, the product EF is defined. In particular,

$$EF = \begin{bmatrix} 1 & 0 & 2 \\ 3 & 4 & 5 \\ -2 & 7 & 9 \end{bmatrix}\begin{bmatrix} 11 \\ 3 \\ 4 \end{bmatrix} = \begin{bmatrix} 1 \cdot 11 + 0 \cdot 3 + 2 \cdot 4 \\ 3 \cdot 11 + 4 \cdot 3 + 5 \cdot 4 \\ (-2) \cdot 11 + 7 \cdot 3 + 9 \cdot 4 \end{bmatrix}$$

$$= \begin{bmatrix} 11 + 0 + 8 \\ 33 + 12 + 20 \\ -22 + 21 + 36 \end{bmatrix} = \begin{bmatrix} 19 \\ 65 \\ 35 \end{bmatrix}.$$

When using matrices, one may have entries that are variables, such as x and y.

EXAMPLE Given $D = \begin{bmatrix} 3 & 1 \\ -2 & 4 \end{bmatrix}$, $X = \begin{bmatrix} x \\ y \end{bmatrix}$, and $H = \begin{bmatrix} 7 \\ 9 \end{bmatrix}$, find $DX + H$.

Solution $DX + H = \begin{bmatrix} 3 & 1 \\ -2 & 4 \end{bmatrix}\begin{bmatrix} x \\ y \end{bmatrix} + \begin{bmatrix} 7 \\ 9 \end{bmatrix} = \begin{bmatrix} 3x + y \\ -2x + 4y \end{bmatrix} + \begin{bmatrix} 7 \\ 9 \end{bmatrix}$

$= \begin{bmatrix} 3x + y + 7 \\ -2x + 4y + 9 \end{bmatrix}$. ■

Stacking Equations. Matrices have many applications in mathematics. Among other things, they are often used to "stack" equations. In particular, the following two equations,

$$2x + 3y = 4$$

and

$$3x - 4y = 6,$$

are equivalent to the following matrix equation

$$AX = B,$$

where $A = \begin{bmatrix} 2 & 3 \\ 3 & -4 \end{bmatrix}$, $X = \begin{bmatrix} x \\ y \end{bmatrix}$, and $B = \begin{bmatrix} 4 \\ 6 \end{bmatrix}$.

EXAMPLE Write the following three equations in the equivalent matrix form $AX = B$ (i.e., find the matrices A, X, and B):

$$x - 7y + 5z = 2$$
$$-x + 2y + 3z = 6$$
$$2x + y - z = 1$$

Solution The matrix A is the matrix of coefficients. The matrix X is a column with the variables x, y, and z. The matrix B is a column with the constants on the right side of the equations. Thus,

$$A = \begin{bmatrix} 1 & -7 & 5 \\ -1 & 2 & 3 \\ 2 & 1 & -1 \end{bmatrix}, X = \begin{bmatrix} x \\ y \\ z \end{bmatrix}, \text{ and } B = \begin{bmatrix} 2 \\ 6 \\ 1 \end{bmatrix}.$$

Using these three matrices, it is easy to check that $AX = B$ becomes

$$\begin{bmatrix} x & -7y & +5z \\ -x & +2y & +3z \\ 2x & +y & -z \end{bmatrix} = \begin{bmatrix} 2 \\ 6 \\ 1 \end{bmatrix}. \quad ■$$

The **identity matrices** of sizes 2×2 and 3×3 are defined as

$$\begin{bmatrix} 1 & 0 \\ 0 & 1 \end{bmatrix} \text{ and } \begin{bmatrix} 1 & 0 & 0 \\ 0 & 1 & 0 \\ 0 & 0 & 1 \end{bmatrix},$$

respectively. In general, the $k \times k$ identity matrix, denoted by I (or I_k when one wants to emphasize the size), has ones going down the main diagonal and zeros off the main diagonal. Thus, the entries a_{ij} of the $k \times k$ identity I are given by

$$a_{ij} = \begin{cases} 1 & \text{if } i = j \\ 0 & \text{if } i \neq j \end{cases}.$$

Whenever I is the $k \times k$ identity matrix and AI is defined, the product will be A (i.e., $AI = A$). Similarly, when the product IA is defined, the product will be A (i.e., $IA = A$).

EXAMPLE Given $A = \begin{bmatrix} 1 & 2 \\ 3 & 4 \\ 5 & 6 \end{bmatrix}$ and $I = \begin{bmatrix} 1 & 0 \\ 0 & 1 \end{bmatrix}$, verify $AI = A$.

Solution Using the definition of matrix multiplication, one calculates that $AI =$

$$\begin{bmatrix} 1 & 2 \\ 3 & 4 \\ 5 & 6 \end{bmatrix} \begin{bmatrix} 1 & 0 \\ 0 & 1 \end{bmatrix} = \begin{bmatrix} 1 \cdot 1 + 2 \cdot 0 & 1 \cdot 0 + 2 \cdot 1 \\ 3 \cdot 1 + 4 \cdot 0 & 3 \cdot 0 + 4 \cdot 1 \\ 5 \cdot 1 + 6 \cdot 0 & 5 \cdot 0 + 6 \cdot 1 \end{bmatrix} = \begin{bmatrix} 1 & 2 \\ 3 & 4 \\ 5 & 6 \end{bmatrix} = A, \text{ as desired.}$$

Transpose. The **transpose** A^T of the $k \times r$ matrix A is an $r \times k$ matrix formed by interchanging the rows and columns of the original matrix A. Thus, the first row of A becomes the first column of A^T, the second row of A becomes the second column of A^T, etc.

EXAMPLE Given $A = \begin{bmatrix} 1 & 2 & 4 \\ 1.5 & -9 & \pi \end{bmatrix}$ and $B = \begin{bmatrix} 2 & 4 \\ 6 & 8 \\ 10 & 12 \end{bmatrix}$, find $3A^T + B$.

Solution Notice that interchanging the rows and columns of A yields $A^T = \begin{bmatrix} 1 & 1.5 \\ 2 & -9 \\ 4 & \pi \end{bmatrix}$. Thus,

$$3A^T + B = 3 \begin{bmatrix} 1 & 1.5 \\ 2 & -9 \\ 4 & \pi \end{bmatrix} + \begin{bmatrix} 2 & 4 \\ 6 & 8 \\ 10 & 12 \end{bmatrix} = \begin{bmatrix} 3 & 4.5 \\ 6 & -27 \\ 12 & 3\pi \end{bmatrix} + \begin{bmatrix} 2 & 4 \\ 6 & 8 \\ 10 & 12 \end{bmatrix}$$

$$= \begin{bmatrix} 5 & 8.5 \\ 12 & -19 \\ 22 & 12 + 3\pi \end{bmatrix}.$$

Selected matrix rules.

1. $A + B = B + A$	Commutative law of addition
2. $(A + B) + C = A + (B + C)$	Associative law of addition
3. $(AB)C = A(BC)$	Associative law of multiplication
4. $A(B + C) = AB + AC$	Left distributive law
5. $(A + B)C = AC + BC$	Right distributive law
6. $(A^T)^T = A$	Double transpose rule
7. $(AB)^T = B^T A^T$	Reverse product transpose rule
8. $k(AB) = (kA)B = A(kB)$	A mixed scalar and matrix
	multiplication rule

EXERCISES 5.3

1. Find $\begin{bmatrix} 3 & 5 \\ 0 & 1 \end{bmatrix} + \begin{bmatrix} 2 & 1 \\ 3 & 1 \end{bmatrix}$.

2. Find $\begin{bmatrix} 5 & 5 \\ 0 & 6 \end{bmatrix} - \begin{bmatrix} 1 & 1 \\ 3 & 1 \end{bmatrix}$.

3. Find $2\begin{bmatrix} 3 & 7 \\ 2 & 1 \end{bmatrix} - 3\begin{bmatrix} 0 & 1 \\ 3 & 1 \end{bmatrix}$.

4. Find $\begin{bmatrix} 3 & 5 \\ 0 & 1 \end{bmatrix}\begin{bmatrix} 2 & 1 \\ 3 & 1 \end{bmatrix}$.

5. Find $\begin{bmatrix} 5 & 2 \\ 0 & 1 \end{bmatrix}\begin{bmatrix} 2 & 3 \\ 7 & 4 \end{bmatrix}$.

6. Find $\begin{bmatrix} 2 & 9 \\ 4 & 5 \end{bmatrix}^T$.

7. Find $\begin{bmatrix} -1 & 3 \\ \sqrt{2} & -7 \end{bmatrix}^T$.

8. Let $A = \begin{bmatrix} 5 & 6 \\ -1 & 0 \end{bmatrix}$. Find (a) $(7 \cdot A)^T$ and (b) $7 \cdot (A^T)$.

9. Let $E = \begin{bmatrix} 1 & 2 \\ 3 & 4 \end{bmatrix}$ and $F = \begin{bmatrix} -1 & 7 \\ 2 & 0 \end{bmatrix}$. Find (a) $(E + 2 \cdot F)^T$ and (b) $E^T + 2 \cdot (F^T)$.

10. Find $5\begin{bmatrix} 1 & 3 & 5 \\ -1 & 0 & 2 \end{bmatrix}^T$.

11. Let $A = \begin{bmatrix} 1 & 0 \\ 3 & 2 \end{bmatrix}$ and $B = \begin{bmatrix} 0 & 2 \\ 3 & 4 \end{bmatrix}$. Find (a) $(AB)^T$ and (b) $B^T A^T$.

12. Find $\begin{bmatrix} 1 & 0 & 0 \\ 2 & 3 & 1 \\ 0 & 2 & 0 \end{bmatrix}\begin{bmatrix} 1 & 0 & 2 \\ 0 & 1 & 3 \\ 2 & -3 & -1 \end{bmatrix}$.

13. Find $\begin{bmatrix} 0 & 1 & 2 \\ 2 & 3 & 1 \\ 0 & 5 & 0 \end{bmatrix} \begin{bmatrix} 3 & 1 & 0 \\ 0 & -1 & 2 \\ 1 & 4 & 0 \end{bmatrix}$.

14. Find two 3×3 **matrices** C **and** D **such that** $CD \neq DC$.

15. Find $\begin{bmatrix} 1 & 2 \\ 5 & -7 \end{bmatrix} \begin{bmatrix} 3 \\ 9 \end{bmatrix}$.

16. Find $\begin{bmatrix} 3 & 0 & 1 \\ 3 & 2 & -1 \\ -1 & 7 & 6 \end{bmatrix} \begin{bmatrix} 1 & 3 \\ 5 & -1 \\ 4 & -2 \end{bmatrix}$.

17. Find $\begin{bmatrix} 1 & 2 \\ 5 & -7 \end{bmatrix} \begin{bmatrix} x \\ y \end{bmatrix}$.

18. Find $\begin{bmatrix} 1 & 3 & 0 \\ 2 & 5 & -1 \\ -3 & 4 & 6 \end{bmatrix} \begin{bmatrix} x \\ y \\ z \end{bmatrix}$.

19. Given $C = \begin{bmatrix} 1 & 1 \\ 2 & -2 \end{bmatrix}$, $X = \begin{bmatrix} x \\ y \end{bmatrix}$, **and** $D = \begin{bmatrix} 3 \\ -2 \end{bmatrix}$, **find the values of** x **and** y **if the following matrix equation holds:** $CX = D$.

20. If $X = \begin{bmatrix} x \\ y \end{bmatrix}$, **then** X **represents the vector** $\overrightarrow{OP}$ **with the tail at the origin and the head at the point** $P = (x, y)$.
  a. Find the matrix product $X^T X$.
  b. Give a geometric interpretation of $\sqrt{X^T X}$.

21. Find $\begin{bmatrix} a & b \\ c & d \end{bmatrix} \begin{bmatrix} e & f \\ g & h \end{bmatrix}$.

22. Find $\begin{bmatrix} a & b & c \end{bmatrix} \begin{bmatrix} d & e \\ f & g \\ h & i \end{bmatrix}$.

23. Let $A = \begin{bmatrix} a & b \\ c & d \end{bmatrix}$, $B = \begin{bmatrix} e & f \\ g & h \end{bmatrix}$, **and assume that** k **is a real number. Show that all three of** $k(AB)$, $(kA)B$, **and** $A(kB)$ **are equal. This verifies the general formula** $k(AB) = (kA)B = A(kB)$ **for the case where both** A **and** B **are** 2×2 **matrices.**

24. Let $A = \begin{bmatrix} a & b & c \end{bmatrix}$, $B = \begin{bmatrix} d & e \\ f & g \\ h & i \end{bmatrix}$, **and assume that** k **is a real number. Show that all three of** $k(AB)$, $(kA)B$, **and** $A(kB)$ **are equal. This verifies the general formula** $k(AB) = (kA)B = A(kB)$ **for the case where** A **is a** 1×3 **matrix and** B **is a** 3×2 **matrix.**

Classroom Discussion 5.3.1

College students often feel that the reverse product transpose rule $(AB)^T = B^T A^T$ is a misprint and that the $B^T A^T$ should be written in with the A^T term first rather than second. To convince students this cannot be correct, point out that if A is

3×7 and B is 7×5, then A^T must be 7×3 and B^T is 5×7. Hence, $B^T A^T$ is defined, but $A^T B^T$ is not defined since the length of A^T does not equal the height of B^T. Of course, just because $(AB)^T \neq A^T B^T$ does not prove that $(AB)^T = B^T A^T$. A classroom discussion could center around the question of what would be an acceptable proof that $(AB)^T = B^T A^T$ for some special case such as A and B both 2×2. Some students will probably be able to come up with a general proof once they understand the proof when the two matrices are each 2×2. ◆

5.4 TRANSFORMATIONS USING MATRICES

In Chapter 3, we discussed transformations of the form $F(x, y) = (ax + h, by + k)$. The Connected Mathematics Project unit *Stretching and Shrinking* includes some studies of such functions in the seventh grade (compare Figure 3.4.3).

Translations. Any translation of the xy-plane can be represented in function form by $F(x, y) = (x + h, y + k)$. The origin $(0, 0)$ is translated to (h, k), and each point is moved h units parallel to the x-axis and k units parallel to the y-axis. Thus, the point with coordinates (x, y) is moved to the point with coordinates $(x + h, y + k)$. Using x' and y' to denote the new x and y coordinates after the point (x, y) has been moved, one has $(x, y) \rightarrow (x', y')$, where $x' = x + h$ and $y' = y + k$. When studying translations of the xy-plane, one needs to understand how they are represented in different ways. For comparison, the translation that takes the origin $(0, 0)$ to the point (h, k) is represented here in the following three ways: (1) in function form, (2) equation form, and (3) matrix form.

Translations of the xy-plane

Function form: $F(x, y) = (x + h, y + k)$

Equation form: $\begin{cases} x' = x + h \\ y' = y + k \end{cases}$

Matrix form: $X' = X + H$ where $X' = \begin{bmatrix} x' \\ y' \end{bmatrix}$, $X = \begin{bmatrix} x \\ y \end{bmatrix}$, and $H = \begin{bmatrix} h \\ k \end{bmatrix}$.

Stretching/Shrinking Parallel to an Axis. The transformation $F(x, y) = (2x, y)$ takes all of the points on the y-axis to themselves. A point off the y-axis is mapped to a point at the same height (i.e., the image point has same y-coordinate) but at twice the distance from the y-axis. Thus, the point $(2, 3)$ goes to the point $(4, 3)$. This transformation is a stretching parallel to the x-axis by a factor of 2. One can think of a rubber sheet that is pasted to a desktop along the y-axis and then stretched parallel to the x-axis in such a way that points move by a factor of 2 away from the y-axis. The transformation $G(x, y) = (x, 2y)$ holds the x-axis fixed and is a stretching by a factor of 2 parallel to the y-axis. The transformation $H(x, y) = (3x, \frac{y}{4})$ stretches by a factor of 3 in the x-direction and at the same time shrinks by a factor of 4 in the y-direction.

The transformation $F(x, y) = (ax, by)$ with $a > 0$ and $b > 0$ stretches (or shrinks) by the factor a in the x-direction and by a factor of b in the y-direction. The origin $(0, 0)$ is mapped to itself. This transformation is a dilation with the center at the origin exactly when $a = b$. Figure 5.4.1 illustrates some transformations of the form $F(x, y) = (ax, by)$ as well as the translation taking (x, y) to $(x + 4, y + 2)$.

Transformations Taking Lines to Lines.   In the rest of this section, we consider transformations that take lines to lines. They can be represented in function form, equation form, or matrix form.

Transformations of the *xy*-plane taking lines to lines

Function form: $F(x, y) = (ax + by + h, cx + dy + k)$, where $ad - bc \neq 0$.

Equation form: $\begin{cases} x' = ax + by + h \\ y' = cx + dy + k, \end{cases}$ where $ad - bc \neq 0$.

Matrix form: $X' = AX + H$, where $\begin{cases} X' = \begin{bmatrix} x' \\ y' \end{bmatrix}, \quad A = \begin{bmatrix} a & b \\ c & d \end{bmatrix}, \quad X = \begin{bmatrix} x \\ y \end{bmatrix}, \\ H = \begin{bmatrix} h \\ k \end{bmatrix}, \quad \text{and} \quad ad - bc \neq 0. \end{cases}$

If $A = \begin{bmatrix} a & b \\ c & d \end{bmatrix}$, then the **determinant** of A may be denoted by either $\det(A)$ or by $\begin{vmatrix} a & b \\ c & d \end{vmatrix}$ and is defined as the quantity $ad - bc$.

$$\det(A) = \begin{vmatrix} a & b \\ c & d \end{vmatrix} = ad - bc \qquad \text{(determinant of 2} \times \text{2)}$$

Geometric Interpretation of 2 × 2 Determinant.   In general, the determinant may be positive, negative, or zero. One geometric interpretation of the determinant is in terms of area. If one sets P and Q to be the respective rows of A [i.e., $P = (a, b)$ and $Q = (c, d)$], then there is a parallelogram determined with vertices at the origin $O = (0, 0)$, $P = (a, b)$, $S = (a + c, b + d)$, and $Q = (c, d)$. The area of this parallelogram will be the absolute value of the determinant of A. In other words, $\text{Area}(OPSQ) = |\det(A)| = |ad - bc|$ (compare Figure 5.4.2).

The transformation $F: R^2 \rightarrow R^2$ given by $F(x, y) = (ax + by + h, cx + dy + k)$ and in matrix form as previous by $X' = AX + H$ takes a figure W of area K to figure $F(W)$ of area $|\det(A)| \cdot K$.

If $F: R^2 \rightarrow R^2$ is represented by $X' = AX + H$, then area $[F(W)] = |\det(A)| \cdot \text{area}(W)$.

Since lines will be mapped to lines, it follows that triangles will be mapped to triangles. More generally, since lines are mapped to lines, it follows that *n*-gons

Section 5
Key Concepts

Transformations and Translations (pp. 291–293)
A transformation is a change in a figure's shape, size, or location.
Transformations can be described using coordinates.

A translation is a transformation that slides a figure to a new
location. The image is congruent to the original figure.

Example The original trapezoid is translated
4 units to the right and 2 units up.

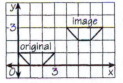

Original: (x, y)
  Image: $(x + 4, y + 2)$

Transformations that Stretch or Squash (pp. 294–295)
You can stretch or squash a figure horizontally by multiplying the
x-coordinates of all its points by the same factor, or vertically by
multiplying the y-coordinates of all its points by the same factor.
A factor greater than one stretches. A factor between zero
and one squashes.

When you stretch or squash a figure horizontally and vertically by the
same factor, the image is similar to the original figure. Figures are
similar if they are the same shape. but not necessarily the same size.

Examples

horizontal stretch   vertical squash   similar to original

30 Key Concepts Question Choose the letters of the transformation
that result in an image similar to the original figure. Explain.

A. $\left(\frac{1}{5}x, \frac{1}{5}y\right)$ **B.** $(x + 3, y - 1)$ **C.** $(4x, 3y)$

FIGURE 5.4.1 In the top example of this figure, the authors have used (x, y) for the original and
$(x + 4, y + 2)$ for the image. This is the transformation $F(x, y) = (x + 4, y + 2)$. Seventh-grade
students are studying transformations of the form $F_1(x, y) = (x + h, y + k)$ and of the form
$F_2(x, y) = (ax, by)$. Reproduced from page 299 of *Book 2* in the Math Thematics grade 7 materials.

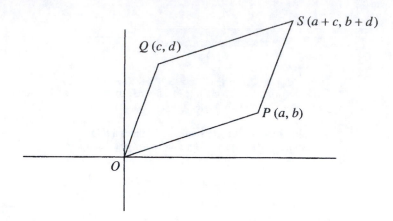

FIGURE 5.4.2 The area of the parallelogram *OPSQ* determined by the rows (a, b) and (c, d) of the matrix $A = \begin{bmatrix} a & b \\ c & d \end{bmatrix}$ is equal to the absolute value of the determinant of *A*. Also, a similar statement holds for the parallelogram determined by the columns of *A*.

are mapped to *n*-gons. Furthermore, parallel lines will be mapped to parallel lines. However, distances and angles will, in general, not be preserved.

If one is given

$$\begin{bmatrix} x' \\ y' \end{bmatrix} = \begin{bmatrix} 3 & 1 \\ 0 & 2 \end{bmatrix} \begin{bmatrix} x \\ y \end{bmatrix} + \begin{bmatrix} 2 \\ 0 \end{bmatrix},$$

then

$$\begin{bmatrix} 3 & 1 \\ 0 & 2 \end{bmatrix} \begin{bmatrix} 0 \\ 0 \end{bmatrix} + \begin{bmatrix} 2 \\ 0 \end{bmatrix} = \begin{bmatrix} 2 \\ 0 \end{bmatrix},$$

$$\begin{bmatrix} 3 & 1 \\ 0 & 2 \end{bmatrix} \begin{bmatrix} 1 \\ 0 \end{bmatrix} + \begin{bmatrix} 2 \\ 0 \end{bmatrix} = \begin{bmatrix} 5 \\ 0 \end{bmatrix},$$

and

$$\begin{bmatrix} 3 & 1 \\ 0 & 2 \end{bmatrix} \begin{bmatrix} 0 \\ 1 \end{bmatrix} + \begin{bmatrix} 2 \\ 0 \end{bmatrix} = \begin{bmatrix} 3 \\ 2 \end{bmatrix}$$

yield that $F(0,0) = (2,0)$, $F(1,0) = (5,0)$, and $F(0,1) = (3,2)$.

Thus, if Δ is the triangle with vertices $(0,0), (1,0)$, and $(0,1)$, then $F(\Delta)$ is the triangle with vertices $(2,0), (5,0)$, and $(3,2)$. Note that area $(\Delta) = \frac{1}{2}$, $\det(A) = 3 \cdot 2 + 0 \cdot 1 = 6$ and area $[F(\Delta)] = \frac{1}{2}(\text{base}) \cdot (\text{height}) = \frac{1}{2} \cdot 3 \cdot 2 = 3$. Hence, one has that area $[F(\Delta)]$ is equal to $|\det(A)| \cdot \text{area}(\Delta)$ (compare Figure 5.4.3).

If A is an $n \times n$ matrix, then for all i, j with $1 \le i, j \le n$, the ij^{th} **minor** M_{ij} of A is the $(n - 1) \times (n - 1)$ determinant formed by deleting the ith row and j^{th} column of the original determinant of A. Clearly, A has n^2 minors.

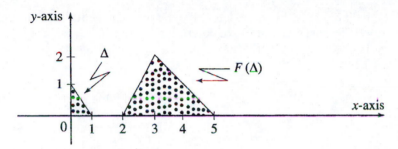

FIGURE 5.4.3 The mapping represented by $\begin{bmatrix} x' \\ y' \end{bmatrix} = \begin{bmatrix} 3 & 1 \\ 0 & 2 \end{bmatrix} \begin{bmatrix} x \\ y \end{bmatrix} + \begin{bmatrix} 2 \\ 0 \end{bmatrix}$ takes the triangle Δ with area $\frac{1}{2}$ to the triangle $F(\Delta)$ with area $3 = 6 \cdot \frac{1}{2} = \det(A) \cdot \text{area}(\Delta)$. If W is a region in the xy-plane, then area $[F(W)] = |\det(A)| \cdot \text{area}(W)$.

EXAMPLE Let $A = \begin{bmatrix} 2 & 0 & 3 \\ 1 & 4 & 5 \\ 7 & 8 & 6 \end{bmatrix}$. Find both M_{12} and M_{33}.

Solution   To find M_{12}, delete the first row and the second column of A to obtain

$$M_{12} = \begin{vmatrix} 1 & 5 \\ 7 & 6 \end{vmatrix} = 1 \cdot 6 - 5 \cdot 7 = -29.$$

Similarly, delete the third row and third column of A to obtain

$$M_{33} = \begin{vmatrix} 2 & 0 \\ 1 & 4 \end{vmatrix} = 2 \cdot 4 - 1 \cdot 0 = 8. \qquad \blacksquare$$

The ij^{th} **cofactor** A_{ij} of a square matrix A is defined by $A_{ij} = (-1)^{i+j} M_{ij}$. Hence,

$$A_{ij} = \begin{cases} M_{ij} & \text{if } i + j \text{ is even} \\ -M_{ij} & \text{if } i + j \text{ is odd} \end{cases}.$$

EXAMPLE Let $A = \begin{bmatrix} 2 & 0 & 3 \\ 1 & 4 & 5 \\ 7 & 8 & 6 \end{bmatrix}$. Find both A_{12} and A_{33}.

Solution   This is the same 3×3 matrix A used in the previous example where M_{12} and M_{33} were found to be -29 and 8, respectively. Thus,

$$A_{12} = (-1)^{1+2} M_{12} = -M_{12} = -(-29) = 29$$

and

$$A_{33} = (-1)^{3+3} M_{33} = +M_{33} = 8. \qquad \blacksquare$$

Definition of $n \times n$ Determinant.   If A is an $n \times n$ matrix, then, just as in the 2×2 case, the determinant may be positive, negative, or zero. For determinants that are larger than 2×2, one may define the determinant inductively using an expansion method. The usual definition of determinant, given here, is in terms of expansion by minors of the first row of A; however, one may expand by any row or column.

$$\det(A) = \sum_{j=1}^{j=n} (-1)^{1+j} a_{1j} M_{1j} = \sum_{j=1}^{j=n} a_{1j} A_{1j} \quad \text{(definition of } n \times n \text{ determinant)}$$

If $n = 3$, then $A = \begin{bmatrix} a_{11} & a_{12} & a_{13} \\ a_{21} & a_{22} & a_{23} \\ a_{31} & a_{32} & a_{33} \end{bmatrix}$ and the determinant of A can be denoted

by either $\det(A)$ or $\begin{vmatrix} a_{11} & a_{12} & a_{13} \\ a_{21} & a_{22} & a_{23} \\ a_{31} & a_{32} & a_{33} \end{vmatrix}$. Using the previous definition of determinant,

one has that

$$\det(A) = \det\left(\begin{bmatrix} a_{11} & a_{12} & a_{13} \\ a_{21} & a_{22} & a_{23} \\ a_{31} & a_{32} & a_{33} \end{bmatrix}\right) = \begin{vmatrix} a_{11} & a_{12} & a_{13} \\ a_{21} & a_{22} & a_{23} \\ a_{31} & a_{32} & a_{33} \end{vmatrix} = a_{11}M_{11} - a_{12}M_{12} + a_{13}M_{13}$$

$$= a_{11}\begin{vmatrix} a_{22} & a_{23} \\ a_{32} & a_{33} \end{vmatrix} - a_{12}\begin{vmatrix} a_{21} & a_{23} \\ a_{31} & a_{33} \end{vmatrix} + a_{13}\begin{vmatrix} a_{21} & a_{22} \\ a_{31} & a_{32} \end{vmatrix}.$$

Hence,

$$\begin{vmatrix} a_{11} & a_{12} & a_{13} \\ a_{21} & a_{22} & a_{23} \\ a_{31} & a_{32} & a_{33} \end{vmatrix} = a_{11}M_{11} - a_{12}M_{12} + a_{13}M_{13}$$

$$\begin{vmatrix} a_{11} & a_{12} & a_{13} \\ a_{21} & a_{22} & a_{23} \\ a_{31} & a_{32} & a_{33} \end{vmatrix} = a_{11}(a_{22}a_{33} - a_{23}a_{32}) - a_{12}(a_{21}a_{33} - a_{23}a_{31})$$
$$+ a_{13}(a_{21}a_{32} - a_{22}a_{31})$$

EXAMPLE Given $A = \begin{bmatrix} 1 & 2 & 3 \\ 0 & 1 & -4 \\ 2 & 0 & -3 \end{bmatrix}$, find $\det(A)$.

Solution

$$\det\left(\begin{bmatrix} 1 & 2 & 3 \\ 0 & 1 & -4 \\ 2 & 0 & -3 \end{bmatrix}\right) = \begin{vmatrix} 1 & 2 & 3 \\ 0 & 1 & -4 \\ 2 & 0 & -3 \end{vmatrix} = 1 \cdot M_{11} - 2 \cdot M_{12} + 3 \cdot M_{13}$$

$$= 1 \cdot \begin{vmatrix} 1 & -4 \\ 0 & -3 \end{vmatrix} - 2 \cdot \begin{vmatrix} 0 & -4 \\ 2 & -3 \end{vmatrix} + 3 \cdot \begin{vmatrix} 0 & 1 \\ 2 & 0 \end{vmatrix}$$

$$= 1 \cdot [1 \cdot (-3) - (-4) \cdot 0]$$
$$- 2 \cdot [0 \cdot (-3) - (-4) \cdot 2] + 3 \cdot (0 \cdot 0 - 1 \cdot 2)$$
$$= -3 - 16 - 6 = -25. \qquad \blacksquare$$

Inverse matrix. If the matrix A is square (i.e., $n \times n$) and also if $\det(A) \neq 0$, then A has an **inverse** A^{-1}. The inverse is a multiplicative inverse, taking the identity I to be the identity matrix. Thus,

$$AA^{-1} = A^{-1}A = I \qquad [A^{-1} \text{ exists for } A\ n \times n \text{ with } \det(A) \neq 0]$$

The inverse A^{-1} may be calculated with the following formula:

$$A^{-1} = \frac{1}{\det(A)} [A_{ij}]^T \qquad [\text{inverse of } A \text{ given that } A \text{ is square and } \det(A) \neq 0]$$

Here $[A_{ij}]^T$ is an $n \times n$ matrix found by taking the transpose of the matrix formed by using the cofactors A_{ij} of A. This formula for A^{-1} is not a very efficient way to find A^{-1} for large n but is reasonable for small values of n, such as $n = 2$ or 3. Below are the formulas for $n = 2$ and for $n = 3$.

If $A = \begin{bmatrix} a & b \\ c & d \end{bmatrix}$ and if $\det(A) \neq 0$, then $A^{-1} = \begin{bmatrix} a & b \\ c & d \end{bmatrix}^{-1}$ is given by

$$A^{-1} = \frac{1}{\det(A)} \begin{bmatrix} A_{11} & A_{12} \\ A_{21} & A_{22} \end{bmatrix}^T = \frac{1}{\det(A)} \begin{bmatrix} M_{11} & -M_{12} \\ -M_{21} & M_{22} \end{bmatrix}^T$$

$$= \frac{1}{(ad - bc)} \begin{bmatrix} d & -b \\ -c & a \end{bmatrix}.$$

EXAMPLE Let $A = \begin{bmatrix} 2 & 4 \\ 0 & -1 \end{bmatrix}$. Find $\det(A)$, A^{-1}, and $\det(A^{-1})$. Find a simple relationship between $\det(A)$ and $\det(A^{-1})$. Also, verify that $AA^{-1} = I$.

Solution Using the definition of determinant for the 2×2 case, one finds

$$\det(A) = \begin{vmatrix} 2 & 4 \\ 0 & -1 \end{vmatrix} = 2 \cdot (-1) - 4 \cdot 0 = -2.$$

Hence,

$$A^{-1} = \frac{1}{\det(A)} \begin{bmatrix} M_{11} & -M_{12} \\ -M_{21} & M_{22} \end{bmatrix}^T = \frac{1}{(-2)} \begin{bmatrix} -1 & -0 \\ -4 & 2 \end{bmatrix}^T$$

$$= \frac{1}{(-2)} \begin{bmatrix} -1 & -4 \\ 0 & 2 \end{bmatrix} = \begin{bmatrix} 1/2 & 2 \\ 0 & -1 \end{bmatrix},$$

which yields

$$\det(A^{-1}) = \begin{vmatrix} 1/2 & 2 \\ 0 & -1 \end{vmatrix} = (1/2) \cdot (-1) - 2 \cdot 0 = -1/2.$$

Clearly, $\det(A^{-1}) = \dfrac{1}{\det(A)}$.

Finally,

$$AA^{-1} = \begin{bmatrix} 2 & 4 \\ 0 & -1 \end{bmatrix} \begin{bmatrix} 1/2 & 2 \\ 0 & -1 \end{bmatrix}$$

$$= \begin{bmatrix} 2 \cdot (1/2) + 4 \cdot 0 & 2 \cdot 2 + 4 \cdot (-1) \\ 0 \cdot (1/2) + (-1) \cdot 0 & 0 \cdot 2 + (-1) \cdot (-1) \end{bmatrix} = \begin{bmatrix} 1 & 0 \\ 0 & 1 \end{bmatrix} = I,$$

which verifies that $AA^{-1} = I$, as desired. ∎

If $A = \begin{bmatrix} a_{11} & a_{12} & a_{13} \\ a_{21} & a_{22} & a_{23} \\ a_{31} & a_{32} & a_{33} \end{bmatrix}$ and $\det(A) \neq 0$, then

$$A^{-1} = \begin{bmatrix} a_{11} & a_{12} & a_{13} \\ a_{21} & a_{22} & a_{23} \\ a_{31} & a_{32} & a_{33} \end{bmatrix}^{-1} = \frac{1}{\det(A)} \begin{bmatrix} A_{11} & A_{12} & A_{13} \\ A_{21} & A_{22} & A_{23} \\ A_{31} & A_{32} & A_{33} \end{bmatrix}^{T}$$

$$= \frac{1}{\det(A)} \begin{bmatrix} A_{11} & A_{21} & A_{31} \\ A_{12} & A_{22} & A_{32} \\ A_{13} & A_{23} & A_{33} \end{bmatrix} = \frac{1}{\det(A)} \begin{bmatrix} M_{11} & -M_{21} & M_{31} \\ -M_{12} & M_{22} & -M_{32} \\ M_{13} & -M_{23} & M_{33} \end{bmatrix}.$$

EXAMPLE Given $A = \begin{bmatrix} 3 & 0 & 0 \\ 2 & -1 & 1 \\ 1 & 1 & 0 \end{bmatrix}$, find A^{-1} and verify that $AA^{-1} = I$.

Solution First calculate the determinant as follows:

$$\det(A) = a_{11}M_{11} - a_{12}M_{12} + a_{13}M_{13} = 3 \cdot (-1) - 0 \cdot M_{12} + 0 \cdot M_{13} = -3.$$

Using

$$A^{-1} = \frac{1}{\det(A)} \begin{bmatrix} M_{11} & -M_{12} & M_{13} \\ -M_{21} & M_{22} & -M_{23} \\ M_{31} & -M_{32} & M_{33} \end{bmatrix}^{T},$$

one obtains

$$A^{-1} = \frac{1}{(-3)} \begin{bmatrix} -1 & -(-1) & 3 \\ -(0) & 0 & -(3) \\ 0 & -(3) & -3 \end{bmatrix}^{T} = \frac{1}{(-3)} \begin{bmatrix} -1 & 0 & 0 \\ 1 & 0 & -3 \\ 3 & -3 & -3 \end{bmatrix}$$

$$= \begin{bmatrix} 1/3 & 0 & 0 \\ -1/3 & 0 & 1 \\ -1 & 1 & 1 \end{bmatrix}.$$

Notice that

$$AA^{-1} = \begin{bmatrix} 3 & 0 & 0 \\ 2 & -1 & 1 \\ 1 & 1 & 0 \end{bmatrix} \begin{bmatrix} 1/3 & 0 & 0 \\ -1/3 & 0 & 1 \\ -1 & 1 & 1 \end{bmatrix} = \begin{bmatrix} 1 & 0 & 0 \\ 0 & 1 & 0 \\ 0 & 0 & 1 \end{bmatrix},$$

which verifies that $AA^{-1} = I$, as desired.   ∎

EXERCISES 5.4

1. Find $\begin{vmatrix} 2 & 3 \\ 5 & 9 \end{vmatrix}$.

2. Find $\begin{vmatrix} 1 & -4 \\ -3 & 7 \end{vmatrix}$.

3. Find $\begin{bmatrix} 1 & 4 \\ 5 & 18 \end{bmatrix}^{-1}$.

4. Find $\begin{bmatrix} 2 & 13 \\ -2 & -12 \end{bmatrix}^{-1}$.

5. Find $\begin{bmatrix} 1 & 0 & 0 \\ 2 & 3 & 0 \\ 2 & 4 & 1 \end{bmatrix}^{-1}$.

6. Find $\begin{bmatrix} 1 & 4 & 5 \\ 2 & 3 & 0 \\ 0 & 4 & 0 \end{bmatrix}^{-1}$.

7. Let $A = \begin{bmatrix} 3 & 0 \\ 0 & 1 \end{bmatrix}$, $H = \begin{bmatrix} 1 \\ 2 \end{bmatrix}$, and let $F: R^2 \to R^2$ be represented by $X' = AX + H$.
   a. Find X', given that $X = \begin{bmatrix} 5 \\ 7 \end{bmatrix}$.
   b. Find $F(5, 7)$.
   c. Find $F(2, 3)$.

8. Let $A = \begin{bmatrix} 1 & 2 \\ 0 & 1 \end{bmatrix}$, $H = \begin{bmatrix} 0 \\ 1 \end{bmatrix}$, and let $F: R^2 \to R^2$ be represented by $X' = AX + H$. Let the isosceles triangle ΔKLM have vertices $K = (0,0)$, $L = (2,0)$, and $M = (1,1)$.
   a. Find the image of triangle ΔKLM (i.e., find the vertices of $\Delta K'L'M'$).
   b. Is the image $\Delta K'L'M'$ also isosceles?

9. Let $A = \begin{bmatrix} 2 & 0 \\ 1 & 1 \end{bmatrix}$, $H = \begin{bmatrix} 5 \\ 5 \end{bmatrix}$, and let $F: R^2 \to R^2$ be represented by $X' = AX + H$.
   a. Represent this transformation in the form

$$x' = ax + by + h$$
$$y' = cx + dy + k$$

In other words, find the parameters a, b, c, d, h, and k.

b. Let R be the square with vertices $(0,0)$, $(2,0)$, $(2,2)$, and $(0,2)$. Find the image of R under this F.

c. What kind of quadrilateral is $F(R)$?

d. Find $\det(A)$.

e. Find the ratio $\dfrac{\text{area}[F(R)]}{\text{area}(R)}$.

10. Consider the transformation $G\colon R^2 \to R^2$ represented by the following two equations:

$$x' = -x + 7y + 1$$
$$y' = 3x + 2y - 5$$

a. Find matrices A and H such that this transformation G is represented in the form $X' = AX + H$.

b. If W is a region in the xy-plane with positive area, find the ratio $\dfrac{\text{area}[G(W)]}{\text{area}(W)}$.

11. Let $B = \begin{bmatrix} 2 & 1 \\ 3 & 5 \end{bmatrix}$.

a. Find $\det(B)$.

b. Find $\det(B^T)$ and compare this answer to $\det(B)$.

c. Find B^2 (i.e., find the product BB).

d. Find $\det(B^2)$ and compare this answer to $\det(B)$.

e. Find B^{-1} and verify your work by showing that $BB^{-1} = I$.

f. Find $\det(B^{-1})$ and compare this answer to $\det(B)$.

g. List at least three conjectures one might make after doing parts (a) to (f).

12. Let $A = \begin{bmatrix} 1 & 3 & 4 \\ 0 & 2 & 1 \\ 0 & 0 & 1 \end{bmatrix}$.

a. Find $4A$.

b. Find A^{-1} and verify that your answer satisfies both $AA^{-1} = I$ and $A^{-1}A = I$.

c. Find $\det(A^{-1})$ and compare this answer to $\det(A)$.

d. Find $\det(4A)$ and compare this answer to $\det(A)$.

13. Let $A = \begin{bmatrix} a & b \\ c & d \end{bmatrix}$ and $B = \begin{bmatrix} e & f \\ g & h \end{bmatrix}$. Verify that $\det(AB) = \det(A) \cdot \det(B)$ for the special case where A and B are 2×2 matrices by showing that both $\det(AB)$ and $\det(A) \cdot \det(B)$ yield the same answer.

14. Find examples of 2×2 matrices C and D such that one has $\det(C + D) \neq \det(C) + \det(D)$.

Classroom Discussion 5.4.1

Students often feel the definition of *matrix multiplication* is very unnatural. The definition has to do with composition of functions. If $F_1(x,y) = (a_{11}x + a_{12}y, a_{21}x + a_{22}y)$ and $F_2(x,y) = (b_{11}x + b_{12}y, b_{21}x + b_{22}y)$, then F_1 is represented by $X' = AX$ and F_2 by $X' = BX$, where A is the matrix given by

$A = \begin{bmatrix} a_{11} & a_{12} \\ a_{21} & a_{22} \end{bmatrix}$, and B is the matrix given by $B = \begin{bmatrix} b_{11} & b_{12} \\ b_{21} & b_{22} \end{bmatrix}$. Let $F_3(x, y) =$ $F_2 \circ F_1(x, y) = (c_{11}x + c_{12}y, c_{21}x + c_{22}y)$. Find each c_{ij} in terms of the elements of the matrices A and B. Then find the matrix products AB and BA. Which of the products AB or BA has the c_{ij} as elements? After doing this exploration, students should be able to write a paper motivating the definition of matrix multiplication. This paper should be at least one page long. ◆

Classroom Discussion 5.4.2

For a good classroom discussion, have students conjecture the relation between $\det(kA)$ and $\det(A)$ where k is a real number and A is an $n \times n$ matrix. If they have difficulty discovering the relation, ask them to consider the special case where A is the $n \times n$ identity matrix. They should quickly see that $\det(I_n) = 1$ and $\det(kI_n) = k^n$. ◆

5.5 ISOMETRIES AND ORTHOGONAL MATRICES

A matrix A is said to be **orthogonal** if its transpose is equal to its inverse (i.e., $A^T = A^{-1}$).

$$A^T = A^{-1} \text{ (definition of orthogonal matrix)}$$

Note that orthogonal matrices must be square and must have a nonzero determinant. The easiest example of an orthogonal $n \times n$ matrix is the $n \times n$ identity matrix I. Clearly, $I = I^T = I^{-1}$. Note that many matrices are square and have a nonzero determinant yet fail to be orthogonal. For example, $A = \begin{bmatrix} 1 & 0 \\ 0 & 2 \end{bmatrix}$ is 2×2 and has a nonzero determinant, yet $A^T \neq A^{-1}$ since $A^T = A = \begin{bmatrix} 1 & 0 \\ 0 & 2 \end{bmatrix}$ and

$A^{-1} = \begin{bmatrix} 1 & 0 \\ 0 & 1/2 \end{bmatrix}$.

The next theorem guarantees that the determinant of an orthogonal matrix is either $+1$ or -1. The proof uses two important results for square matrices. The determinant of the product is the product of the determinants [i.e., $\det(AB) = \det(A) \cdot \det(B)$] and the determinant of the transpose is equal to the determinant of the original matrix [i.e., $\det(A^T) = \det(A)$].

$\det(AB) = \det(A) \cdot \det(B)$	(for square matrices, the determinant of the product equals the product of the determinants)
$\det(A^T) = \det(A)$	(for square matrices, the determinant of the transpose equals the determinant of the original matrix)

Theorem 5.5.1. *If A is an orthogonal matrix, then* $\det(A) = \pm 1$.

Proof. From the definition of orthogonal, one has $A^T = A^{-1}$. Hence, using $AA^{-1} = I$, one obtains $AA^T = I$. Taking the determinant of each side yields

$$\det(AA^T) = \det(I).$$

Hence,

$$\det(A) \cdot \det(A^T) = \det(I),$$

which yields

$$[\det(A)]^2 = 1.$$

Taking the general (i.e., $\pm$) square root of the previous equation yields

$$\det(A) = \pm 1,$$

as desired.

Rotations about the Origin.   A transformation $F : R^n \rightarrow R^n$ is said to be a **proper rotation holding the origin fixed** if it has a matrix representation of the form $X' = AX$, where A is an $n \times n$ orthogonal matrix with $\det(A) = +1$.

Here $X' = \begin{bmatrix} x'_1 \\ \dots \\ x'_n \end{bmatrix}$ and $X = \begin{bmatrix} x_1 \\ \dots \\ x_n \end{bmatrix}$ are $n \times 1$ matrices, where $R^n = \{(x_1, x_2, \dots,$

$x_n) \mid$ each $x_i \in R^1\}$. A transformation $F: R^n \rightarrow R^n$ is called an **improper (or orientation-reversing) rotation holding the origin fixed** if it has a matrix representation of the form $X' = AX$, where A is an $n \times n$ orthogonal matrix with $\det(A) = -1$. Naturally, middle school materials generally only use the word *rotation* when referring to proper rotations.

A proper rotation about the origin in the Euclidean plane by angle θ is given by $F(x, y) = (x \cos\theta - y \sin\theta, x \sin\theta + y \cos\theta)$, see Figure 5.5.1. Here the given angle θ is measured counterclockwise from the x-axis. The corresponding matrix representation for this rotation is

$$\begin{bmatrix} x' \\ y' \end{bmatrix} = \begin{bmatrix} \cos\theta & -\sin\theta \\ \sin\theta & \cos\theta \end{bmatrix} \begin{bmatrix} x \\ y \end{bmatrix}.$$

Using the identity $\sin^2(\theta) + \cos^2(\theta) = 1$, it is easily checked that $\begin{vmatrix} \cos\theta & -\sin\theta \\ \sin\theta & \cos\theta \end{vmatrix} = 1$.

EXAMPLE Let $X' = AX$ represent a counterclockwise rotation of $45°$ about the origin. Find the matrix A and find the image of $(1, 0)$ under this rotation.

Solution   Since $\theta = 45°$, $A = \begin{bmatrix} \cos 45° & -\sin 45° \\ \sin 45° & \cos 45° \end{bmatrix} = \begin{bmatrix} 1/\sqrt{2} & -1/\sqrt{2} \\ 1/\sqrt{2} & 1/\sqrt{2} \end{bmatrix}.$

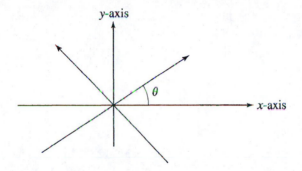

FIGURE 5.5.1 The proper rotation of the Euclidean plane about the origin through an angle θ is represented by the matrix equation $X' = AX$, where $A = \begin{bmatrix} \cos\theta & -\sin\theta \\ \sin\theta & \cos\theta \end{bmatrix}$. Points are rotated counterclockwise by angle θ. When θ is negative, then a rotation by angle θ is the same as a rotation by an amount $|\theta|$ in the clockwise direction.

Thus, $X' = AX$ becomes

$$\begin{bmatrix} x' \\ y' \end{bmatrix} = \begin{bmatrix} 1/\sqrt{2} & -1/\sqrt{2} \\ 1/\sqrt{2} & 1/\sqrt{2} \end{bmatrix} \begin{bmatrix} x \\ y \end{bmatrix} = \begin{bmatrix} x/\sqrt{2} - y/\sqrt{2} \\ x/\sqrt{2} + y/\sqrt{2} \end{bmatrix}.$$

Hence, the image of the point $(1,0)$ is $(1/\sqrt{2}, 1/\sqrt{2})$ since using $x = 1$ and $y = 0$ yields

$$\begin{bmatrix} x' \\ y' \end{bmatrix} = \begin{bmatrix} 1/\sqrt{2} & -1/\sqrt{2} \\ 1/\sqrt{2} & 1/\sqrt{2} \end{bmatrix} \begin{bmatrix} 1 \\ 0 \end{bmatrix} = \begin{bmatrix} 1/\sqrt{2} \\ 1/\sqrt{2} \end{bmatrix}. \qquad \blacksquare$$

An improper rotation about the origin of the Euclidean plane may be thought of as reflection across a line through the origin followed by a rotation about the origin by an angle θ. If one chooses to reflect across the x-axis and then rotate, the improper rotation will have a matrix representation of the form $X' = AX$, where $A = \begin{bmatrix} \cos\theta & \sin\theta \\ \sin\theta & -\cos\theta \end{bmatrix}$. Note that this matrix has a determinant equal to -1.

Letting $\theta = 0$ yields a simple reflection across the x-axis with corresponding matrix

$A = \begin{bmatrix} \cos 0 & \sin 0 \\ \sin 0 & -\cos 0 \end{bmatrix} = \begin{bmatrix} 1 & 0 \\ 0 & -1 \end{bmatrix}$. Hence, reflection across the x-axis is given by

$\begin{bmatrix} x' \\ y' \end{bmatrix} = \begin{bmatrix} 1 & 0 \\ 0 & -1 \end{bmatrix} \begin{bmatrix} x \\ y \end{bmatrix} = \begin{bmatrix} x \\ -y \end{bmatrix}$. Representing this with equations yields

$$x' = x$$

$$y' = -y.$$

Each rigid motion of the Euclidean plane may be represented as a proper rotation or improper rotation about the origin followed by a translation. Thus, the rigid motions of the plane are exactly those transformations $F: R^2 \rightarrow R^2$ that have matrix representations of the form

$$X' = AX + H,$$

where

$$X' = \begin{bmatrix} x' \\ y' \end{bmatrix}, A^T = A^{-1}, X = \begin{bmatrix} x \\ y \end{bmatrix}, \text{ and } H = \begin{bmatrix} h \\ k \end{bmatrix}.$$

Similarly, the rigid motions of Euclidean three-dimensional space are exactly those transformations $F: R^3 \rightarrow R^3$ that have matrix representations of the form

$$X' = AX + H,$$

where

$$X' = \begin{bmatrix} x' \\ y' \\ z' \end{bmatrix}, A^T = A^{-1}, X = \begin{bmatrix} x \\ y \\ z \end{bmatrix}, \text{ and } H = \begin{bmatrix} h \\ k \\ m \end{bmatrix}.$$

Furthermore, similar results hold for n-dimensional Euclidean space.

EXERCISES 5.5

1. The transformation $F(x, y) = (-x, -y)$ is a rotation of the xy-plane about the origin by $180°$. Find the matrix A such that this transformation is represented by $X' = AX$.

2. The transformation represented by $X' = BX$, where $B = \begin{bmatrix} 1/\sqrt{2} & -1/\sqrt{2} \\ 1/\sqrt{2} & 1/\sqrt{2} \end{bmatrix}$ is a rotation of the xy-plane about the origin by $45°$. Show that $B^4 = -I$, where I is the 2×2 identity matrix. Explain how this exercise is related to the previous exercise.

3. Let $F: R^2 \rightarrow R^2$ be a reflection across the line $y = x$. Find a matrix C such that this transformation is represented by $X' = CX$.

4. Let $F_1: R^3 \rightarrow R^3$ be represented by $X' = AX$, where $A = \begin{bmatrix} 0 & -1 & 0 \\ 1 & 0 & 0 \\ 0 & 0 & 1 \end{bmatrix}$ and

$X = \begin{bmatrix} x \\ y \\ z \end{bmatrix}$. This transformation F_1 is a rotation by $90°$ about the z-axis. In particular, F_1 holds the z-axis fixed and rotates points on the positive x-axis to points on the positive y-axis at the same time each point on the positive y-axis is rotated to a point on the negative x-axis. Let $F_2: R^3 \rightarrow R^3$ be represented by

$X' = BX$, where $B = \begin{bmatrix} 1 & 0 & 0 \\ 0 & 0 & -1 \\ 0 & 1 & 0 \end{bmatrix}$. Then F_2 is a rotation by $90°$ about the x-axis.

a. Where are points on the positive y-axis rotated to by F_2?
b. Calculate the product AB.
c. Calculate the product BA.
d. Which of AB and BA represents $F_2 \circ F_1$ and which represents $F_1 \circ F_2$?

e. Do rotations by 90° about the z-axis and by 90° about the x-axis commute?

f. Is this problem related to anything covered in Chapter 4? Explain.

Classroom Discussion 5.5.1

Recall that a (proper) rotation $F: R^n \to R^n$ holding the origin fixed has a matrix representation of the form $X' = AX$, where A is an $n \times n$ orthogonal matrix with $\det(A) = +1$ and $X = \begin{bmatrix} x_1 \\ \cdots \\ x_n \end{bmatrix}$. Many students have difficulty accepting that one can have dimensions larger than three. Of course, once they accept dimension four, then higher dimensions are easier to accept. In the case of $n = 4$, one usually sets $x = x_1$, $y = x_2$, $z = x_3$, and $w = x_4$. To get them to develop some intuition about Euclidean four-dimensional space, ask them to describe, in words, some possible rotations in R^4. For example, first ask them to describe the rotation in the plane, R^2, represented by $X' = AX$, where $A = \begin{bmatrix} 0 & -1 \\ 1 & 0 \end{bmatrix}$, then ask them to describe the rotation in four-dimensional space, R^4, given by $X' = BX$, where $B = \begin{bmatrix} 0 & -1 & 0 & 0 \\ 1 & 0 & 0 & 0 \\ 0 & 0 & 1 & 0 \\ 0 & 0 & 0 & 1 \end{bmatrix}$. If students have difficulty, then ask them to find where the unit (i.e., length one) vectors along the coordinate axes go under these rotations. ◆

5.6 ISOMETRIES AND COMPLEX NUMBERS

Complex numbers can be used to study isometries of the Euclidean plane. Recall that the imaginary unit i is a square root of -1. Thus,

$$i^2 = -1.$$

A **complex number** is one of the form $a + bi$, where a and b are real numbers. The real part is a and the imaginary part is b. Either or both of these parts are allowed to be zero. When $a = 0$, the complex number is said to **pure imaginary**. When $b = 0$, the complex number is said to be **real**. Thus, both pure imaginary numbers and real numbers are special cases of complex numbers. The customary notation used for complex numbers is z. When one is considering two or more complex numbers, one can use subscript notation such as z_1, z_2, etc., to denote complex numbers that may be different. If z_1 and z_2 are complex numbers given by

$$z_1 = a + bi$$

and

$$z_2 = c + di,$$

their **sum** is found by adding corresponding real parts and corresponding imaginary parts.

$$z_1 + z_2 = (a + bi) + (c + di) = (a + c) + (b + d)i \qquad \text{(addition)}$$

Their **difference** is found by taking the difference of the real parts and the difference of the imaginary parts.

$$z_1 - z_2 = (a + bi) - (c + di) = (a - c) + (b - d)i \quad \text{(subtraction)}$$

Thus,

$$(3 + 4i) + (-1 + 2i) = 2 + 6i$$

and

$$(5 + 7i) - (4 - i) = 1 + 8i.$$

The **product** of $z_1 = a + bi$ and $z_2 = c + di$ is defined as $(ac - bd) + (ad + bc)i$.

$$z_1 \cdot z_2 = (ac - bd) + (ad + bc)i \quad \text{(multiplication)}$$

The previous definition is motivated by the following equations, using $i^2 = -1$:

$$z_1 \cdot z_2 = (a + bi) \cdot (c + di) = a \cdot (c + di) + (bi) \cdot (c + di)$$

$$z_1 \cdot z_2 = ac + adi + bci + bdi^2 = ac + adi + bci + bd(-1)$$

$$z_1 \cdot z_2 = (ac - bd) + (ad + bc)i$$

Using $z_1 = 2 + 3i$ and $z_2 = -4 + 2i$, one finds

$$(2 + 3i) \cdot (-4 + 2i) = [2 \cdot (-4) - 3 \cdot 2] + [2 \cdot 2 + 3 \cdot (-4)]i = -14 - 8i.$$

The **conjugate** of $z = a + bi$ is denoted by $= \bar{z}$ and is obtained from z by changing the sign of the imaginary part. Thus, $\bar{z} = a - bi$.

$$\overline{(a + bi)} = a - bi \quad \text{(conjugate)}$$

Hence,

$$\overline{(9 + 10i)} = 9 - 10i.$$

Using xy-coordinates for R^2, each point (x, y) may be identified with the complex number $z = x + iy$. This clearly yields a $1 - 1$ onto correspondence with the complex numbers and the points of the usual xy-plane. Hence, each complex number may be thought of as a point of the xy-plane and vice versa.

$$x + iy \Leftrightarrow (x, y)$$
$$1 \Leftrightarrow (1, 0)$$
$$i \Leftrightarrow (0, 1)$$

Notice that if one multiplies a complex number $z = x + iy$ by its conjugate $\bar{z} = x - iy$, then one obtains the real number $x^2 + y^2$.

$$z \cdot \bar{z} = (x + yi) \cdot (x - yi) = [x \cdot x - (y) \cdot (-y)] + [x \cdot (-y) + x \cdot y]i = x^2 + y^2$$

The **absolute value** or **norm** $|z|$ of a complex number $z = x + iy$ is defined as $\sqrt{x^2 + y^2}$, which is the square root of the product $z \cdot \bar{z}$.

$$|z| = \sqrt{z \cdot \bar{z}} = \sqrt{x^2 + y^2} \qquad \text{(absolute value)}$$

Since one may think of the complex number $z = x + iy$ as identified with the point (x, y), the absolute value $|z|$ represents the Euclidean length of the segment from the origin $(0, 0)$ to the point (x, y). The angle θ this line segment makes with the positive x-axis is said to be the **argument** of z, which is denoted by $\arg(z)$.

$$\theta = \arg(z) = \tan^{-1}\left(\frac{y}{x}\right) \quad \text{assuming } x \neq 0.$$

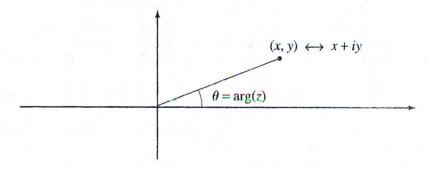

Fractions that have denominators with nonzero imaginary parts can be changed to get rid of the i in the denominator by multiplying both numerator and denominator by the conjugate of the denominator. Hence,

$$\frac{2 + 3i}{3 + 4i} = \frac{(2 + 3i)(3 - 4i)}{(3 + 4i)(3 - 4i)} = \frac{[6 - (-12)] + (-8 + 9)i}{[9 - (-16)] + (12 - 12)i} = \frac{18}{25} + \frac{i}{25}.$$

Some rules for complex numbers are:

$z_1 + z_2 = z_2 + z_1$	Commutative law of addition
$z_1 + (z_2 + z_3) = (z_1 + z_2) + z_3$	Associative law of addition
$z + 0 = 0 + z = z$	Existence of additive identity
$z + (-z) = (-z) + z = 0$	Existence of additive inverse
$z_1 \cdot z_2 = z_2 \cdot z_1$	Commutative law of multiplication
$z_1 \cdot (z_2 \cdot z_3) = (z_1 \cdot z_2) \cdot z_3$	Associative law of multiplication
$z \cdot 1 = 1 \cdot z = z$	Existence of multiplicative identity
$z \cdot z^{-1} = z^{-1} \cdot z = 1$	For $z \neq 0$, there exists a multiplicative inverse
$z_1 \cdot (z_2 + z_3) = z_1 \cdot z_2 + z_1 \cdot z_3$	Distributive law (of multiplication over addition)
$\bar{\bar{z}} = z$	Double conjugate rule

$\overline{(z_1 + z_2)} = \overline{z_1} + \overline{z_2}$ Conjugate of sum is sum of conjugates

$\overline{(z_1 \cdot z_2)} = \overline{z_1} \cdot \overline{z_2}$ Conjugate of product is product of conjugates

$\overline{\left(\dfrac{z_1}{z_2}\right)} = \dfrac{\overline{z_1}}{\overline{z_2}}$ Conjugate of ratio is ratio of conjugates

$\overline{z^{-1}} = (\overline{z})^{-1}$ For $z \neq 0$, the inverse and conjugate commute

$|\overline{z}| = |z|$ Absolute value of conjugate is absolute value of original

$|z_1 \cdot z_2| = |z_1| \cdot |z_2|$ Absolute value of product is product of absolute values

When using **polar coordinates** (r, θ), one lets r denote the distance of (x, y) from the origin $(0, 0)$ and lets θ denote the angle the segment from the origin to (x, y) makes with the positive x-axis, see Figure 5.6.1. If $z = x + iy$, then $r = |z| = \sqrt{x^2 + y^2}$.

Recall from Chapter 2 that the Euler number of a polyhedron is given by $F - E + V$, where F is the number of faces, E is the number of edges, and V is the number of vertices. Euler made many contributions to mathematics. Among many other things, he studied a number that is now denoted by e in his honor. The number e is the base of the natural logarithms. Like π, the number e is an irrational number. An approximate value of e, correct to five decimal places, is given by 2.71828.

$$e \approx 2.71828 \quad \text{(approximate value of the number } e\text{)}$$

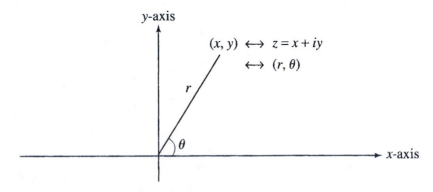

FIGURE 5.6.1 The complex number $z = x + iy$ corresponds to the point (x, y). Set $r = |z| = \sqrt{x^2 + y^2}$. If $\theta = \arg(z)$ is the angle the segment from the origin to (x, y) makes with the positive x-axis, the polar coordinates of (x, y) are given by (r, θ). One makes the identifications $z = x + iy \leftrightarrow (x, y) \leftrightarrow (r, \theta)$.

Euler linked trigonometry to complex numbers via the number e. **Euler's Identity** is given by $e^{i\theta} = \cos(\theta) + i\sin(\theta)$.

$$e^{i\theta} = \cos(\theta) + i\sin(\theta) \qquad \text{(Euler's Identity)}$$

The conjugate of $e^{i\theta}$ is $e^{-i\theta}$ since $\cos(-\theta) = \cos(\theta)$ and $\sin(-\theta) = -\sin(\theta)$ yield $e^{-i\theta} = e^{i(-\theta)} = \cos(-\theta) + i\sin(-\theta) = \cos(\theta) - i\sin(\theta)$. Also, the absolute value of $e^{i\theta}$ equals 1 since

$$|e^{i\theta}| = \sqrt{[\cos(\theta) + i\sin(\theta)] \cdot [\cos(\theta) - i\sin(\theta)]}$$

$$= \sqrt{\cos^2(\theta) + \sin^2(\theta)} = \sqrt{1} = 1.$$

In particular, for any real value of θ (which may be measured in degrees or radians), the complex number $e^{i\theta}$ represents a point on the unit circle centered at the origin that corresponds to a radius, making an angle θ with the positive x-axis, see Figure 5.6.2.

Euler's identity can be used to find the value of e^z for any complex number z. If $z = x + iy$, then $e^z = e^{(x+iy)} = e^x \cdot e^{iy} = e^x \cdot [\cos(y) + i\sin(y)]$.

Given a complex number $z = x + iy$, then $z = x + iy \leftrightarrow (x, y) \leftrightarrow (r, \theta)$ as illustrated in Figure 5.6.1, where $r = |z| = \sqrt{x^2 + y^2}$ and where θ is the angle the segment from $(0, 0)$ to (x, y) makes with the positive x-axis. Polar and cartesian coordinates are related by

$$x = r\cos(\theta)$$

$$y = r\sin(\theta).$$

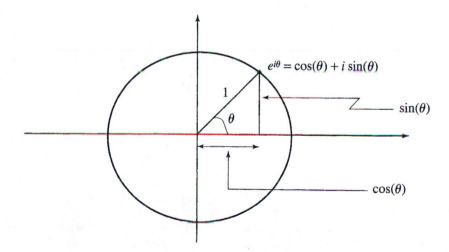

FIGURE 5.6.2 The number $e^{i\theta}$ corresponds to a point on the unit circle that may be found by constructing a radius, making an angle θ with the positive x-axis.

Also, $z = x + iy = r\cos(\theta) + ir\sin(\theta) = r[\cos(\theta) + i\sin(\theta)] = r \cdot e^{i\theta}$. The representation $z = r \cdot e^{i\theta}$ is called the **complex polar representation**. Notice that the r and θ for the complex polar representation are the same r and θ as for ordinary polar coordinates.

EXAMPLE Given $z = 4 + 3i$, find the cartesian representation (x, y), the polar representation (r, θ), and the complex polar representation $r \cdot e^{i\theta}$.

Solution Using $z = 4 + 3i$, one obtains $x = 4$ and $y = 3$. Thus, the cartesian representation is $(4, 3)$. Since $r = \sqrt{x^2 + y^2}$, one obtains $r = \sqrt{4^2 + 3^2} = \sqrt{25} = 5$. Also, $\theta = \tan^{-1}\left(\frac{y}{x}\right)$ yields $\theta = \tan^{-1}(3/4) \approx .6435$ radians (about $36.87°$). Thus, z has a polar representation of $(5, .6435)$, where the angle is only an approximate value. Of course, one may add or subtract multiples of 2π radians ($360°$) to any angle. Thus, the polar representation can be written more generally as $(5, \theta)$, where $\theta \approx .6435 \pm n \cdot 2 \cdot \pi$ radians ($\theta \approx 36.87° \pm n \cdot 360°$). Using 5 for r and using the (approximate) value of .6435 radians for the angle θ, one finds that the complex polar representation is $z \approx 5 \cdot e^{i(.6435)}$. ∎

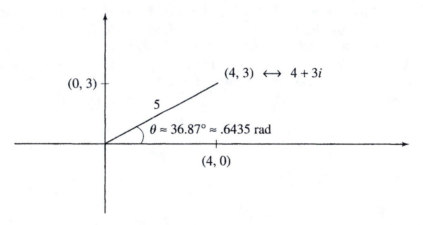

Complex addition may be interpreted as vector addition, and complex subtraction may be interpreted as vector subtraction. In particular, given $z_1 = x_1 + iy_1$ and $z_2 = x_2 + iy_2$, the sum $z_1 + z_2 = (x_1 + iy_1) + (x_2 + iy_2) = (x_1 + x_2) + i(y_1 + y_2)$ corresponds to the vector sum of the vectors from the origin to (x_1, y_1) and (x_2, y_2), respectively.

$$\begin{bmatrix} x_1 \\ y_1 \end{bmatrix} + \begin{bmatrix} x_2 \\ y_2 \end{bmatrix} = \begin{bmatrix} x_1 + x_2 \\ y_1 + y_2 \end{bmatrix}$$

This sum is illustrated in Figure 5.6.3.

Of course, the difference $z_1 - z_2 = (x_1 - x_2) + i(y_1 - y_2)$ corresponds to the vector difference

$$\begin{bmatrix} x_1 \\ y_1 \end{bmatrix} - \begin{bmatrix} x_2 \\ y_2 \end{bmatrix} = \begin{bmatrix} x_1 - x_2 \\ y_1 - y_2 \end{bmatrix}.$$

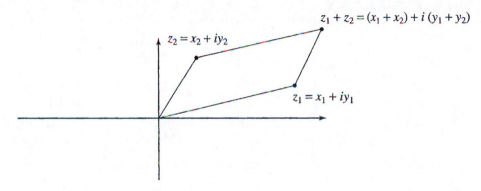

FIGURE 5.6.3 Addition of the complex number $z_1 = x_1 + iy_1 \leftrightarrow (x_1, y_1) \leftrightarrow \begin{bmatrix} x_1 \\ y_1 \end{bmatrix}$ and the complex number $z_2 = x_2 + iy_2 \leftrightarrow (x_2, y_2) \leftrightarrow \begin{bmatrix} x_2 \\ y_2 \end{bmatrix}$ corresponds to the vector addition $\begin{bmatrix} x_1 \\ y_1 \end{bmatrix} + \begin{bmatrix} x_2 \\ y_2 \end{bmatrix} = \begin{bmatrix} x_1 + x_2 \\ y_1 + y_2 \end{bmatrix}$.

Taking the conjugate of $z = x + iy$ yields $\bar{z} = x - iy$. Thus, the effect of $f(z) = \bar{z}$ is to take the point (x, y) to the point $(x, -y)$, which geometrically corresponds to reflection across the x-axis, see Figure 5.6.4.

> Complex addition $\leftrightarrow$ vector addition.
> Complex subtraction $\leftrightarrow$ vector subtraction.
> Taking the conjugate $\leftrightarrow$ reflection across the x-axis.

To obtain a geometric interpretation of complex multiplication,

$$z_1 \cdot z_2 = (x_1 + iy_1) \cdot (x_2 + iy_2)$$
$$= (x_1 \cdot x_2 - y_1 \cdot y_2) + i(x_1 \cdot y_2 + x_2 \cdot y_1),$$

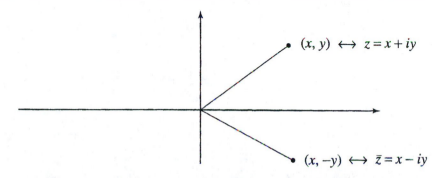

FIGURE 5.6.4 The mapping $f(z) = \bar{z}$ reflects the xy-plane across the x-axis. The point (x, y) is taken to the point $(x, -y)$. Using matrix notation, this corresponds to $\begin{bmatrix} x' \\ y' \end{bmatrix} = \begin{bmatrix} 1 & 0 \\ 0 & -1 \end{bmatrix} \begin{bmatrix} x \\ y \end{bmatrix}$.

it is convenient to switch to complex polar notation. Using $z_1 = r_1 e^{i\theta_1}$ and $z_2 = r_2 e^{i\theta_2}$, one finds

$$z_1 \cdot z_2 = (r_1 \cdot e^{i\theta_1}) \cdot (r_2 \cdot e^{i\theta_2}) = r_1 \cdot r_2 \cdot e^{i\theta_1} \cdot e^{i\theta_2} = r_1 \cdot r_2 \cdot e^{i\theta_1 + i\theta_2}$$

$$= r_1 \cdot r_2 \cdot e^{i(\theta_1 + \theta_2)}.$$

Thus, if $z = z_1 \cdot z_2 = r \cdot e^{i\theta}$, then $r = r_1 \cdot r_2$ and $\theta = \theta_1 + \theta_2$. Hence, the absolute value r of the product is the product $r_1 \cdot r_2$ of absolute values, and the argument θ of the product is the sum $\theta_1 + \theta_2$ of the arguments.

$$|z_1 \cdot z_2| = |z_1| \cdot |z_2|$$

$$\arg(z_1 \cdot z_2) = \arg(z_1) + \arg(z_2)$$

In particular, multiplication of z by $e^{i\theta}$ rotates z about the origin by an angle θ for every z. Furthermore, multiplying z by a real number h yields a dilation with dilation factor h, which is centered at the origin.

Multiplication by $e^{i\theta}$ $\leftrightarrow$ rotation about origin by angle θ.

Multiplication by a real number h $\leftrightarrow$ dilation about origin by factor h.

The information given in Figure 5.6.5 can be summarized by the following:

$$R_{O,\theta} \quad \leftrightarrow \quad e^{i\theta} \quad \leftrightarrow \quad \begin{bmatrix} \cos(\theta) & -\sin(\theta) \\ \sin(\theta) & \cos(\theta) \end{bmatrix}.$$

For example, to rotate counterclockwise by $90°$ (i.e., $\pi/2$ radians), one just multiplies z by $e^{i(\pi/2)} = \cos(\pi/2) + i\sin(\pi/2) = 0 + i \cdot 1 = i$. Thus, $f(z) = i \cdot z$ is the rotation by $90°$ about the origin $O = (0,0)$. In particular, to find where $(3, 4)$ goes,

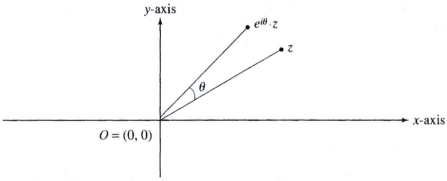

Figure 5.6.5 The mapping $e^{i\theta} \cdot z$ rotates points of the xy-plane by the angle θ (measured in radians) about the origin. Thus, $R_{O,\theta}(z) = e^{i\theta} \cdot z$. In terms of matrix multiplication, this corresponds to the matrix equation $\begin{bmatrix} x' \\ y' \end{bmatrix} = \begin{bmatrix} \cos(\theta) & -\sin(\theta) \\ \sin(\theta) & \cos(\theta) \end{bmatrix} \begin{bmatrix} x \\ y \end{bmatrix}$.

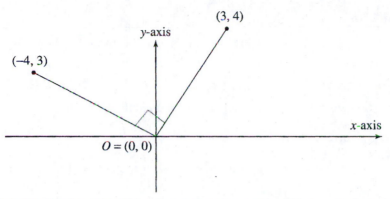

Figure 5.6.6 The transformation $f(z) = i \cdot z$ rotates the xy-plane $90°$ (counterclockwise) about the origin. In particular, the point $(3, 4)$ is mapped to the point $(-4, 3)$. More generally, the point (x, y) is mapped to the point $(-y, x)$. Using matrix notation, this rotation is represented by $\begin{bmatrix} x' \\ y' \end{bmatrix} = \begin{bmatrix} \cos(90°) & -\sin(90°) \\ \sin(90°) & \cos(90°) \end{bmatrix} \begin{bmatrix} x \\ y \end{bmatrix} = \begin{bmatrix} 0 & -1 \\ 1 & 0 \end{bmatrix} \begin{bmatrix} x \\ y \end{bmatrix} = \begin{bmatrix} -y \\ x \end{bmatrix}$.

one sets $z = 3 + 4i$ and calculates $f(3 + 4i) = i \cdot (3 + 4i) = -4 + 3i$. Hence, the point $(3, 4)$ is mapped by this rotation to the point $(-4, 3)$ (compare Figure 5.6.6). More generally, setting $z = x + iy$, one finds that $f(z) = i \cdot z = i \cdot (x + iy) = -y + ix$. Hence, this rotation takes the general point (x, y) to the point $(-y, x)$.

Translations of the xy-plane are even easier using complex numbers. The translation taking the origin $(0, 0)$ to the point (h, k) is given by $F(x, y) = (x + h, y + k)$. With complex numbers one first sets $c = h + ik$ and then defines $f(z) = z + c$. Hence, $f(x + iy) = (x + iy) + (h + ik) = (x + h) + i(y + k)$, see Figure 5.6.7.

> Addition of complex number $c = h + ik \leftrightarrow$ translation taking origin to (h, k).

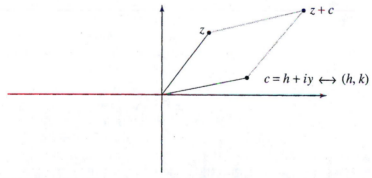

Figure 5.6.7 The mapping $f(z) = z + c$, where $c = h + ik$ is a translation of the xy-plane taking the origin to the point (h, k) and the general point (x, y) to the point $(x + h, y + k)$. Using matrix notation, this corresponds to $\begin{bmatrix} x' \\ y' \end{bmatrix} = \begin{bmatrix} x \\ y \end{bmatrix} + \begin{bmatrix} h \\ k \end{bmatrix}$.

EXERCISES 5.6

1. Let $z_1 = 4 + 5i$ and $z_2 = 2 + 3i$. Simplify the following (i.e., express as $x + iy$):

   a. $z_1 + z_2$

   b. $z_1 \cdot z_2$

   c. $13 \cdot \dfrac{z_1}{z_2}$

2. Let $z_3 = 1 + 2i$ and $z_4 = 2 + 2i$. Find the following:

   a. $|z_3|$

   b. $|z_4|$

   c. $|(z_3)^2 \cdot (z_4)^2|$

3. In each case, find the matrix A such that the following may be expressed in the form $X' = AX$, where $X' = \begin{bmatrix} x' \\ y' \end{bmatrix}$ and $X = \begin{bmatrix} x \\ y \end{bmatrix}$.

   a. $f(z) = e^{i\pi} \cdot z$

   b. $f(z) = -i \cdot z$

   c. $f(z) = e^{i(5\pi/2)} \cdot z$

CHAPTER 5 REVIEW

Vectors, matrices, and transformations are all of increasing importance in the study of mathematics and in applications of mathematics to other fields. Vectors have been studied in two ways. They were first presented as oriented line segments where one identifies all oriented line segments that point in the same direction and have the same length. In the coordinate approach they are thought of as columns of numbers where the numbers are the coordinates of the head of an oriented line segment with the tail at the origin. If $O = (0,0)$ is the origin and $P = (x, y)$ is a point in the xy-plane, then the vector $\overrightarrow{OP}$ with the tail at O and the head at P is represented by the column $\begin{bmatrix} x \\ y \end{bmatrix}$. Using the oriented line segment approach, vector addition is accomplished by the head-to-tail method. Using the coordinate representation method, vector addition is accomplished by adding corresponding coordinates. Thus, x-coordinates are added together to obtain the final x-coordinate, and y-coordinates are added together to obtain the final y-coordinate.

One can use vectors to represent many physical quantities that have both magnitude and direction associated with them. If one pushes or pulls on an object, then the force exerted on the object involves both the magnitude of the push or pull and the direction of the push or pull. Also, velocity and acceleration are vectors. Newton's famous second law is $\overrightarrow{F} = m \cdot \overrightarrow{a}$, where $\overrightarrow{F}$ is a force vector applied to an object of mass m, and $\overrightarrow{a}$ is the acceleration vector of the object that results from the applied force vector. Note that a force applied in a certain direction yields an acceleration in the same direction. If you put an ice cube on a smooth flat surface and push it north, it will accelerate toward the north. If you push a bigger ice cube having a larger mass m with the same force, it will accelerate less but will still go toward the north. If you push these ice cubes harder, then in each case they will accelerate more than they did originally.

Matrices are introduced as rectangular arrays of numbers and have many uses in the social and behavioral sciences as well as in mathematics and science.

They are particularly important in the study of geometric transformations. If $X = \begin{bmatrix} x \\ y \end{bmatrix}$, $X' = \begin{bmatrix} x' \\ y' \end{bmatrix}$, $A = \begin{bmatrix} a & b \\ c & d \end{bmatrix}$, $H = \begin{bmatrix} h \\ k \end{bmatrix}$, and $\det(A) \neq 0$, then the transformation of the xy-plane given by

$$X' = AX + H$$

takes lines to lines. Furthermore, any transformation of the xy-plane that takes lines to lines can be written in this form. The condition that $\det(A) \neq 0$ is very important since otherwise the image will not be the entire xy-plane. If $\det(A) = 0$, the image is either a single line or a single point. A transformation of the form $X' = AX + H$ does not necessarily preserve distance or angles or area. However, it will take parallel lines to parallel lines. If W is a subset of the xy-plane with nonzero area and with image W', then one has that the absolute value of the determinant of A will be the ratio of the area of the image W' to the area of the original set W.

$$|\det(A)| = \frac{\text{area}(W')}{\text{area}(W)} \quad \leftrightarrow \quad \text{area}(W') = |\det(A)| \cdot \text{area}(W)$$

Imaginary numbers arise naturally from the quadratic formula but were not really taken very seriously for several centuries. They were thought to be truly imaginary and of little or no value but are now known to be of real value in many ways. Combining the real number x with the imaginary number iy results in the complex number $x + iy$. If one wants a "geometric" way to think of this complex number, one just identifies it with the point in the xy-plane with coordinates (x, y). Thus, imaginary numbers and more generally complex numbers have a "concrete" mathematical realization. Complex numbers play particularly important roles in physics, engineering, and mathematics. Among other things, they can be used to define important transformations of the xy-plane. Rotations and translations of the xy-plane take very simple forms using complex numbers.

Selected formulas:

Addition of column vectors:
$$\begin{bmatrix} x_1 \\ y_1 \end{bmatrix} + \begin{bmatrix} x_2 \\ y_2 \end{bmatrix} = \begin{bmatrix} x_1 + x_2 \\ y_1 + y_2 \end{bmatrix}$$

Addition of matrices:
$$\begin{bmatrix} a_{11} & a_{12} & \cdots \\ a_{21} & a_{22} & \cdots \\ \cdots & \cdots & \cdots \end{bmatrix} + \begin{bmatrix} b_{11} & b_{12} & \cdots \\ b_{21} & b_{22} & \cdots \\ \cdots & \cdots & \cdots \end{bmatrix}$$
$$= \begin{bmatrix} a_{11} + b_{11} & a_{12} + b_{12} & \cdots \\ a_{21} + b_{21} & a_{22} + b_{22} & \cdots \\ \cdots & \cdots & \cdots \end{bmatrix}$$

Scalar multiplication:
$$a \begin{bmatrix} b_{11} & b_{12} & \cdots \\ b_{21} & b_{22} & \cdots \\ \cdots & \cdots & \cdots \end{bmatrix} = \begin{bmatrix} ab_{11} & ab_{12} & \cdots \\ ab_{21} & ab_{22} & \cdots \\ \cdots & \cdots & \cdots \end{bmatrix}$$

Matrix multiplication: $A B = [a_{ij}] [b_{ij}] = \left[\sum_{h=1}^{h=k} a_{ih} b_{hj}\right]$

Determinant: $\begin{vmatrix} a & b \\ c & d \end{vmatrix} = ad - bc,$

$\begin{vmatrix} a_{11} & a_{12} & a_{13} \\ a_{21} & a_{22} & a_{23} \\ a_{31} & a_{32} & a_{33} \end{vmatrix} = a_{11}M_{11} - a_{12}M_{12} + a_{13}M_{13}$

Inverse matrix: $AA^{-1} = A^{-1}A = I (A$ is $n \times n$ and $\det(A) \neq 0)$

Complex addition: $(x_1 + iy_1) + (x_2 + iy_2) = (x_1 + x_2)$
$+ i(y_1 + y_2)$

The square of i is -1: $i^2 = -1$

Complex multiplication: $(x_1 + iy_1) \cdot (x_2 + iy_2) = (x_1 x_2 - y_1 y_2)$
$+ i(x_1 y_2 + x_2 y_1)$

Conjugate: $\overline{(x + iy)} = x - iy$

Absolute value: $|x + iy| = \sqrt{x^2 + y^2}$

Euler's identity: $e^{i\theta} = \cos(\theta) + i \sin(\theta)$

CHAPTER 5 REVIEW EXERCISES

1. Let $O = (0,0)$ and $A = (2,3)$. If $\overrightarrow{OB} = 5 \cdot \overrightarrow{OA}$, find the coordinates of the point B.

2. Let $O = (0,0)$ and $E = (2,4)$. If $\overrightarrow{OF} = \left(\frac{1}{2}\right) \cdot \overrightarrow{OE}$, find the coordinates of F.

3. Let $O = (0,0)$ and $C = (\pi, 1)$. If $\overrightarrow{OD} = (-3) \cdot \overrightarrow{OC}$, find the coordinates of D.

4. Let $O = (0,0)$ and $H = (6,12)$. If $\overrightarrow{OK} = \left(-\frac{1}{3}\right) \cdot \overrightarrow{OH}$, find the coordinates of K.

5. Given $O = (0,0)$ and $J = (2,2)$, find the norm $|\overrightarrow{OJ}|$.

6. Given $O = (0,0)$ and $L = (3,4)$, find the norm $|\overrightarrow{OL}|$.

7. Given $O = (0,0)$ and $S = (x, y)$, find the norm $|\overrightarrow{OS}|$.

8. Given $O = (0,0)$ and $W = (a, 2a)$, find the norm $|\overrightarrow{OW}|$.

9. Let $O = (0,0), X = (1,0)$, and $Y = (1,2)$.

   a. Find the coordinates of the tail (i.e., beginning point) of the vector $\overrightarrow{OX} + \overrightarrow{XY}$.

   b. Find the coordinates of the head (i.e., ending point) of the vector $\overrightarrow{OX} + \overrightarrow{XY}$.

10. Let $|\overrightarrow{AB}| = 6$ and $|\overrightarrow{BC}| = 8$. Find $|\overrightarrow{AC}|$ assuming that the lines $\overleftrightarrow{AB}$ and $\overleftrightarrow{BC}$ are perpendicular.

11. Let $|\overrightarrow{EF}| = 10$ and $|\overrightarrow{FG}| = 13$.

    a. What is the maximum possible value of $|\overrightarrow{EF} + \overrightarrow{FG}|$?

    b. What is the minimum possible value of $|\overrightarrow{EF} + \overrightarrow{FG}|$?

12. Let $|\overrightarrow{RS}| = 10$ and $|\overrightarrow{ST}| = 13$.

    a. What is the maximum possible value of $|\overrightarrow{RS} - \overrightarrow{ST}|$?

    b. What is the minimum possible value of $|\overrightarrow{RS} \pm \overrightarrow{ST}|$?

13. $|\overrightarrow{LM}| = 5$ and $|\overrightarrow{PQ}| = 5$.

    a. What is the maximum possible value of $|\overrightarrow{LM} + \overrightarrow{PQ}|$?

    b. What is the maximum possible value of $|\overrightarrow{LM} - \overrightarrow{PQ}|$?

    c. What is the minimum possible value of $|\overrightarrow{LM} + \overrightarrow{PQ}|$?

    d. What is the minimum possible value of $|\overrightarrow{LM} - \overrightarrow{PQ}|$?

14. If $\overrightarrow{OA} = \begin{bmatrix} 5 \\ 9 \end{bmatrix}$ and $\overrightarrow{OB} = \begin{bmatrix} 1 \\ -2 \end{bmatrix}$, find $\overrightarrow{OA} + \overrightarrow{OB}$.

15. If $\overrightarrow{OC} = \begin{bmatrix} 14 \\ -5 \end{bmatrix}$ and $\overrightarrow{OD} = \begin{bmatrix} 4 \\ 2 \end{bmatrix}$, find $\overrightarrow{OC} + \overrightarrow{OD}$.

16. If $\overrightarrow{OE} = \begin{bmatrix} 9 \\ 5 \end{bmatrix}$ and $\overrightarrow{OF} = \begin{bmatrix} \sqrt{3} \\ \pi \end{bmatrix}$, find $\overrightarrow{OE} - \overrightarrow{OF}$.

17. If $\overrightarrow{OA} = \begin{bmatrix} 0 \\ 9 \end{bmatrix}$ and $\overrightarrow{OB} = \begin{bmatrix} 1 \\ 10 \end{bmatrix}$, find $\overrightarrow{OA} - \overrightarrow{OB}$.

18. If $\overrightarrow{OC} = \begin{bmatrix} -1 \\ -2 \end{bmatrix}$ and $\overrightarrow{OD} = \begin{bmatrix} 2 \\ 0 \end{bmatrix}$, find $4 \cdot \overrightarrow{OC} - 5 \cdot \overrightarrow{OD}$.

19. If $P = (-1, 4)$ and $Q = (1, -3)$, find $\overrightarrow{PQ}$ (express your answer as a column vector).

20. If $R = (-2, -1)$ and $S = (-3, 4)$, find $\overrightarrow{RS}$ (express your answer as a column vector).

21. Find $\begin{bmatrix} 1 & 0 \\ -1 & 2 \end{bmatrix} + \begin{bmatrix} 3 & 5 \\ 7 & 10 \end{bmatrix}$.

22. Find $\begin{bmatrix} 0 & -7 \\ 2 & 1 \end{bmatrix} - \begin{bmatrix} 3 & -9 \\ \pi & 4 \end{bmatrix}$.

23. Find $4 \begin{bmatrix} 0 & 1 \\ 3 & 2 \end{bmatrix} + 2 \begin{bmatrix} 1 & 2 \\ 5 & 0 \end{bmatrix}$.

24. Find $\begin{bmatrix} 3 & 5 \\ 0 & 1 \end{bmatrix} - 2 \begin{bmatrix} 1 & 2 \\ 3 & 5 \end{bmatrix}$.

25. Find $\begin{bmatrix} 0 & -1 \\ 2 & 1 \end{bmatrix} \begin{bmatrix} 1 & 5 \\ 3 & 2 \end{bmatrix}$.

26. Find $\begin{bmatrix} 2 & 1 \\ 3 & 0 \end{bmatrix} \begin{bmatrix} -1 & -4 \\ 5 & 2 \end{bmatrix}$.

27. Find $2 \begin{bmatrix} 1 & 2 \\ 7 & 4 \end{bmatrix}^T + 3 \begin{bmatrix} 3 & -2 \\ 1 & 5 \end{bmatrix}^T$.

28. Find $2 \begin{bmatrix} 0 & 1 & 2 \\ 3 & 4 & 5 \end{bmatrix}^T + 3 \begin{bmatrix} 1 & 0 & 2 \\ 3 & 7 & 9 \end{bmatrix}^T$.

29. Find $\begin{bmatrix} 0 & 1 & 0 \\ 1 & 3 & 2 \\ 1 & 3 & 1 \end{bmatrix} \begin{bmatrix} 2 & 1 & 0 \\ 1 & 1 & 0 \\ -4 & -1 & 2 \end{bmatrix}$.

30. Find $\begin{bmatrix} 2 & 0 & 1 \\ 5 & 0 & -2 \\ -3 & 1 & 0 \end{bmatrix} \begin{bmatrix} 1 & 2 & 3 \\ 4 & -1 & -2 \\ -3 & 0 & 0 \end{bmatrix}$.

31. Find $\begin{bmatrix} 2 & 3 \\ 4 & -5 \end{bmatrix} \begin{bmatrix} x \\ y \end{bmatrix}$.

32. Find $\begin{vmatrix} 1 & 2 \\ 4 & 3 \end{vmatrix}$.

33. Find $\begin{vmatrix} 1 & 6 \\ -2 & -7 \end{vmatrix}$.

34. Find $\begin{bmatrix} 2 & 3 \\ -4 & -5 \end{bmatrix}^{-1}$.

35. Find $\begin{bmatrix} 1 & -2 \\ 3 & -9 \end{bmatrix}^{-1}$.

36. Let $A = \begin{bmatrix} a & b \\ c & d \end{bmatrix}$ and $B = \begin{bmatrix} e & f \\ g & h \end{bmatrix}$. Verify that $(A + B)^T = A^T + B^T$ for the special case of 2×2 matrices by showing that both $(A + B)^T$ and $A^T + B^T$ yield the same answer.

37. Let $A = \begin{bmatrix} a & b \\ c & d \end{bmatrix}$ and $B = \begin{bmatrix} e & f \\ g & h \end{bmatrix}$. Verify the reverse transpose rule [i.e. prove $(AB)^T = B^T A^T$] for the special case when A and B are 2×2 matrices by showing that both $(AB)^T$ and $B^T A^T$ yield the same answer.

38. Let $A = \begin{bmatrix} a & b \\ c & d \end{bmatrix}$ and assume $ad - bc \neq 0$. Verify that $B = \frac{1}{ad-bc} \begin{bmatrix} d & -b \\ -c & a \end{bmatrix}$ is actually equal to A^{-1}, showing that $AB = I$, where I is the 2×2 identity matrix.

39. Let $A = \begin{bmatrix} a & b \\ c & d \end{bmatrix}$, $B = \begin{bmatrix} e & f \\ g & h \end{bmatrix}$, and $C = \begin{bmatrix} i & j \\ k & l \end{bmatrix}$. Verify the left distributive law [i.e. prove $A(B + C) = AB + AC$] for the special case of A, B, and C all 2×2 matrices by showing that both $A(B + C)$ and $AB + AC$ yield the same answer.

40. Let $A = \begin{bmatrix} a & b \\ c & d \end{bmatrix}$, $B = \begin{bmatrix} e & f \\ g & h \end{bmatrix}$, and $C = \begin{bmatrix} i & j \\ k & l \end{bmatrix}$. Verify the right distributive law [i.e., prove $(A + B)C = AC + BC$] for the special case of A, B, and C all 2×2 matrices by showing that both $(A + B)C$ and $AC + BC$ yield the same answer.

41. Let $A = \begin{bmatrix} a & b \\ c & d \end{bmatrix}$, let h be a real number, and let $n = 2$. Verify that $\det(hA) = h^n \det(A)$ for the $n \times n$ matrix A in the special case of $n = 2$, by showing that both $\det(hA)$ and $h^2 \det(A)$ yield the same answer.

42. Let $A = \begin{bmatrix} 0 & 1 \\ 1 & 0 \end{bmatrix}$, $H = \begin{bmatrix} 1 \\ 2 \end{bmatrix}$, and let $F: R^2 \rightarrow R^2$ be represented by $X' = AX + H$.

　a. Let $\triangle BCD$ be the triangle with vertices $B = (2,3)$, $C = (6,3)$, and $D = (4,3)$. Find the image $\triangle B'C'D'$ of $\triangle BCD$.

　b. Find $\det(A)$.

　c. Find the ratio $\dfrac{\text{area}(\triangle B'C'D')}{\text{area}(\triangle BCD)}$.

43. Let $K = \begin{bmatrix} 3 & 2 \\ 0 & 1 \end{bmatrix}$, $L = \begin{bmatrix} 3 \\ 4 \end{bmatrix}$, and let $F: R^2 \to R^2$ be represented by $X' = KX + L$.

   a. Let ΔEFG be the triangle with vertices $E = (-2, 0)$, $F = (-2, 2)$, and $G = (6, 0)$. Find the image $\Delta E'F'G'$ of ΔEFG.

   b. Find $\det(K)$.

   c. Find the ratio $\dfrac{\text{area}(\Delta E'F'G')}{\text{area}(\Delta EFG)}$.

44. Let $F: R^2 \to R^2$ be the transformation that is represented by $X' = AX$, where $A = \begin{bmatrix} 1 & 1 \\ 0 & 1 \end{bmatrix}$. Let ΔOPQ have vertices given by $O = (0, 0)$, $P = (1, 0)$, and $Q = (0, 1)$.

   a. Find the image $\Delta O'P'Q'$ of ΔOPQ.

   b. Graph both $\Delta O'P'Q'$ and ΔOPQ.

   c. Find $\det(A)$.

   d. Find the ratio $\dfrac{\text{area}(\Delta O'P'Q')}{\text{area}(\Delta OPQ)}$.

45. Let $z_1 = 2 - 3i$ and $z_2 = -1 + 2i$. Simplify the following (i.e., express as $x + iy$):

   a. $z_1 + z_2$

   b. $2z_1 - 3z_2$

   c. $z_1 \cdot z_2$

   d. $\dfrac{z_1}{z_2}$

46. Let $z_1 = 4 - 2i$ and $z_2 = 1 - 5i$. Simplify the following (i.e., express as $x + iy$):

   a. $z_1 + z_2$

   b. $2z_1 - 3z_2$

   c. $z_1 \cdot z_2$

   d. $\dfrac{z_1}{z_2}$

47. Let $z_1 = 3 - i$ and $z_2 = -1 + 2i$. Find the following:

   a. $|z_1|$

   b. $|z_2|$

   c. $|z_1 \cdot z_2|$

   d. $\left| \dfrac{z_1}{z_2} \right|$

48. Let $z_1 = 1 - i$ and $z_2 = -1 + i$. Find the following:

   a. $|z_1|$

   b. $|z_2|$

   c. $|z_1 \cdot z_2|$

   d. $\left| \dfrac{z_1}{z_2} \right|$

49. Simplify the following (i.e., express as $x + iy$):

   a. $e^{i\pi}$

   b. $e^{i\left(\frac{\pi}{2}\right)}$

50. Simplify the following (i.e., express as $x + iy$):

   a. $e^{i\left(\frac{\pi}{4}\right)}$

   b. $3e^{i\left(\frac{5\pi}{2}\right)}$

51. Each of the following represents rotation by some angle θ $(0 < \theta < 360°)$ about the origin. In each case, state the value of the angle θ.

    a. $z' = iz$

    b. $z' = e^{i\left(\frac{\pi}{2}\right)}z$

    c. $z' = \left(\dfrac{1}{\sqrt{2}} + \dfrac{i}{\sqrt{2}}\right)z$

52. Each of the following represents rotation by some angle θ $(0 < \theta < 360°)$ about the origin. In each case, state the value of the angle θ.

    a. $z' = e^{i\pi}z$

    b. $z' = e^{i\left(\frac{3\pi}{2}\right)}z$

    c. $z' = \left(-\dfrac{1}{\sqrt{2}} + \dfrac{i}{\sqrt{2}}\right)z$

RELATED READING FOR CHAPTER 5

Billstein, R., and J. Williamson. *Math Thematics: Book 2.* Evanston, IL: McDougal Littell, 1999.

Coxford, A. F., and Z. Usiskin. *Geometry: A Transformational Approach.* River Forest, IL: Laidlaw Brothers, 1971.

Day, R., et al. *Navigating through Geometry in Grades 9–12.* Reston, VA: National Council of Teachers of Mathematics, 2001.

Dayoub, I. M., and J. W. Lott. *Geometry: Constructions and Transformations.* Menlo Park, CA: Dale Seymour Publications, 1997.

Kay, D. C. *College Geometry: A Discovery Approach.* 2nd ed. New York: Addison Wesley Longman, 2001.

Lappan, G., et al. *Stretching and Shrinking.* Connected Mathematics Project. Needham, MA: Pearson/Prentice-Hall, 2004.

Smart, J. R. *Modern Geometries.* 4th ed. Pacific Grove, CA: Brooks/Cole Publishing, 1994.

Three Other Geometries

6.1 TAXICAB GEOMETRY
6.2 SPHERICAL GEOMETRY
6.3 HYPERBOLIC GEOMETRY

In this chapter we study three different geometries, taxicab geometry, spherical geometry, and hyperbolic geometry. In **taxicab geometry**, angles and lines are the same as in the Euclidean plane. However, there is a fundamental difference in how distance is measured. In order to find the taxicab distance between the two points (x_1, y_1) and (x_2, y_2) in the xy-plane, one adds the distance $|x_1 - x_2|$ between their x-components to the distance $|y_1 - y_2|$ between their y-components. Taxicab geometry is relatively simple and is an easy-to-understand alternative to Euclidean geometry. It demonstrates that Euclidean geometry is not the only geometry that can be understood using the human mind. Taxicab geometry has many features that are the same as in Euclidean geometry and many features that are different. One very important way in which taxicab geometry differs from Euclidean geometry is that in taxicab geometry SAS does not necessarily imply congruent triangles. Taxicab geometry shows that SAS does not follow from the other axioms and postulates of Euclidean geometry.

> Taxicab geometry $\Rightarrow$ Euclid's "proof" of SAS is not valid.

Spherical geometry was studied in ancient times but was considered Euclidean because spheres were studied as subsets of Euclidean three-dimensional space. Recall from Chapter 1 that given a sphere, a great circle was defined as the intersection of the sphere with a plane through the sphere's center. The great circles are the largest circles on the sphere and are the geodesics of the sphere. These geodesics play the role of lines in spherical geometry. A small circle on the sphere is defined as a circle formed by the intersection of the sphere with a plane that does not pass through the sphere's center. In Section 6.2, we study spherical geometry with special emphasis on the Earth, which is very nearly a sphere. We define spherical coordinates (ρ, θ, ϕ) and provide their relationships to cartesian coordinates (x, y, z) in R^3. We define latitude and longitude for the Earth, which are related to spherical coordinates. Given the latitudes and longitudes of two points on the Earth, one can find the spherical distance between the points. This spherical distance is often called the *great circle distance*. It is the minimal flying distance between the two points.

> Spherical geometry $\Rightarrow$ great circle routes minimize flying distance.

Two-dimensional **hyperbolic geometry** shows that the parallel postulate is not something that one must accept in geometry. In fact, the existence of Euclidean geometry is equivalent to the existence of hyperbolic geometry. Thus, if you accept the existence of Euclidean geometry, you must accept the existence of hyperbolic geometry and vice versa. For many years people have wondered what the three-dimensional universe looks like if one "fixes an instant in time." The overall structure of the three-dimensional universe is still uncertain; however, there are models that have been studied as real possibilities. One such possibility is three-dimensional hyperbolic space. Although this book will not consider this model, two-dimensional hyperbolic geometry is studied in Section 6.3. Nicholai Lobachevski (*luh buh chayf skee*) (1793–1856) published the first work on hyperbolic geometry in 1829. At approximately the same time, two other mathematicians, John Bolyai (*boh lyoy*) (1802–1860) and Carl Friedrich Gauss (*gous*) (1777–1855), developed the foundations of hyperbolic geometry working independently of each other and of Lobachevski. Thus the first geometry to be considered non-Euclidean was developed by three different individuals working independently. In hyperbolic geometry, one replaces Playfair's postulate with a nonequivalent alternative. One assumes in two-dimensional hyperbolic geometry that if a point P is off a line ℓ, then there are always at least two parallels to ℓ through P. In fact, replacing Playfair's postulate with the postulate that there are at least two parallels to ℓ through P when P is off ℓ yields a geometry where each point P off ℓ actually has an infinite number of parallels to ℓ containing it.

> Hyperbolic geometry $\Rightarrow$ there are alternatives to the parallel postulate.

6.1   TAXICAB GEOMETRY

Taxicab geometry was devised by Herman Minkowski (*ming kuf skee*) (1864–1909). Among other things, taxicab geometry is interesting because the SAS axiom is not valid. Thus, taxicab geometry illustrates the need in Euclidean geometry for either the SAS axiom or some other axiom that is equivalent to the SAS axiom.

Taxicab Points, Lines, and Angles.   In taxicab geometry, the set of points is the usual *xy*-plane. The lines are the usual straight lines of the *xy*-plane, and the angles are measured the same in Euclidean geometry.

Taxicab points: ordinary ordered pairs (x, y) in the xy-plane.
Taxicab lines: ordinary (Euclidean) lines in the xy-plane.
Taxicab angles: have usual (Euclidean) measures.

Taxicab Distance. The fundamental difference from Euclidean geometry is the way in which distance is measured. If points $P = (x_1, y_1)$ and $Q = (x_2, y_2)$ are points in R^2, then recall that the Euclidean distance between these same two points is denoted by either PQ or by $d_E(P, Q)$ and is, given by $\sqrt{(x_1 - x_2)^2 + (y_1 - y_2)^2}$.

$$d_E[(x_1, y_1), (x_2, y_2)] = \sqrt{(x_1 - x_2)^2 + (y_1 - y_2)^2} \text{ (Euclidean distance)}$$

On the other hand, the **taxicab distance** $d_T(P, Q)$ between P and Q is given by $|x_1 - x_2| + |y_1 - y_2|$.

$$d_T(P, Q) = d_T[(x_1, y_1), (x_2, y_2)] = |x_1 - x_2| + |y_1 - y_2| \text{ (taxicab distance)}$$

EXAMPLE Given $P = (2, 3)$ and $Q = (7, 6)$, find both the taxicab distance and the Euclidean distance between these two points.

Solution

$$d_T[(2, 3), (7, 6)] = |2 - 7| + |3 - 6| = 5 + 3 = 8.$$

$$d_E[(2, 3), (7, 6)] = \sqrt{(2 - 7)^2 + (3 - 6)^2} = \sqrt{25 + 9} = \sqrt{34} \approx 5.83. \quad \blacksquare$$

The name *taxicab geometry* comes from the concept of a city with perpendicular streets. Think of one set of streets as being parallel to the x-axis and the other set of streets as being parallel to the y-axis. If one wishes to take a taxicab from point P to point Q, then the cab must stay on the streets. Thus, the taxicab might first go parallel to one axis, say the x-axis, for a distance $|x_1 - x_2|$ corresponding to the amount the x-coordinates of the two points differ. Then the cab might make a turn (at S in Figure 6.1.1) and go parallel to the y-axis for a distance $|y_1 - y_2|$ corresponding to the amount the y-coordinates of the two points differ. The taxicab distance between P and Q would be the sum of these two quantities $d_T(P, Q) = |x_1 - x_2| + |y_1 - y_2|$. In contrast, Euclidean distance $d_E(P, Q) = PQ$ is measured along the Euclidean (and taxicab) straight line segment from P to Q. Notice that the taxicab distance is always longer than or the same as the Euclidean distance. The two distances are equal exactly when the two points lie on a line parallel to one of the coordinate axes.

$$d_E(P, Q) \leq d_T(P, Q) \text{ (Euclidean } \leq \text{ taxicab)}$$

Of course, in studying taxicab geometry, one may use Euclidean geometry as an aid since many of the objects of taxicab geometry, such as points, lines, and angles,

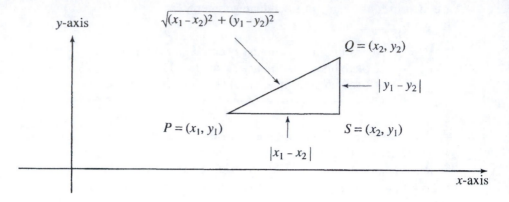

FIGURE 6.1.1 Let $P = (x_1, y_1)$, $Q = (x_2, y_2)$, and $S = (x_2, y_1)$. If both the x-coordinates and y-coordinates of the two points P and Q differ, then $\triangle PSQ$ will be a right triangle as shown here. The taxicab distance from P to Q is given by $d_T(P, Q) = |x_1 - x_2| + |y_1 - y_2|$. This distance comes from the Euclidean distance from P to S added to the Euclidean distance from S to Q. For comparison, the Euclidean distance PQ from P to Q will be the Euclidean length of the hypotenuse of the triangle $\triangle PSQ$.

are the same. Since lines are the same in these two geometries, triangles are also the same. Furthermore, the angle sum of a triangle in taxicab geometry is one straight angle (i.e., 180° or π radians), just as in Euclidean geometry. On the other hand, circles for the two geometries are different. This follows since circles are defined using distance, and Euclidean distance is different from taxicab distance. Thus, given a point O and radius $r > 0$, the Euclidean circle C_E and the taxicab circle C_T with center O and radius r are given, respectively, by $C_E = \{P | d_E(O, P) = r\}$ and $C_T = \{P | d_T(O, P) = r\}$.

$$C_E = \{P | d_E(O, P) = r\} \text{ (Euclidean circle)}$$
$$C_T = \{P | d_T(O, P) = r\} \text{ (taxicab circle)}$$

If the coordinates of the center O are given by (a, b) and the variable point P on the corresponding circles has coordinates (x, y), then using $O = (a, b)$ and $P = (x, y)$, the equations for the two circles are as follows:

$$(x - a)^2 + (y - b)^2 = r^2 \text{ (Euclidean circle } C_E)$$

and

$$|x - a| + |y - b| = r \text{ (taxicab circle } C_T).$$

A taxicab circle is illustrated in Figure 6.1.2.

Congruent Segments and Triangles.   Two segments $\overline{PQ}$ and $\overline{RS}$ are **congruent in the Euclidean sense**, denoted as $\overline{PQ} \cong_E \overline{RS}$, if they have equal Euclidean lengths. They are **congruent in the taxicab sense**, denoted as $\overline{PQ} \cong_T \overline{RS}$, if they have equal

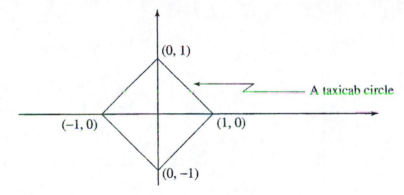

FIGURE 6.1.2 The taxicab circle of radius one centered at the origin $O = (0,0)$ is a square with vertices at $(1,0)$, $(0,1)$, $(-1,0)$ and $(0,-1)$. It is shaped like a baseball diamond and centered at the origin. In fact, all taxicab circles are squares with sides of slope ± 1. Thus, taxicab circles appear as diamond shapes when the x and y axes are graphed as horizontal and vertical lines, respectively.

taxicab lengths.

$$\overline{PQ} \cong_E \overline{RS} \text{ iff } d_E(P,Q) = d_E(R,S).$$
$$\overline{PQ} \cong_T \overline{RS} \text{ iff } d_T(P,Q) = d_T(R,S).$$

Two triangles in taxicab geometry are congruent if all three corresponding angles are congruent and if each pair of the three pairs of corresponding sides have the same taxicab length.

Two triangles $\triangle ABC$ and $\triangle DEF$ are **congruent taxicab triangles** with sides in the indicated order, written $\triangle ABC \cong_T \triangle DEF$, if and only if all six of the following are valid.

1. $\overline{AB} \cong_T \overline{DE}$ 4. $\angle A \cong \angle D$
2. $\overline{AC} \cong_T \overline{DF}$ 5. $\angle B \cong \angle E$
3. $\overline{BC} \cong_T \overline{EF}$ 6. $\angle C \cong \angle F$

The order of the vertices is very important for congruence in taxicab geometry, just as in Euclidean geometry. Notice that since angles are measured the same in taxicab geometry as these are in Euclidean geometry, the congruence of angles in number 4 could have been written as $\angle A \cong_T \angle D$ instead of as $\angle A \cong \angle D$. Similarly, 5 and 6 could have been written as $\angle B \cong_T \angle E$ and $\angle C \cong_T \angle F$, respectively.

The next example shows that two triangles may be congruent in the Euclidean sense and not congruent in the taxicab sense.

EXAMPLE Let $A = (\sqrt{2},0)$, $B = (0,\sqrt{2})$, $C = (0,0)$, $A' = (3,1)$, $B' = (5,1)$, and $C' = (4,0)$. Investigate the congruence of the two triangles $\triangle ABC$ and $\triangle A'B'C'$ in both Euclidean geometry and taxicab geometry.

Solution   The triangles are illustrated here.

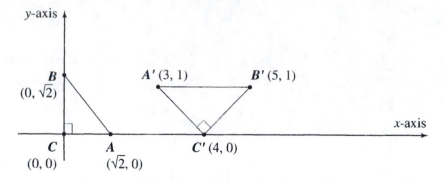

Using Euclidean geometry, it is easy to check that

$$m(\angle A) = m(\angle A') = m(\angle B) = m(\angle B') = 45°,$$

$$m(\angle C) = m(\angle C') = 90°,$$

$$d_E(A, C) = d_E(B, C) = d_E(A', C') = d_E(B', C') = \sqrt{2},$$

and

$$d_E(A, B) = d_E(A', B') = 2.$$

Thus, $\triangle ABC \cong_E \triangle A'B'C'$ (i.e., the triangles are congruent in Euclidean geometry). However,

$$d_T(A, C) = |\sqrt{2} - 0| + |0 - 0| = \sqrt{2}$$

and

$$d_T(A', C') = |3 - 4| + |1 - 0| = 2$$

yield

$$d_T(A, C) \neq d_T(A', C').$$

Hence, these two triangles are *not* congruent in taxicab geometry.   ∎

The following example shows that two triangles may be congruent simultaneously in both the Euclidean and taxicab senses.

EXAMPLE   Let $D = (3, 0)$, $E = (0, 4)$, $F = (0, 0)$, $D' = (-8, -4)$, $E' = (-5, -8)$, and $F' = (-5, -4)$. Investigate the congruence of the two triangles $\triangle DEF$ and $\triangle D'E'F'$ in both Euclidean geometry and taxicab geometry.

Solution   The triangles are illustrated here.

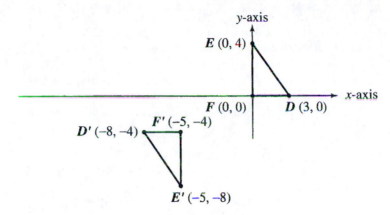

Clearly, $m(\angle F) = m(\angle F') = 90°$. The Euclidean lengths of the legs are easily seen to be $d_E(D,F) = d_E(D',F') = 3$ and $d_E(E,F) = d_E(E',F') = 4$. Furthermore, $d_E(D,F) = d_E(D',F') = \sqrt{3^2 + 4^2} = \sqrt{25} = 5$. Thus, using SSS for Euclidean geometry, one finds that $\triangle DEF \cong_E \triangle D'E'F'$ (i.e., the triangles are congruent in Euclidean geometry). This implies that all of the corresponding angles are congruent both in the Euclidean sense and in the taxicab sense.

One may use $d_T[(x_1,y_1),(x_2,y_2)] = |x_1 - x_2| + |y_1 - y_2|$ to check the taxicab lengths of the sides.

$$d_T(D,F) = |3 - 0| + |0 - 0| = 3,$$
$$d_T(D',F') = |-8 - (-5)| + |-4 - (-4)| = 3,$$
$$d_T(E,F) = |0 - 0| + |4 - 0| = 4,$$
$$d_T(E',F') = |-5 - (-5)| + |-8 - (-4)| = 4,$$
$$d_T(D,E) = |3 - 0| + |0 - 4| = 7,$$
$$d_T(D',E') = |-8 - (-5)| + |-4 - (-8)| = 7.$$

Thus, $\overline{DF} \cong_T \overline{D'F'}$, $\overline{EF} \cong_T \overline{E'F'}$, and $\overline{DE} \cong_T \overline{D'E'}$. Since all of the corresponding angles are congruent and all of the corresponding sides are congruent in the taxicab sense, the two triangles satisfy $\triangle DEG \cong_T \triangle D'E'F'$ (i.e., the triangles are taxicab congruent).   ■

EXERCISES 6.1

1. Let $R = (2,3)$ and $S = (5,7)$. Find
   a. The Euclidean distance RS from R to S [i.e., find $d_E(R,S)$].
   b. The taxicab distance $d_T(R,S)$ from R to S.
2. Let $P = (2,2)$, $Q = (4,4)$, and $S = (4,2)$.
   a. Find the taxicab angles of the triangle $\triangle PSQ$
   b. Check that for Euclidean geometry the Pythagorean theorem is satisfied. That is, check that $[d_E(P,S)]^2 + [(d_E(S,Q)]^2 = [d_E(P,Q)]^2$.
   c. Does the formula from the Pythagorean theorem hold using taxicab distances? That is, does $[d_T(P,S)]^2 + [d_T(S,Q)]^2$ equal $[d_T(P,Q)]^2$?

3. Let $A = (0,0)$, $B = (2,0)$, and $C = (1,1)$.
   a. Find the taxicab angles of the triangle $\triangle ABC$.
   b. Find the Euclidean lengths, of the sides [i.e., find $d_E(A,B), d_E(B,C), d_E(C,A)$].
   c. Find the taxicab lengths of the sides [i.e., find $d_T(A,B), d_T(B,C), d_T(C,A)$].
   d. Would you say that this is an equilateral triangle in taxicab geometry? Why or why not?
   e. Would you say that this is an equiangular triangle in taxicab geometry? Why or why not?

4. Let $A' = (0,0)$, $B' = \left(\frac{1}{2}, \frac{3}{2}\right)$, and $C' = \left(\frac{3}{2}, \frac{1}{2}\right)$.
   a. Find the Euclidean lengths of the sides of this triangle [i.e., find $d_E(A',B')$, $d_E(B',C')$, $d_E(C',A')$].
   b. Find the taxicab lengths of the sides of $\triangle A'B'C'$ [i.e., find $d_T(A',B')$, $d_T(B',C')$, $d_T(C',A')$].
   c. Does $\triangle A'B'C'$ have any right angles? Why or why not?

5. Does SSS imply congruent triangles in taxicab geometry? (Hint: Do exercises 3 and 4 before doing this exercise.)

6.2 SPHERICAL GEOMETRY

As in previous chapters, let $R^3 = \{(x,y,z)|x,y,z \text{ are real}\}$. Here x, y, z are cartesian coordinates. Of course, x, y, and z are "length" coordinates along the respective axes. **Spherical coordinates** (ρ, θ, ϕ) for Euclidean three-dimensional space have ρ as a "length" coordinate and both θ and ϕ as angle coordinates. The length coordinate ρ represents distance from the origin $O = (0,0,0)$. Thus, $\rho = \sqrt{x^2 + y^2 + z^2}$ and clearly $0 \le \rho < \infty$. The angle ϕ is the angle that the positive z-axis makes with the vector from the origin to the point (x,y,z). Measured in degrees, the angle ϕ satisfies $0° \le \phi \le 180°$, and using radians, the angle ϕ satisfies $0 \le \phi \le \pi$. To obtain the angle θ, the segment having endpoints at the origin O and at the point (x,y,z) is projected onto the xy-plane by moving each point of the segment parallel to the z-axis. The resulting segment lies in the xy-plane and has endpoints at the origin O and at the point $(x,y,0)$. The length of this segment in the xy-plane is given by $r = \sqrt{x^2 + y^2}$, and the angle θ is an angle measured in the xy-plane between this segment and the positive x-axis. Notice that $0 \le r < \infty$, and, using degrees, the angle θ satisfies $0° \le \theta < 360°$. Of course, using radians, the angle θ satisfies $0 \le \theta < 2\pi$. In effect, the r and θ are polar coordinates in the xy-plane.

From Figure 6.2.1 it follows that $\sin(\phi) = $ (opposite)/(hypotenuse) $= r/\rho$ and $\cos(\phi) = $ (adjacent)/(hypotenuse) $= z/\rho$. Hence,

$$r = \rho \sin(\phi) \text{ and } z = \rho \cos(\phi).$$

Clearly, θ is the same as the angle θ used in polar coordinates and studied in Section 5.6. Since some readers may have skipped Section 5.6, we will again discuss polar coordinates. They are illustrated in Figure 6.2.2. For **polar coordinates** (r, θ) in

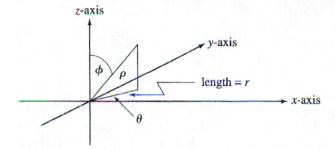

FIGURE 6.2.1 Let a point have cartesian coordinates (x, y, z) and let it have (ρ, θ, ϕ) as spherical coordinates. Then ρ represents the distance of the point from the origin, and ϕ is the angle the segment from the origin to (x, y, z) makes with the positive z-axis. Projecting this segment to the xy-plane, one obtains a projected segment of length r, and this projected segment makes an angle θ with the positive x-axis.

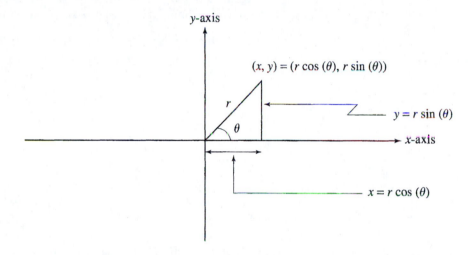

FIGURE 6.2.2 Polar coordinates (r, θ) are related to xy-coordinates by $x = r \cos(\theta)$ and $y = r \sin(\theta)$. Here r is the distance from the origin in the xy-plane, and the angle θ is the angle that the positive x-axis makes with the vector from the origin to the point (x, y) in the xy-plane.

the xy-plane, one has

$$
\begin{array}{ll}
x = r \cos(\theta) & r = \sqrt{x^2 + y^2} \\
y = r \sin(\theta) & \tan(\theta) = \frac{y}{x} \text{ (if } x \neq 0)
\end{array}
\qquad \text{(polar coordinates)}
$$

EXAMPLE Given the point P in the plane with cartesian coordinates $x = 6$, $y = 8$, find the polar coordinates (r, θ) of P.

Solution Using $\tan(\theta) = \frac{y}{x} = \frac{8}{6} = \frac{4}{3}$, one finds that $\theta = \tan^{-1}\left(\frac{4}{3}\right)$ is approximately either $53.13°$ or $233.13°$. In radians, these angles are approximately $.9273$ and

4.069, respectively. Since the point P with coordinates $x = 6$, $y = 8$ lies in the first quadrant, one can eliminate the value of $233.13°$ (4.066 radians). Hence, one obtains $\theta \approx 53.13°$. Furthermore, $r = \sqrt{x^2 + y^2} = \sqrt{6^2 + 8^2} = \sqrt{100} = 10$. Thus, the polar coordinates of P are (approximately) given by $(r, \theta) \approx (10, 53.13°)$ using degrees. ■

EXAMPLE   Given a point Q with polar coordinates $r = 1$, $\theta = 45°$, find the cartesian coordinates (x, y) of Q.

Solution   Clearly, $x = r\cos(\theta) = 1 \cdot \cos(45°) = \frac{1}{\sqrt{2}} \approx .707$ and $y = r\sin(\theta) = 1 \cdot \sin(45°) = \frac{1}{\sqrt{2}} \approx .707$. Thus, the cartesian coordinates of this point are given by $\left(\frac{1}{\sqrt{2}}, \frac{1}{\sqrt{2}}\right) \approx (.707, .707)$. ■

Using $r = \rho\sin(\phi)$, $x = r\cos(\theta)$, $y = r\sin(\theta)$, and $z = \rho\cos(\phi)$ yields the following equations relating cartesian coordinates (x, y, z) to spherical coordinates (ρ, θ, ϕ).

> Relating xyz-coordinates to $\rho\theta\phi$-coordinates.
> $$x = \rho\sin(\phi)\cos(\theta)$$
> $$y = \rho\sin(\phi)\sin(\theta)$$
> $$z = \rho\cos(\phi)$$

Spherical Distance.   Let $P = (x_1, y_1, z_1)$ and $Q = (x_2, y_2, z_2)$ be two points on the sphere of radius ρ about the origin O. If P and Q are distinct and not diametrically opposed, there will be exactly one great circle containing both of them. If P and Q are diametrically opposed, there will be an infinite number. In any case, one can take a fixed plane containing P, Q, and the origin and intersect the sphere with this plane to get a great circle containing P and Q. The **spherical distance** $d_S(P, Q)$ is defined to be the length of the shortest arc of the great circle joining P and Q. If $m(\angle POQ)$ is the measure of the angle $\angle POQ$ in radians, then the definition of radian measure yields $m(\angle POQ) = (\text{length of arc})/(\text{radius}) = d_S(P, Q)/\rho$. Hence, $d_S(P, Q) = \rho \cdot m(\angle POQ)$ [for $m(\angle POQ)$ in radians].

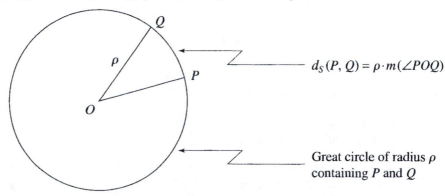

$$d_S(P, Q) = \rho \cdot m(\angle POQ)$$

Great circle of radius ρ containing P and Q

The **dot product** $P \cdot Q$ of $P = (x_1, y_1, z_1)$ and $Q = (x_2, y_2, z_2)$ is defined by $P \cdot Q = (x_1, y_1, z_1) \cdot (x_2, y_2, z_2) = x_1 x_2 + y_1 y_2 + z_1 z_2$.

$$(x_1, y_1, z_1) \cdot (x_2, y_2, z_2) = x_1 x_2 + y_1 y_2 + z_1 z_2 \qquad \text{(dot product)}$$

EXAMPLE Find the dot product of $P = (1, 2, 3)$ and $Q = (0, -1, 4)$.

Solution Using the formula for the dot product, one finds that $P \cdot Q = (1) \cdot (0) + (2) \cdot (-1) + (3) \cdot (4) = 10$. ∎

Thus, using the formula $P \cdot Q = |P| \cdot |Q| \cdot \cos(\angle POQ)$, one obtains (assuming P and Q are distinct from the origin)

$$m(\angle POQ) = \cos^{-1}\left(\frac{P \cdot Q}{|P| \cdot |Q|}\right) = \cos^{-1}\left(\frac{x_1 x_2 + y_1 y_2 + z_1 z_2}{\sqrt{x_1^2 + y_1^2 + z_1^2} \cdot \sqrt{x_2^2 + y_2^2 + z_2^2}}\right).$$

Now using the assumption that P and Q lie on the sphere of radius ρ centered at the origin, one has $|P| = |Q| = \rho$ and $d_S(P, Q) = \rho \cdot m(\angle POQ)$. Hence, one has $d_S(P, Q) = \rho \cdot \cos^{-1}\left(\frac{x_1 x_2 + y_1 y_2 + z_1 z_2}{\rho^2}\right)$.

$$d_S((x_1, y_1, z_1), (x_2, y_2, z_2)) = \rho \cdot \cos^{-1}\left(\frac{x_1 x_2 + y_1 y_2 + z_1 z_2}{\rho^2}\right)$$

(spherical distance)

This formula is often applied to Earth where $\rho \approx 4{,}000$ miles, but it also works for spheres with any radius centered at the origin. When using the spherical distance formula, always express the value of the inverse cosine in radians.

EXAMPLE Let $P = (1, 1, \sqrt{2})$ and $Q = (0, -1, \sqrt{3})$ be two points on the sphere with radius $\rho = 2$ centered at the origin O. Find their spherical distance.

Solution The formula for spherical distance yields

$$d_S(P, Q) = d_S[(1, 1, \sqrt{2}), (0, -1, \sqrt{3})]$$

$$= 2 \cdot \cos^{-1}\left(\frac{(1) \cdot (0) + (1) \cdot (-1) + (\sqrt{2}) \cdot (\sqrt{3})}{2^2}\right)$$

Thus,

$$d_S(P, Q) = 2 \cdot \cos^{-1}\left(\frac{-1 + \sqrt{6}}{4}\right) \approx 2 \cdot \cos^{-1}\left(\frac{1.4495}{4}\right)$$

$$\approx 2 \cdot \cos^{-1}(.3623) \approx 2.4. \quad ∎$$

Of course, in the previous formula, $\cos^{-1}$ must be found in terms of radians, and one uses the **principal value** of the $\cos^{-1}$. Thus,

$$0 \le \cos^{-1}\left(\frac{x_1 x_2 + y_1 y_2 + z_1 z_2}{\sqrt{x_1^2 + y_1^2 + z_1^2} \cdot \sqrt{x_2^2 + y_2^2 + z_2^2}}\right) \le \pi.$$

The formulas

$$
\begin{array}{ll}
x_1 = \rho \sin(\phi_1)\cos(\theta_1) & x_2 = \rho \sin(\phi_2)\cos(\theta_2) \\
y_1 = \rho \sin(\phi_1)\sin(\theta_1) & y_2 = \rho \sin(\phi_2)\sin(\theta_2) \\
z_1 = \rho \cos(\phi_1) & z_2 = \rho \cos(\phi_2)
\end{array}
$$

yield the following formula for the spherical distance $d_S(P,Q)$ when P and Q are on the surface of Earth and have spherical angular coordinates given by θ_1, ϕ_1, and θ_2, ϕ_2, respectively:

$$d_S(P,Q) = \rho \cdot \cos^{-1}[\sin(\phi_1)\cos(\theta_1)\sin(\phi_2)\cos(\theta_2)$$
$$+ \sin(\phi_1)\sin(\theta_1)\sin(\phi_2)\sin(\theta_2) + \cos(\phi_1)\cos(\phi_2)]$$

Places on Earth have traditionally been catalogued by latitude and longitude using degrees, minutes, and seconds. An angular measure of 1 minute is denoted by $1'$, and an angular measure of 1 second is denoted by $1''$. Here, 60 minutes equals 1 degree and 60 seconds equals 1 minute.

$$1° = 60'$$
$$1' = 60''$$

For example, the latitude of New York is $40°40'$ N (i.e., 40 degrees and 40 minutes north of the equator), which is approximately equal to $40.667°$ N since 40 minutes is approximately $.667°$.

$$40' = (40') \cdot \left(\frac{1°}{60'}\right) = \left(\frac{40}{60}\right)° = \left(\frac{2}{3}\right)° \approx .667°$$

Latitude and Longitude.   Take a sphere of radius $\rho = 4,000$ miles as a model of Earth, then imagine that one travels to this sphere's center and measures angles. Think of having cartesian coordinates (x, y, z) with origin centered at the Earth's center and with the North Pole on the positive z-axis. Then the xy-plane is the plane of Earth's equator. The **latitude** of a point on Earth's surface is given by measuring up (north) or down (south) from the plane of the equator, see Figure 6.2.3. The **North Pole** $(0,0,\rho)$ has latitude $90°$ N and the **South Pole** $(0,0,-\rho)$ has latitude $90°$ S. To find the latitude of a point P that is neither the North Pole nor the South Pole, take the plane H_P containing P, the North Pole and the South Pole. This plane H_P contains the origin $O = (0,0,0)$ since the origin lies on the line joining the North Pole and South Pole. Also, this plane H_P intersects the sphere in a great circle with two arcs joining the North Pole and the South Pole. Of course, the point P lies on one of these arcs. The arc of the great circle from the North Pole to the South Pole,

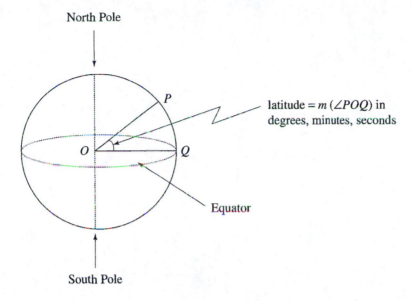

FIGURE 6.2.3 Latitude can be thought of as the measure (in degrees) of an angle with the vertex at Earth's center. The angle $\angle POQ$ is measured up (N) or down (S) from the equator. The plane containing the point P, the North Pole and South Pole is denoted by H_P. In the diagram, the plane H_P is drawn as the plane of the page. The point Q is one of the two points where the plane H_P intersects the equator. The point Q is on that arc of the great circle from the North Pole to the South Pole that contains P. This arc is said to be the longitudinal line determined by the point P. A longitudinal line is half of a great circle with one endpoint at the North Pole and the other endpoint at the South Pole.

which contains P, is said to be the **longitudinal line** determined by P. Let the point Q be the intersection of the arc containing P and the equator. The point Q is the closest point on the equator to the point P. The latitude angle is the measure in degrees (with possibly minutes, and/or seconds) of the angle $\angle POQ$. This angle is in this plane H_P, is measured in degrees, and is given in degrees N or S depending on if P is in the Northern or Southern Hemisphere.

When the above point P is on the equator, then $P = Q$ and $m(\angle POQ) = 0°$. Thus, the latitude of P is given by both $0°$ N and equivalently by $0°$ S. In geography, the word *parallel* is used to describe the set of points at some fixed latitude. Thus, points of the Earth with latitude $45°$ N are said to be on the 45th (northern) parallel. Note for $0° < \sigma < 90°$, the σth northern parallel is a circle in the Northern Hemisphere that lies in a plane parallel to the equatorial plane. Since $\sigma \neq 0°$, this circle is not a great circle and must have a radius less than ρ. In fact, the radius of the northern (or southern) σ^{th} parallel is given by $\rho \cos(\sigma)$.

Knowing the latitude of a point allows the calculation of the angle ϕ used in spherical coordinates. Recall that ϕ is the angle that the segment $\overline{OP}$ makes with the positive z-axis. Thus, relating ϕ to the latitude $\sigma°$ (N or S) yields

$$\phi = 90° - \sigma° \quad \text{for latitude } \sigma°\text{N}$$
$$\phi = 90° + \sigma° \quad \text{for latitude } \sigma°\text{S}.$$

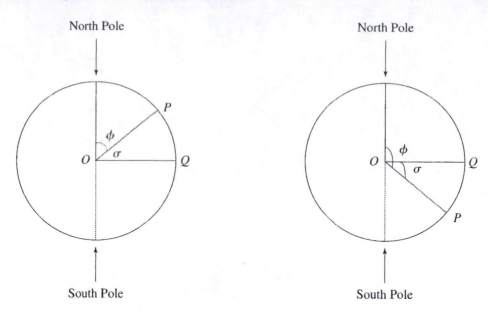

Consider the longitudinal line passing through Greenwich, England. This longitudinal line is taken as the zero longitudinal line and may be designated as either $0°$ E or $0°$ W. Thus, Paris is east and New York is west. Consider the cartesian coordinates with origin centered at Earth's center, the positive z-axis containing the North Pole and the positive x-axis containing the point on the equator where the longitudinal line containing Greenwich, England, intersects the equator. The y-axis is then in the equatorial plane with the positive y-axis pointing into the Eastern Hemisphere and the negative y-axis pointing into the Western Hemisphere. Longitudinal angles are measured in the equatorial plane (i.e., xy-plane), see Figure 6.2.4.

Spherical Triangles.   The great circles on a sphere play the role of lines in spherical geometry. Hence collinear sets in spherical geometry are sets that lie on a single great circle. A subset of that that does not lie on any single great circle is said to be a **noncollinear** set in this geometry. If three points are noncollinear, then it is easy to see that no two of them can be diametrically opposed. A spherical triangle has three sides, and each side of the spherical triangle is an arc lying on a different great circle. Furthermore, the three arcs of great circles that make up the spherical triangle are all shorter than a semicircle of a great circle (i.e., each side is less than half the length of a great circle).

A spherical triangle separates the sphere into two pieces just as an ordinary triangle in the Euclidean plane separates the plane into an interior and exterior. In the case of a spherical triangle, the sphere less the spherical triangle has two pieces that cannot be joined by a curve on the sphere without crossing over the spherical triangle. The piece with the smaller area is called the **interior**, and the piece with the larger area is called the **exterior**.

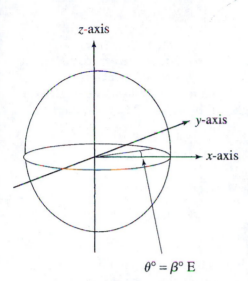

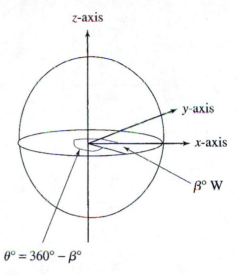

$\theta° = \beta° \text{ E}$ $\qquad\qquad\qquad$ $\theta° = 360° - \beta°$

FIGURE 6.2.4 Longitudinal angles are measured in the *xy*-plane using degrees and go from 0° to 180° in either the east direction or west direction. The angle θ that is used in spherical coordinates (and polar coordinates) agrees with an eastern longitude. Thus, $\theta° = \beta°$ E. A longitude of $\beta°$ W corresponds to $\theta° = 360° - \beta°$ W.

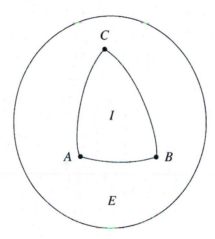

FIGURE 6.2.5 The spherical triangle $\triangle ABC$ separates the sphere into two pieces. The interior *I* has the smaller surface area and the exterior *E* has the larger. The exterior looks like the surface of a ball with a piece cut out. The interior looks like the "triangular" piece that has been cut out.

Angle Sums in Spherical Geometry.   Angles in spherical geometry are measured in tangent planes using Euclidean rays that are tangent to curves. Thus, in Figure 6.2.5, the two great circular arcs joining A to B and joining A to C have tangent rays making an angle in the tangent plane to the sphere at the point A. The measure $m(\angle BAC)$ is then defined to be the measure of the angle between the two rays in the tangent plane at A. Notice that for the spherical triangle $\triangle ABC$, there will be two angles at the vertex A, the interior angle at A that points into the interior I of $\triangle ABC$, and the complete (or total) exterior angle at A that points to the exterior E of the triangle. The interior angle has measure $m(\angle BAC)$. The complete exterior angle at A has measure given by $360° - m(\angle BAC)$ if one measures angles in degrees or by $2 \cdot \pi - m(\angle BAC)$ if one measures angles with radians. Thus, the **angle sum** $AS(\triangle ABC)$ of the (interior) of the triangle $\triangle ABC$ is

$$AS(\triangle ABC) = m(\angle BAC) + m(\angle ABC) + m(\angle BCA).$$

Of course, the **angle sum of the exterior** of $\triangle ABC$ using radians will be

$$AS_{\text{exterior}}(\triangle ABC) = 6 \cdot \pi - [m(\angle BAC) + m(\angle ABC) + m(\angle BCA)]$$
$$AS_{\text{exterior}}(\triangle ABC) = 6 \cdot \pi - AS(\triangle ABC).$$

All spherical triangles have an angle sum strictly greater than that of a straight angle (i.e., greater than $180°$ using degrees or greater than π using radians). The amount greater than that of a straight angle is called the **excess** and is denoted by $\varepsilon(\triangle ABC)$. Using radian measure one has the following formula for the (interior) excess $\varepsilon(\triangle ABC) = AS(\triangle ABC) - \pi$.

$$\boxed{\varepsilon(\triangle ABC) = AS(\triangle ABC) - \pi \quad \text{(excess using radian measure)}}$$

For the exterior of $\triangle ABC$, one has a corresponding excess given by

$$\varepsilon_{\text{exterior}}(\triangle ABC) = AS_{\text{exterior}}(\triangle ABC) - \pi.$$

Using this equation and $AS_{\text{exterior}}(\triangle ABC) = 6 \cdot \pi - AS(\triangle ABC)$, one obtains

$$\varepsilon_{\text{exterior}}(\triangle ABC) = 5 \cdot \pi - AS(\triangle ABC) \quad \text{(radian measure)}.$$

Notice that if one adds the excess of $\triangle ABC$ to the excess of its exterior, one obtains $4 \cdot \pi$. For the special case of a sphere of radius one unit, this is the surface area of the sphere:

$$\varepsilon(\triangle ABC) + \varepsilon_{\text{exterior}}(\triangle ABC) = 4 \cdot \pi \quad \text{(radian measure)}$$

In Euclidean geometry, all triangles have the same angle sum (i.e., π radians), and thus the angle sum of a triangle is not related to its area. On the other hand, in spherical geometry, the angle sum of a triangle is related in a fundamental way to its angle sum. In particular, if one is doing spherical geometry on a sphere of radius ρ, then the area of a given spherical triangle is ρ^2 times its excess, where the excess is measured using radians.

$$\boxed{\text{area}(\triangle ABC) = \varepsilon(\triangle ABC) \cdot \rho^2 \quad \text{(excess measured in radians)}}$$

For the area of the exterior of $\triangle ABC$ one obtains the corresponding formula

$$\boxed{\text{area}_{\text{exterior}}(\triangle ABC) = \varepsilon_{\text{exterior}}(\triangle ABC) \cdot \rho^2 \quad \text{(excess in radians)}}$$

Using $\varepsilon(\Delta ABC) + \varepsilon_{exterior}(\Delta ABC) = 4 \cdot \pi$ and the fact that the surface area of the sphere is given by the sum of the area of ΔABC and the area of the exterior of ΔABC, one obtains the usual formula $4 \cdot \pi \cdot \rho^2$ for the (surface) area of a sphere of radius ρ.

EXAMPLE A spherical triangle ΔABC on Earth is found to have angles with measures of $m(\angle A) = 60°, m(\angle B) = 60°$, and $m(\angle C) = 120°$. Find the excess in degrees, the excess in radians, and the area of this spherical triangle in square miles. Use 4,000 miles for the radius of Earth.

Solution For degrees, one may use $\varepsilon(\Delta ABC) = m(\angle A) + m(\angle B) + m(\angle C) - 180°$ to obtain

$$\varepsilon(\Delta ABC) = 60° + 60° + 120° - 180° = 60°.$$

Converting degrees to radians, one finds that

$$\varepsilon(\Delta ABC) = (60°) \cdot \left(\frac{\pi \text{ radians}}{180°}\right) = \frac{\pi}{3}\text{radians}.$$

The formula for area of a spherical triangle is area(ΔABC) $= \varepsilon(\Delta ABC) \cdot \rho^2$. Thus, using $\rho = 4{,}000$ miles yields

$$\text{area}(\Delta ABC) = \left(\frac{\pi}{3}\right) \cdot (4{,}000 \text{ mi.})^2 \approx 16{,}755{,}000 \text{ mi}^2. \qquad \blacksquare$$

Isometries of Spherical Geometry. Given a sphere K of radius ρ, an isometry of this sphere is a $1 - 1$ mapping of this sphere onto itself such that spherical distance is preserved. In other words, an isometry of K is a $1 - 1$ onto mapping $F: K \rightarrow K$ such that for all $P, Q \in K$, the following equality holds:

$$d_S(P, Q) = d_S[F(P), F(Q)].$$

In studying spheres, it is important to use known facts about circles. If one is given a circle of radius ρ and the length L of a chord, then one may calculate the central angle θ subtended by the chord by using the law of cosines.

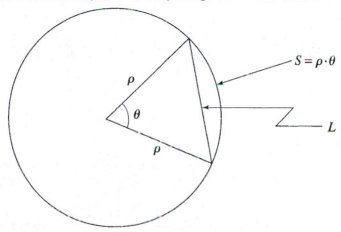

The law of cosines applied to this triangle with two sides of length ρ and the side opposite the angle θ of length L yields

$$L^2 = \rho^2 + \rho^2 - 2 \cdot \rho^2 \cdot \cos(\theta).$$

Hence,

$$\theta = \cos^{-1}\left(\frac{2 \cdot \rho^2 - L^2}{2 \cdot \rho^2}\right).$$

If the angle θ is measured in radians, then the arc subtended by this angle has length

$$S = \rho \cdot \theta = \rho \cdot \cos^{-1}\left(\frac{2 \cdot \rho^2 - L^2}{2 \cdot \rho^2}\right).$$

Conversely, given a circle of radius ρ and an arc of length S, then the central angle θ in radians is $\theta = \frac{S}{\rho}$. Knowing the angle θ, the length L of the chord corresponding to this arc can be found using $L^2 = \rho^2 + \rho^2 - 2 \cdot \rho^2 \cdot \cos(\theta)$.

The previous shows that given the radius ρ of a circle, the length L of a chord determines the length S of the corresponding arc, conversely, the length S of an arc determines the length L of its chord. Applying this equivalence of L and S to great circles on a sphere K shows that a $1 - 1$ onto mapping $F: K \to K$ preserves spherical distance, that is, $d_S(P, Q) = d_S[F(P), F(Q)]$ for all $P, Q \in K$, iff it also preserves Euclidean distance, that is, $PQ = F(P)F(Q)$ for all $P, Q \in K$. Using this information, we show that the isometries of a sphere $K \subset R^3$ may be thought of as isometries of R^3, which take K onto itself. Hence, one may use knowledge of Euclidean isometries of R^3 to study the isometries of spherical geometry. Each isometry of a sphere $K \subset R^3$ extends to an isometry of R^3. Of course, an isometry taking a sphere $K \subset R^3$ onto itself must extend to an isometry of R^3, which takes the center of the sphere to itself.

Using xyz-coordinates, let K be the sphere of radius ρ centered at the origin of R^3.

$$K = \{(x, y, z) | x^2 + y^2 + z^2 = \rho^2\}$$

The isometries of K extend to isometries of R^3, which have matrix representations given by orthogonal matrices. Thus, each isometry of the sphere K with center at the origin corresponds to an isometry $F: R^3 \to R^3$ of xyz-space taking the origin to the origin that may be represented by

$$X' = AX$$

where $X' = \begin{bmatrix} x' \\ y' \\ z' \end{bmatrix}, X = \begin{bmatrix} x \\ y \\ z \end{bmatrix}$, and A is an orthogonal matrix (i.e., $A^{-1} = A^T$).

Isometry of a sphere centered at the origin $\Leftrightarrow$ orthogonal matrix A.

For example, reflection across the yz-plane is given by $F(x, y, z) = (-x, y, z)$, which corresponds to $X' = AX$, where

$$A = \begin{bmatrix} -1 & 0 & 0 \\ 0 & 1 & 0 \\ 0 & 0 & 1 \end{bmatrix} \quad \text{(reflection across } yz\text{-plane)}.$$

Rotation about the z-axis by an angle θ measured in the xy-plane is given by $X' = AX$, where

$$A = \begin{bmatrix} \cos(\theta) & -\sin(\theta) & 0 \\ \sin(\theta) & \cos(\theta) & 0 \\ 0 & 0 & 1 \end{bmatrix} \quad \text{(rotation about } z\text{-axis by angle } \theta).$$

The previous angle θ is measured counterclockwise in the xy-plane, starting with the positive x-axis. For example, rotation by $90°$ moves points on the positive x-axis onto the positive y-axis. Using $\sin(90°) = 1$ and $\cos(90°) = 0$, one finds that for rotation about the z-axis by $90°$, the previous matrix A is given by

$$\begin{bmatrix} 0 & -1 & 0 \\ 1 & 0 & 0 \\ 0 & 0 & 1 \end{bmatrix} \quad \text{(rotation about } z\text{-axis by } \theta = 90°).$$

For rotations about the x-axis, one measures the angle of rotation θ in the yz-plane. Here positive angles correspond to rotating points on the positive y-axis toward the positive z-axis. Rotation about the x-axis by an angle θ measured in the yz-plane is given by $X' = BX$, where

$$B = \begin{bmatrix} 1 & 0 & 0 \\ 0 & \cos(\theta) & -\sin(\theta) \\ 0 & \sin(\theta) & \cos(\theta) \end{bmatrix} \quad \text{(rotation about } x\text{-axis by angle } \theta).$$

Using $\sin(90°) = 1$ and $\cos(90°) = 0$, one finds that for rotation about the x-axis by $90°$, the above matrix B is given by

$$\begin{bmatrix} 1 & 0 & 0 \\ 0 & 0 & -1 \\ 0 & 1 & 0 \end{bmatrix} \quad \text{(rotation about } x\text{-axis by } \theta = 90°).$$

Composition of Isometries. Just as in Chapter 5, composition of functions represented in matrix form can be related to matrix operations. In particular, let $F: R^3 \rightarrow R^3$ and $G: R^3 \rightarrow R^3$ be isometries of R^3, taking the origin to the origin represented by orthogonal matrices A and B, respectively.

Then F is represented by $X' = AX$, and G is represented by $X' = BX$, where

$$X' = \begin{bmatrix} x' \\ y' \\ z' \end{bmatrix} \quad X = \begin{bmatrix} x \\ y \\ z \end{bmatrix}, \text{ and A and B are orthogonal matrices (i.e., } A^{-1} = A^T \text{ and}$$

$B^{-1} = B^T).$

The composition of F followed by G (i.e., $G \circ F$) is represented by $X' = CX$, where $C = BA$. Similarly, the composition of G followed by F (i.e., $F \circ G$) is represented by $X' = DX$, where $D = AB$. Of course, matrix multiplication is not commutative in general. Thus, it is no surprise that the composition of two isometries of a sphere will, in general, fail to commute.

$$F: R^3 \to R^3 \Leftrightarrow X' = AX.$$

$$G: R^3 \to R^3 \Leftrightarrow X' = BX.$$

$$G \circ F: R^3 \to R^3 \Leftrightarrow X' = BAX.$$

$$F \circ G: R^3 \to R^3 \Leftrightarrow X' = ABX.$$

EXERCISES 6.2

1. Assume the radius of Earth is 4,000 miles.
   a. Find the circumference of a great circle on Earth.
   b. Find the length of a longitudinal line.
2. The 30th north parallel is a circle in a plane that is parallel to the equatorial plane and is smaller than a great circle on Earth.
   a. Find the radius of the 30th north parallel.
   b. Find the circumference of the 30th north parallel.
3. Using spherical distance:
   a. Find how far each point of the 30th north parallel is from the North Pole.
   b. Find how far each point of the 30th north parallel is from the South Pole.
4. Using Euclidean distance:
   a. Find how far each point of the 30th north parallel is from the North Pole.
   b. Find how far is each point of the 30th north parallel is from the South Pole.
5. A spherical triangle $\triangle ABC$ is given on the surface of Earth. Point A is the North Pole. Point B is located on the equator and has latitude and longitude given by $0°$ N and $0°$ E. The point C is located on the equator and has latitude and longitude given by $0°$ N and $1°$ W.
   a. Find the angle sum $AS(\triangle ABC)$ of this triangle in degrees.
   b. Find the excess $\varepsilon(\triangle ABC)$ in radians.
   c. Find the area in square miles of this spherical triangle.
6. The latitude and longitude of Paris are given by $48°51'$ N and $2°21'$ E. The latitude and longitude of New York are given by $40°40'$ N and $73°58'$ W.
   a. Find the cartesian coordinates (x, y, z) for Paris ($=$ point P) using the xyz-coordinates discussed in this section.
   b. Find the cartesian coordinates (x, y, z) for New York ($=$ point Q).
   c. Find the spherical distance $d_S(P, Q)$ from Paris to New York.

Classroom Connection 6.2.1

An intuitive understanding of spherical distance is something that should be given to students before college. To show middle or high school students an experimental way to approximate the spherical distance from London to New York, take a globe of Earth with some given radius ρ_0 and circumference $C_0 = 2\pi\rho_o$. Use string to measure the distance on the globe from London to New York. The string should be pulled tight, with one end on London and one end on New York. Put the string down flat next to a ruler and measure the length of string that connected London and New York. Let this length be L_0. If the string is used a second time to find the circumference C_o of the globe, then one has the following ratio:

$$\frac{d_S\,(\text{London, New York})}{\text{radius of earth}} = \frac{L_0}{\rho_0} = \frac{2\pi L_0}{2\pi\rho_0} = \frac{2\pi L_0}{C_o}.$$

Using 4,000 miles for the radius of Earth, this yields

$$d_S\,(\text{London, New York}) = (4{,}000\text{ miles}) \cdot \left(\frac{L_0}{\rho_0}\right) = (4{,}000\text{ miles}) \cdot \left(\frac{2\pi L_0}{C_0}\right). \quad \blacklozenge$$

6.3 HYPERBOLIC GEOMETRY

The postulates for hyperbolic geometry are the same as for Euclidean geometry with one exception. The parallel postulate (i.e., Euclid's fifth postulate) is replaced with an alternative that is often called the **hyperbolic parallel postulate**, see Figure 6.3.1. Just as with Euclidean geometry, hyperbolic geometry can be done in higher dimensions. In this section, we consider only two-dimensional hyperbolic geometry.

Hyperbolic Parallel Postulate

Given a point P off a line ℓ in the hyperbolic plane, there are at least two lines containing P that are parallel to ℓ.

Upper Half-Plane Model.   There are several models of the hyperbolic plane. One of the most important of these is known as the **Poincaré upper half-plane model**, which was extensively studied by the mathematician Poincaré (*Poin-car-a*) (1854–1912), see Figure 6.3.1. This model is called the (Poincaré) upper half-plane model because the points used in this model, lie in the upper half of the usual xy-plane. Let H denote the points of this model.

$$H = \{(x,y)|y > 0\} \quad \text{(points of this model)}$$

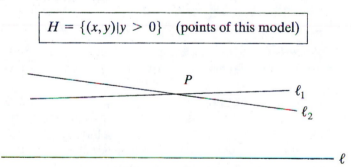

FIGURE 6.3.1 If the point P is off the line ℓ in the hyperbolic plane, then the hyperbolic parallel postulate guarantees that there are at least two lines ℓ_1 and ℓ_2 that contain P and are parallel to ℓ. Here lines are parallel to ℓ if they do not intersect ℓ. It is not hard to prove that if ℓ_1 and ℓ_2 are parallel to ℓ and pass through P, then there are an infinite number of lines through P that are parallel to ℓ.

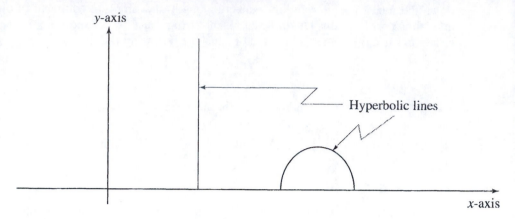

FIGURE 6.3.2 The points of the Poincaré upper half-plane model are the points in the upper half of the usual *xy*-plane (i.e., $H = \{(x, y) \mid y > 0\}$). A hyperbolic line is either a Euclidean semicircle with the center on the *x*-axis or the intersection of the upper half-plane *H* with a vertical Euclidean line. The two hyperbolic lines shown here are parallel since they do not intersect. Points of the *x*-axis would not be visible to a two-dimensional hyperbolic person in this model.

In this model, the **hyperbolic lines** are defined as (1) the intersection of the set *H* with Euclidean circles having centers on the *x*-axis and (2) the intersection of the set *H* with Euclidean lines that have undefined slope in the *xy*-plane, see Figure 6.3.2. Hyperbolic lines of type (1) are Euclidean semicircles of the form $\{(x, y) \mid y > 0 \text{ and } (x - a)^2 + y^2 = r^2\}$. Hyperbolic lines of type (2) are Euclidean half lines of the form $\{(x, y) \mid y > 0 \text{ and } x = a\}$. Thus, the hyperbolic lines are given by the Euclidean semicircles $y > 0$ and $(x - a)^2 + y^2 = r^2$ and by the Euclidean rays $y > 0$ and $x = a$.

$$y > 0 \text{ and } (x - a)^2 + y^2 = r^2$$

(hyperbolic lines for this model)

$$y > 0 \text{ and } x = a$$

We discuss hyperbolic distance in this model later in the chapter. Hyperbolic distance is measured quite differently from Euclidean distance, hence, the lines in hyperbolic geometry are quite different from Euclidean lines. When using Euclidean distance, the shortest join of two points lies on a Euclidean line, when using hyperbolic distance, the shortest join of two points lies on a hyperbolic line.

Notice that the half-plane *H* does not include points of the *x*-axis, only points strictly above the *x*-axis. Thus, a two-dimensional hyperbolic person living in this model would not be able to "see" any points of the *x*-axis. To such a person, the points of the *x*-axis would be infinitely far away. Since the *x*-axis is not included in *H*, the hyperbolic lines of the first type are semicircles that are "open" in the sense that they do not include the two points where the Euclidean circle they lie on intersect the *x*-axis. Similarly, a hyperbolic line of the second type does not include the point where the Euclidean line it lies on intersects the *x*-axis.

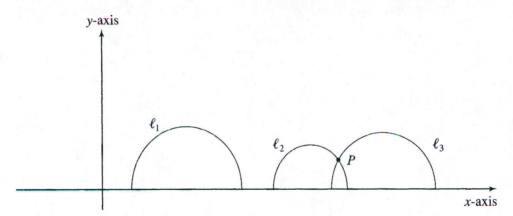

FIGURE 6.3.3 Point P lies off the hyperbolic line ℓ_1. By the hyperbolic parallel postulate, there must be at least two lines passing through P that are parallel to ℓ_1. In particular, ℓ_2 and ℓ_3 are hyperbolic lines parallel to ℓ_1 and pass through P. In fact, there are an infinite number of hyperbolic lines passing through P that are parallel to ℓ_1 (i.e., each has empty intersection with ℓ_1).

The hyperbolic parallel postulate can be illustrated by taking two intersecting hyperbolic lines and then taking a line that has empty intersection with these two. An example is illustrated in Figure 6.3.3.

Angles in Hyperbolic Geometry.   Interestingly, in the Poincaré upper half-plane model, the usual Euclidean angles are equal in measure to the hyperbolic angles. Many models of the hyperbolic plane do not have this feature. Of course, given two intersecting hyperbolic lines, the Euclidean angle between them is defined to be the Euclidean angle between the Euclidean lines tangent to the given hyperbolic lines. This is illustrated in Figure 6.3.4.

> Hyperbolic angles = Euclidean angles (in this model).

Distances in Hyperbolic Geometry.   Unlike angles, distance and area are different for Euclidean geometry and hyperbolic geometry. Let P and Q be points in the upper half-plane. If the second point Q lies directly above the first, then they have the same first coordinate, and the second coordinate y_2 of Q must be larger than the second coordinate y_1 of P. Thus, $P = (a, y_1)$ and $Q = (a, y_2)$, where $y_2 > y_1 > 0$. In this case, the hyperbolic distance $d_H(P, Q)$ is given by

$$d_H(P, Q) = d_H((a, y_1), (a, y_2)) = \ln\left[\frac{y_2}{y_1}\right].$$

> $d_H[(a, y_1), (a, y_2)] = \ln\left(\frac{y_2}{y_1}\right)$ (when one point is directly above the other).

Here, $\ln$ denotes the natural log (i.e., log base e).

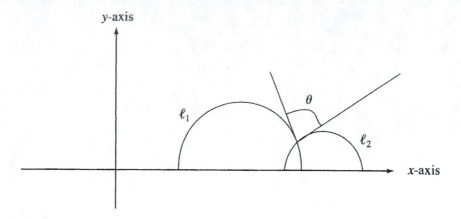

FIGURE 6.3.4 The Euclidean angle formed by the intersection of lines ℓ_1 and ℓ_2 has the same measure θ as the hyperbolic angle formed by this intersection. The angle is measured between the two Euclidean lines tangent to the two Euclidean circles corresponding to the two hyperbolic lines. Euclidean angles and hyperbolic angles are equal in measure in this model.

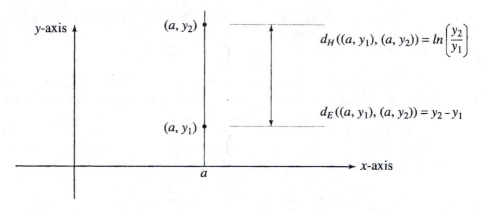

When two points P and Q are given in the upper half-plane with neither point directly above the other, then they have different x-coordinates. To find the center A of the Euclidean circle that contains both of these points and has its center on the x-axis, one intersects the x-axis with the perpendicular bisector of the Euclidean segment joining P and Q. Let the angle that the Euclidean segment $\overline{AP}$ makes with the positive x-axis have measure α, and let the angle that the Euclidean segment $\overline{AQ}$ makes with the positive x-axis have measure β. Assuming $0 < \alpha < \beta$, the hyperbolic distance $d_H(P, Q)$ is given by $d_H(P, Q) = \ln\left(\frac{\csc(\beta) - \cot(\beta)}{\csc(\alpha) - \cot(\alpha)}\right)$.

$$d_H(P, Q) = \ln\left(\frac{\csc(\beta) - \cot(\beta)}{\csc(\alpha) - \cot(\alpha)}\right)$$

(when the points are *not* along a vertical line).

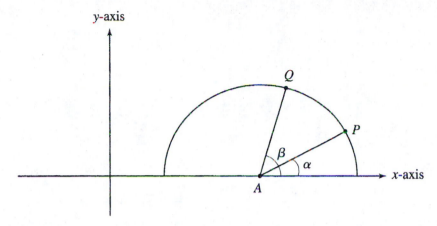

Recall from spherical geometry that a spherical triangle has all three sides on great circles. In hyperbolic geometry, one defines a set of points to be (hyperbolically) noncollinear if they all fail to lie on a single hyperbolic line. Given three hyperbolically noncollinear points A, B, and C, the hyperbolic triangle $\triangle ABC$ is the union of the three hyperbolic line segments joining the points.

A spherical triangle always has angle sum greater than a straight angle. The opposite is true in hyperbolic geometry. In the hyperbolic plane, a hyperbolic triangle always has an angle sum strictly less than one straight angle. The difference is called the **defect** of the triangle and is denoted by def($\triangle ABC$). Using radian measure and recalling that the radian measure of a straight angle is equal to π, one finds that def($\triangle ABC$) = $\pi - m(\angle A) - m(\angle B) - m(\angle C)$.

$$\text{def}(\triangle ABC) = \pi - m(\angle A) - m(\angle B) - m(\angle C) \text{ (defect measured in radians)}$$

The hyperbolic area $A_{\text{hyperbolic}}(\triangle ABC)$ of a triangle $\triangle ABC$ in the hyperbolic plane is given by its defect (measured in radians) $A_{\text{hyperbolic}}(\triangle ABC) = \text{def}(\triangle ABC)$, see Figure 6.3.5.

$$A_{\text{hyperbolic}}(\triangle ABC) = \text{def}(\triangle ABC) \quad (\text{hyperbolic area = defect})$$

EXERCISES 6.3

1. Explain why each hyperbolic triangle in the Poincaré upper half-plane must have hyperbolic area less than π.
2. Consider the Euclidean segment joining $P = (0,1)$ to $Q = (0,7)$. Since P and Q are joined by a hyperbolic line that lies on a Euclidean line, the hyperbolic segment joining the points is the same point set as the Euclidean line joining these segments. However, the hyperbolic length [i.e., $d_H(P,Q)$] of this segment is different from the Euclidean length (i.e., PQ). Also, the hyperbolic midpoint of this segment differs from the Euclidean midpoint.
   a. What is the Euclidean length of this segment?
   b. What is the hyperbolic length of this segment?

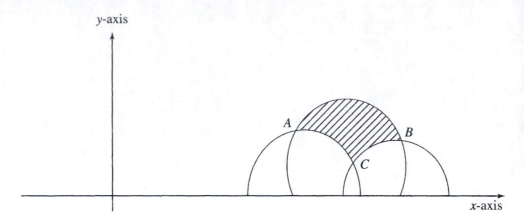

FIGURE 6.3.5 The hyperbolic triangle $\triangle ABC$ has all three sides lying on hyperbolic lines. The hyperbolic area of this hyperbolic triangle will be equal to the defect of the triangle, where the defect is given by $\pi - m(\angle A) - m(\angle B) - m(\angle C)$. Here the angles are measured in radians and are measured pointing into the triangle (i.e., into the shaded region). The angles have the same measure in both Euclidean geometry and hyperbolic geometry.

    c. Which is larger, the Euclidean length or the hyperbolic length?
    d. What is the Euclidean midpoint of the segment from P to Q?
    e. What is the hyperbolic midpoint of the segment from P to Q?
  3. Find the equation of the Euclidean circle that contains the hyperbolic line joining the points $A = (0,4)$ and $B = (4,8)$.
  4. Let ℓ be the hyperbolic line that lies on the y-axis (i.e., $\ell = \{(0,y)|y > 0\}$. There are two hyperbolic lines ℓ_1 and ℓ_2 that pass through the point $C = (0,1) \in \ell$ and are such that each make an angle of $45°$ with the line ℓ. Find the equations of the two Euclidean circles that contain the respective lines ℓ_1 and ℓ_2.

Classroom Discussion 6.3.1

Students often feel that the Poincaré upper half-plane model is not valid because Euclidean geometry is being used to study a non-Euclidean geometry. On a deeper level they may even doubt that hyperbolic geometry really exists. To help students understand that it is permissible to use Euclidean geometry to study other geometries, point out that almost everyone accepts that one can use three-dimensional Euclidean geometry to study spherical geometry. Also mention that the consistency of Euclidean geometry has been shown to be equivalent to the consistency of hyperbolic geometry. If one believes in the existence in one of these geometries, then one is "forced" to believe in the existence of the other. ◆

CHAPTER 6 REVIEW

One way to gain a real insight into what Euclidean geometry really represents is to see what is not Euclidean. There are many geometries besides Euclidean geometry, and in this chapter we presented three of the alternatives. Each is important in certain ways. Taxicab geometry and its higher-dimensional versions have several applications. One key use of taxicab geometry is to provide an example of a space that satisfies all but the SAS axiom. Thus, the existence of taxicab geometry shows that in studying Euclidean geometry one needs to assume the SAS axiom or some equivalent. It demonstrates, very graphically, that not only was Euclid's proof of SAS false, but also no valid proof can be provided without adding another assumption to those of Euclid. The key idea of taxicab geometry is that distances between the two points (x_1, y_1) and (x_2, y_2) are found by adding the absolute value of the difference in the x-coordinates $|x_1 - x_2|$ to the absolute value of the difference in the y-coordinates $|y_1 - y_2|$. In this geometry, SAS, SSS, and the Pythagorean theorem are not valid.

Spherical geometry is the study of geometry on spheres. Since Earth is very close to being a sphere, one uses spherical geometry to study the geometry of the real world when considering points that are not close together. In particular, great circles are used to lay out the paths of plane flights between distant cities since great circles minimize distance. The key thing is that distance between two points on a given sphere of radius ρ is measured along the (or a) shortest curve joining the two points with the requirement that the curve stay on the sphere's surface. If two given points are diametrically opposed, then there will be many such shortest curves, and they will all lie on circles of radius ρ joining the two points. In the usual case of two points that are not diametrically opposed, there will be one shortest arc of the great circle joining these points. The spherical distance between the points is then the length of this shortest arc. Given a collection of three points such that no great circle contains all three of them, there is a spherical triangle determined. It consists of the union of the three shortest arcs of great circles joining the points. It divides the sphere into two parts, and the part with the smallest area is called the *interior* of the spherical triangle. The angle sum of a spherical triangle is always greater than a straight angle (i.e., greater than π radians). The amount, measured in radians, by which the angle sum is greater than π is called the *excess*. The area of a spherical triangle's interior is ρ^2 times the excess. Hence, if the radius is fixed, the angle sum of a spherical triangle determines its area.

Hyperbolic geometry was originally studied synthetically (i.e., without coordinates). Several different coordinate models were devised much later. In each of the coordinate models, one has points represented by ordered pairs of real numbers and lines represented by sets of points. In all cases, the "lines" and "points" must satisfy the hyperbolic parallel postulate. We studied hyperbolic geometry in this chapter by using the Poincaré upper half-plane model. The points of this model are the points in the upper half of the xy-plane. A line in this geometry is the intersection of the upper half-plane with a Euclidean line of the form $x = c$ or is the top half of a Euclidean circle with the center on the x-axis. Three points in this model are said to be noncollinear if there is no hyperbolic line containing all three of them. Three such noncollinear points determine a hyperbolic triangle consisting of the three hyperbolic line segments joining the points. The angle sum of a hyperbolic triangle is always less than a straight angle (i.e., less than π radians). The amount less, measured in radians, is called the *defect*. The area of a hyperbolic triangle is equal to its defect. Thus, just as in spherical geometry, the angle sum of a triangle determines its area.

Selected Formulas.

Euclidean distance: $d_E[(x_1, y_1), (x_2, y_2)] = \sqrt{(x_1 - x_2)^2 + (y_1 - y_2)^2};$

Taxicab distance: $d_T[(x_1, y_1), (x_2, y_2)] = |x_1 - x_2| + |y_1 - y_2|$

Polar coordinates: $x = r\sin(\theta), y = r\cos(\theta)$

Spherical coordinates: $x = \rho\sin(\phi)\cos(\theta)$

$$y = \rho\sin(\phi)\sin(\theta)$$

$$z = \rho\cos(\phi)$$

Dot product: $(x_1, y_1, z_1) \cdot (x_2, y_2, z_2) = x_1 x_2 + y_1 y_2 + z_1 z_2$

Spherical distance: $d_S[(x_1, y_1, z_1), (x_2, y_2, z_2)] = \rho \cdot \cos^{-1}\left(\frac{x_1 x_2 + y_1 y_2 + z_1 z_2}{\rho^2}\right)$

Angle sum: $AS(\triangle ABC) = m(\angle A) + m(\angle B) + m(\angle C)$

Excess of spherical triangle: $\varepsilon(\triangle ABC) = AS(\triangle ABC) - \pi$ (measured in radians)

Area of spherical triangle: $\text{area}(\triangle ABC) = \varepsilon(\triangle ABC) \cdot \rho^2$

Hyperbolic distance: $d_H[(a, y_1), (a, y_2)] = \ln\left(\frac{y_2}{y_1}\right)$ (same x coordinate, $y_2 > y_1$)

$d_H(P, Q) = \ln\left(\frac{\csc(\beta) - \cot(\beta)}{\csc(\alpha) - \cot(\alpha)}\right)$ (points with unequal x-coordinates)

Defect: $\text{def}(\triangle ABC) = \pi - m(\angle A) - m(\angle B) - m(\angle C)$ (in radians)

Area of hyperbolic triangle: $A_{\text{hyperbolic}}(\triangle ABC) = \text{def}(\triangle ABC)$

CHAPTER 6 REVIEW EXERCISES

1. Let $A = (-1, 5)$ and $B = (3, 2)$. Find
   a. The Euclidean distance $d_E(A, B)$ from A to B.
   b. The taxicab distance $d_T(A, B)$ from A to B.

2. Let the right $\triangle OPQ$ have vertices given by $O = (0, 0)$, $P = (1, 0)$ and $Q = (0, \sqrt{3})$.
   a. Find the lengths of the three sides of this triangle in Euclidean geometry.
   b. Find the lengths of the three sides of this triangle in taxicab geometry.
   c. Verify that for Euclidean geometry the Pythagorean theorem is satisfied. That is, check that $[d_E(O, P)]^2 + [d_E(O, Q)]^2 = [d_E(P, Q)]^2$.
   d. Verify that the formula from the Pythagorean theorem does not hold in taxicab geometry. That is, check that $[d_T(O, P)]^2 + [d_T(O, Q)]^2 \neq [d_T(P, Q)]^2$.

3. Let $\triangle ABC$ have vertices given by $A = (-2, 0)$, $B = (0, 2)$, and $C = (0, 0)$. Let $\triangle A'B'C'$ have vertices given by $A' = (3, 3)$, $B' = (5, 3)$, and $C' = (4, 4)$.
   a. Find all of the Euclidean (hence also taxicab) angles of both triangles.
   b. Find the Euclidean lengths of the sides of both triangles.
   c. Find the taxicab lengths of the sides of both triangles.

   d. Do the two triangles satisfy SAS in taxicab geometry?

   e. Are the two triangles congruent in taxicab geometry?

4. Assuming the radius of the moon is 1,080 miles, find the circumference of a great circle on the moon.

5. Assume the radius of Earth is 4,000 miles and the radius of the moon is 1,080 miles.

   a. If a spherical triangle on Earth has an area of 2,000,000 square miles, find its spherical excess.

   b. If a spherical triangle on moon has an area of 2,000,000 square miles, find its spherical excess.

   c. Find the ratio of your above answers (i.e., $\frac{\text{ans}.a}{\text{ans}.b}$).

   d. Let $r_1 = 4,000$ miles be the radius of Earth, and let $r_2 = 1,080$ miles be the radius of the moon. Find $\left(\frac{r_2}{r_1}\right)^2$.

6. a. If a spherical triangle on Earth has an excess of $30°$, find the area of this spherical triangle measured in square miles.

   b. If a spherical triangle on the moon has an excess of $30°$, find the area of this spherical triangle measured in square miles.

   c. Find the ratio of your above answers to part a and b.

7. Sphere 1 has a radius of r_1 and sphere 2 has a radius of r_2. If spherical triangle Δ_1 on sphere 1 and spherical triangle Δ_2 on sphere 2 have the same area, find the ratio of their excesses [i.e., find $\frac{\varepsilon(\Delta_1)}{\varepsilon(\Delta_2)}$].

8. Using Euclidean distance, find how far each point of the 45th north parallel is from the North Pole.

9. Using spherical distance, find how far each point of the 45th north parallel is from the North Pole. Is your answer more or less than the Euclidean distance found in the previous question?

10. The latitude and longitude of city A on Earth are given by $45°0'N$ and $90°0'W$. The latitude and longitude of city B are given by $40°0'S$ and $90°0'W$.

   a. Find the Cartesian coordinates (x, y, z) for city A($=$ point A) using the xyz-coordinates discussed in this section.

   b. Find the Cartesian coordinates (x, y, z) for city B($=$ point B).

   c. Find the spherical distance $d_S(A, B)$.

11. Let $R = (0, 3)$, $S = (0, 9)$, and $T = (0, 27)$ be three points in the Poincaré upper half-plane. Find (a) $d_H(R, S)$, (b) $d_H(S, T)$, and (c) $d_H(R, T)$.

12. Let $A = (0, a)$, $B = (0, b)$, and $C = (0, c)$ be three points in the Poincaré upper half-plane. Assume that $0 < a < b < c$.

   a. Find $d_H(A, B), d_H(B, C)$, and $d_H(A, C)$.

   b. Using logarithms, show that $d_H(A, B) + d_H(B, C) = d_H(A, C)$.

13. Find the equation of the Euclidean circle that contains the hyperbolic line joining the points $A = (-3, 4)$ and $B = (3, 4)$.

14. Find the equation of the Euclidean circle that contains the hyperbolic line joining the points $C = (1, 1)$ and $D = (3, 3)$.

15. Find the hyperbolic distance $d_H(A, B)$ between the two points $A = (-5, 5)$ and $B = (5, 5)$.

RELATED READING FOR CHAPTER 6

Blumenthal, L. *A Modern View of Geometry*. San Francisco, CA: W. H. Freeman, 1961.

Kay, D. C. *College Geometry: A Discovery Approach* 2nd ed., New York: Addison Wesley Longman, 2001.

Smart, J. R. *Modern Geometries* 4th ed., Pacific Grove, CA: Brooks/Cole Publishing, 1994.

Stahl, S. *The Poincaré Half-Plane*. Boston, MA: Jones and Bartlett, 1993.

APPENDIX I

Geometer's Sketchpad

This appendix is a very brief introduction to the Geometer's Sketchpad (GSP). This software is very user friendly and a novice will find it easy to start using this program after a short introduction. Additional information may be found: http://www.keypress.com/sketchpad

TOOLBOX BUTTONS

After opening up a new sketch in the GSP software, the screen appears as shown here with the menus listed across the top and the toolbox on the left.

The toolbox has six buttons.

The *top button* looks like an arrow and is for the *selection arrow tools*.

There are three options for this button. The first option is used for selection of objects, and this appears as an arrow as shown previously. To select an object on the screen, first click on the Selection Arrow Tools button using this first option and then click on the desired object in the sketch. To select several objects, hold down the Shift key while selecting the desired objects. The second option is for rotating selected objects about a fixed point, which is identified using the Mark Center option under the Transform menu. The last option is for dilating selected objects using a center point, which is identified with the Mark Center option.

The *second button* is the *point tool*.

This button is used to define new points. Click on this point tool and then click on a spot of the sketch to place a point there. The screen then shows a very small circle and a second small circle around the first circle. The inner circle represents the point. The outer circle shows that the point has been selected. If a point has not been selected, then it is just shown as a single small circle.

The *third button* is the *compass tool*.

This tool is used to make circles. However, many individuals find it more useful to construct circles by selecting two points and then using a command in the Construct menu. The Construct menu commands will be shown as part of the GSP menus given later.

The *fourth button* is for the *straightedge tools*.

Clicking on the Segment button in the toolbox yields three choices. Use this button to construct segments, rays, or straight lines.

The fifth button is the *text tool*.

Use this button to add text to the sketch and to name various objects in the sketch. Click on this tool and then click on a point of the sketch; start to type at that point.

The sixth button is the *Custom tool* and is used to create new tools. This tool greatly extends the number of activities that may be done with this software. More information on this tool is available at the previously mentioned web page.

MENU OPTIONS

We now cover the items in the menus across the top of the sketch. For the most part, the items are self-explanatory. We briefly discuss each menu.

Use the *File menu* to start new sketches, open previous sketches, save sketches, print sketches, etc.

File Menu Options
New Sketch
Open
Save
Save As...
Close
Document Options...
Page Setup...
Print Preview...
Print...
Quit

Use the *Edit menu* to undo commands, cut, paste, set up action buttons, select objects, and other related operations. The Preferences command is at the menu's bottom. This allows choices of units, color, and text. One may select different angle measures, such as degrees or radians, and one may select different length measures, such as centimeters or inches.

Edit Menu Options
Undo
Redo
Cut
Paste
Clear
Action Buttons
Select All
Select Children
Split/Merge
Edit Definition...
Properties
Preferences

Use the *Display menu* to change the appearance of objects, including line width and color. The Hide Objects command allows one to hide objects, such as lines used to help construct objects. This allows one to "clean up" diagrams, making them much neater, nicer, and more understandable. The Animate command is used to make objects move on the computer screen.

Display Menu Options

Line Width
Color
Text
Hide Objects
Show All Hidden
Show Labels
Trace
Erase Traces
Animate
Increase Speed
Decrease Speed
Stop Animate
Show Motion Controller
Hide Toolbox

The *Construct menu* contains many of the most important GSP commands. The Point on Object command allows one to construct a point on an object. After this point has been constructed, it may be moved around on the given object. If one selects (using the select tool in the toolbox) two points, then the Segment command may be used to construct the segment joining the two points. Rays and lines are constructed in similar fashion. The difference is in selecting the particular line tool from the toolbox. To construct parallel lines, first select a line and a point off the line. To construct a line perpendicular to a given line, select both a line and a point. The point may be located on or off the given line.

Construct Menu Options

Point on Object
Midpoint
Intersection
Segment
Ray
Line
Parallel Line
Perpendicular Line
Angle Bisector
Circle by Center + Point
Arc on Circle
Arc Through 3 Points
Interior
Locus

Use the *Transform menu* to translate, rotate, reflect, and dilate. It is especially useful in illustrating rigid motions and dilations. To rotate about a point, first select a point and then use the Mark Center command to let the computer know what the center of the rotation will be. To dilate, first select a point, then use the Mark Center command to let the computer know what the center of the dilation will be.

Transform Menu Options
Mark Center
Mark Mirror
Mark Angle
Mark Vector
Mark Distance
Translate. . .
Rotate. . .
Dilate. . .
Reflect
Iterate. . .

Use the *Measure menu* to measure angles, slopes, lengths, and areas. It is also used to work with coordinates and has a valuable Calculate command that can combine measured quantities.

Measure Menu Options
Length
Distance
Perimeter
Circumference
Angle
Area
Arc Angle
Arc Length
Radius
Ratio
Calculate
Coordinates
Abscissa (x)
Ordinate (y)
Coordinate Distance
Slope
Equation

Use the *Graph menu* to plot points on a coordinate plane and give the graphs of many common functions.

Graph Menu Options
Define Coordinate System
Mark Coordinate System
Grid Form
Show Grid
Snap Points
Plot Points
New Parameter
New Function
Plot New Function
Derivative
Tabulate
Add Data
Remove Table Data

The *Help menu* contains a great deal of very useful information. Individuals learning how to use the GSP should use it often.

Help Menu Options
Contents
What's New
Elements
Menus
Keyboard
Advanced Topics
About Sketchpad. . .

SIMPLE CONSTRUCTIONS

A beginning GSP user may wish to start by experimenting with the tools in the toolbox and then investigate the menus. It is very easy to do many Euclidean constructions with this software. The following is a simple "hands-on" introduction to the GSP that can be used on an individual level or presented in a classroom with students working on computers that have the GSP software.

Step 1: Line Segments

The easiest way to construct a segment is to click on the straightedge tools, select the Segment option, and then use this tool directly. However, for many involved diagrams, it is better to construct segments using the Construct menu to join points. To do this, start by choosing the point tool. Point and click to make a first and then a second point. Use the arrow selection tool while holding down the Shift key to select both points. Use the Segment option under the Construct menu to construct the segment joining the two points. Now use the arrow selection tool to select the segment by clicking on the segment. Use the Measure menu to find the segment's length and then to find the segment's slope. The length and slope will be displayed on the screen. One may use the arrow selection tool to "grab" one end of the segment by pointing at that end and then holding down one of the mouse buttons. One may then move this end. As the end moves, the length and slope of the segment that is displayed on the screen will change.

Step 2: Parallel and Perpendicular Lines

To construct a line, first construct two points and then select them as described in Step 1. Now use the Line option under the Construct menu. To make a parallel line to the line just constructed, make a point off the line and use the arrow selection tool to select both the point and the line at the same time. Hold down the Shift key when selecting the second of these two objects. Now select the Parallel option under the Construct menu. Use this same type of approach to construct a line that is perpendicular to a given line and that passes through a given point. The given point may be on or off the first line.

Step 3: Measuring Angles

Given three points, select them in some order by using the arrow selection tool while holding down the Shift key. Use the Measure menu to find the measure of the angle determined. The measure will be displayed on the screen. The angle measured will have a vertex at the second of the three points selected. For example, let the points be labeled A, B, and C. To measure the angle $\angle ABC$ with vertex B, one may select the points in either the order A-B-C or in the order C-B-A. If one measures two or more angles and wishes to find the sum of the measures, one begins by finding the measure of each angle, thus obtaining their displayed measures. Select the displayed measures using the arrow selection tool. Now use the Calculate option under the Measure menu to add the different angles together. The Calculate option will open a box that shows the angles selected for addition under the value key. In the box is a keypad for doing the desired additions. This makes it easy to add the measures of the angles in a triangle or, more generally, in any polygon.

Step 4: Polygons

One way to begin constructing a polygon with n sides is to start by using the point tool to make n points. Select them in the desired order using the arrow selection tool and holding down the Shift key. To construct the polygon's sides, use the Segment option under the Construct menu. The interior of the polygon can be obtained as a set and displayed in color by selecting the vertices of the polygon in order and then using the Interior option under the Construct menu. Once the interior is filled in with color, use the arrow selection tool to select the interior, then use the options under the Measure menu to find the polygon's area and perimeter. One may also "grab" the interior of the polygon and move the polygon around rigidly. Given a polygon, one may also use the arrow selection tool to "grab" a single vertex and then deform the polygon by moving that single vertex. If the area and perimeter are displayed, they will change dynamically as the vertex is moved. Measuring the angles of a polygon and using the Calculate option as described in Step 3, one can find the polygon's angle sum and demonstrate the angle sum formula for polygons. This works best with convex polygons since the software measures angles as less than or equal to 180°. Using this software, one obtains 360° for the angle sum of a quadrilateral only when the quadrilateral is convex.

Step 5: Reflections and Rotations

Use the Transform menu to reflect across a line or rotate about a point. Before reflecting across a line, select the line using the arrow selection tool, and then use the Mark Mirror option under the Transform menu. This tells the software that this is the line that the object will be reflected across. After the given line has been marked as the mirror, select some object, such as a triangle, and use the Transform menu to reflect it across the line and get an image triangle. If one reflects a triangle with vertices labeled A, B, and C, then the image triangle will have vertices labeled A', B', and C'. To rotate about a point O, select the point and then use the Mark Center option under the Transform menu to tell the software that this will be the center of the rotation. Once the center is marked, select an object to rotate, and use the Rotate option under the Transform menu to do the rotation.

The following are some relatively simple GSP projects that are good exercises for future teachers. Given some direction and help from their teacher, middle or high school students can do these exercises after they become familiar with basic GSP operations. One can assign several of these projects to future teachers and ask them to turn in hard copy for each project assigned together with several pages describing the project, the mathematical significance of the project, and how they might incorporate this project into a middle or high school classroom.

SOME GSP PROJECTS

1. Illustrate that a transversal to a parallel line creates alternate interior angles of equal measure. Also, illustrate that for parallel lines with a transversal, one has corresponding angles of equal measure and interior angles on the same side that are supplementary.

2. Illustrate that the angle sum of a triangle is 180°. Make this illustration more impressive by moving one vertex of the triangle and finding that the angle sum remains equal to 180°.

3. Illustrate that the perpendicular bisector of a line segment has its points equidistant from the endpoints of the original segment.

4. Illustrate that the perpendicular bisectors of a triangle's sides meet at a common point (the circumcenter). Also, illustrate that this point is the center of the circumcircle by showing that it is equidistant from the triangle's three vertices. One can make this and many of the other projects listed dynamic by moving one vertex of the original triangle.

5. Illustrate that the angle bisectors of a triangle meet at a common point (the incenter). Also illustrate that this point is equidistant from the three sides of the triangle.

6. Illustrate that the altitudes of a triangle meet at a common point (the orthocenter).

7. Illustrate that the medians of a triangle meet at a common point (the centroid). Illustrate that each median splits the original triangle into two triangles of equal area. Also illustrate that the centroid is located two-thirds of the way from each vertex to the side opposite that vertex.

8. Illustrate that if one joins the midpoints of a quadrilateral, the resulting figure is a parallelogram.

9. Illustrate the Pythagorean theorem by showing that the sum of areas of squares determined by the legs is equal to the area of a square determined by the hypotenuse.

10. Illustrate that reflection across a line takes an object, such as a triangle or a polygon, to a congruent object. Using a triangle, one may also illustrate that orientation is reversed under such a reflection.

11. Illustrate that a similarity transformation with some scale factor S takes an object, such as a triangle, to an object that has a perimeter that is S times as large as the original and an area that is S^2 times as large as the original.

APPENDIX II

Euclid's Assumptions

The Elements is a collection of thirteen books on mathematics. Some of the results were due to Euclid, but most of the mathematics in Euclid's *Elements* represent contributions by others.

EUCLID'S DEFINITIONS, POSTULATES, AND AXIOMS

At the start of *Book 1* of Euclid's *Elements* is a list of definitions. Although these definitions are not up to modern mathematical standards, we list them to provide some understanding of the historical development of geometry.

Definitions:

1. A point is that which has no part.
2. A line is breadthless length. [Remark: Euclid uses *line* to mean *curve*.]
3. The extremities of a line are points.
4. A straight line is a line which lies evenly with the points on itself.
5. A surface is that which has length and breadth only.
6. The extremities of a surface are lines.
7. A plane surface is a surface which lies evenly with the straight lines on itself.
8. A plane angle is the inclination to one another of two lines in a plane which meet one another and do not lie in a straight line.
9. When the lines containing the angle are straight, the angle is called rectilinear.
10. When a straight line set up on a straight line makes the adjacent angles equal to one another, each of the equal angles is right, and the straight line standing on the other is called perpendicular to that on which it stands.
11. An obtuse angle is an angle greater than a right angle.
12. An acute angle is an angle less than a right angle.
13. A boundary is that which is an extremity of anything.
14. A figure is that which is contained by any boundary or boundaries.
15. A circle is a plane figure contained by one line such that all the straight lines falling upon it from one point among those lying within the figure are equal to one another.
16. And the point is called the center of the circle.
17. A diameter of the circle is any straight line drawn through the center and terminated in both directions by the circumference of the circle, and such a straight line also bisects the surface.

18. A semicircle is the figure contained by the diameter and the circumference cut off by it, and the center of the semicircle is the same as that of the circle.

19. Rectilinear figures are those which are contained by straight lines, trilateral figures being those contained by three, quadrilateral those contained by four, and multilateral those contained by more than four straight lines.

20. Of trilateral figures, an equilateral triangle is that which has its three sides equal, an isosceles triangle that which has two of its sides alone equal, and a scalene triangle that which has its three sides unequal.

21. Furthermore, of trilateral figures, a right-angled triangle is that which has a right angle, and an obtuse-angled triangle is that which has an obtuse angle, and an acute-angled triangle is that which has its three angles acute.

22. Of quadrilateral figures, a square is that which is both equilateral and right-angled, an oblong is that which is right-angled but not equilateral, a rhombus that which is equilateral but not right-angled, and a rhomboid is that which has its opposite sides and angles equal to one another but is neither equilateral nor right-angled. And let quadrilaterals other than this be called trapezia.

23. Parallel straight lines are straight lines which, being in the same plane, and being produced indefinitely in both directions and do not meet one another in either direction.

Postulates:

1. To draw a straight line from any point to any point.
2. To produce a finite straight line continuously in a straight line.
3. To describe a circle with any center and distance.
4. That all right angles are equal to one another.
5. If a straight line falling on two straight lines make the interior angles on the same side less than two right angles, the two straight lines, if extended indefinitely, meet on that side on which the angles are less than the two right angles.

Common Notions (Axioms):

1. Things which are equal to the same thing are also equal to each other.
2. If equals are added to equals, the wholes are equal.
3. If equals be subtracted from equals, the remainders are equal.
4. Things which coincide with each other are equal to each other.
5. The whole is greater than the part.

In the 1800s it became increasingly apparent that Euclid had not explicitly stated all of his assumptions and that the definitions he used were not mathematically acceptable. Thus, a number of criticisms surfaced, and several axioms have now been added to the list of postulates and common notions that Euclid originally created. The following three axioms are now often assumed, in some form, when studying Euclidean geometry. The first of these is a completeness assumption that, in essence, says that a straight Euclidean line is essentially a copy of the real number system. In particular, the points along a straight Euclidean line are more than essentially just a copy of the rational numbers. The second is the SAS axiom, which says that for triangles, side-angle-side guarantees congruence. The argument that Euclid gave for his proposition stating that SAS yields congruent triangles is simply not a valid argument. The third axiom is the plane separation axiom which, in essence, states that if a line crosses another, then from then on it stays on that side of the line. The reader is warned that different books may state equivalent axioms using different approaches.

Axiom of Completeness

Given a line $\overleftrightarrow{AB}$, let each point $C \in \overleftrightarrow{AB}$ on the same side of A as point B be assigned the positive number AC, which is the Euclidean distance from A to C. Let A be assigned the number zero. Let each point $D \in \overleftrightarrow{AB}$ that is not on the same side of A as B be assigned the negative number $-AD$, where AD represents the distance from A to D. Then this axiom states that the above correspondence between real numbers and points on the line $\overleftrightarrow{AB}$ is 1-1 and onto. In other words, for each real number, there is a unique point determined on the line, and for each point on the line $\overleftrightarrow{AB}$, there is a unique real number determined.

SAS Axiom

Given two triangles with two pairs of sides that are congruent and with the included angles being congruent, the two triangles are congruent.

Plane Separation Axiom

Given a line ℓ in a plane, the points of the plane off ℓ may be divided into two disjoint sets H_1 and H_2 such that if $A \in H_1$ and $B \in H_2$, then the line segment $\overline{AB}$ joining A and B has a point in common with the line ℓ. Also, if A and B are in the same set H_i, then the line segment $\overline{AB}$ has an empty intersection with the line ℓ.

In today's world, all of Euclid's postulates and common notions would be lumped together and called either *postulates* or *axioms*. An **axiom** (or **postulate**) is a statement that is assumed to be true without proof. The words *axiom* and *postulate* are generally used interchangeably.

Euclid's approach is known as the **axiomatic approach**. Basically his idea was to lay down a set of axioms and then prove results based on these assumptions. The modern approach is to take certain words used in the axioms as undefined. In geometry these undefined words include *point* and *line*. Note that when one takes the analytic approach to geometry, then meanings are assigned to words such as *point* and *line*. For example, one can use the xy-plane (i.e., $R^2 = \{(x, y) \mid x, y \in R^1\}$) as a model for the Euclidean plane. Then a point is defined (in this model) as an ordered pair (x, y) of real numbers. Furthermore, a line is defined as a set of points of the following form:

$$\{(x, y) \mid Ax + By + C = 0, \text{ where at least one of } A \text{ or } B \text{ is not zero}\}.$$

Evidently, Euclid had certain reservations about his fifth postulate, which is known as the parallel postulate. In particular, he did not use the fifth postulate in establishing his first twenty-eight results, which he called *propositions*. At some point he needed to use the parallel postulate to obtain important results, such as the angle sum of a triangle being equal to two right angles (i.e., 180°). Many authors tried to prove that the parallel postulate could be established using Euclid's other assumptions, but they were unsuccessful. Eventually, in 1829, a Russian mathematician, Lobachevski, published an article suggesting that an alternative to Euclidean geometry could be constructed using an alternative to the fifth postulate. Lobachevski's alternative was that, in a plane, if one has a point P not on a line ℓ, then there will be more than one parallel to ℓ through P. This was the start of non-Euclidean geometry. The geometry that Lobachevski published as an alternative to Euclidean geometry is called *Lobachevskian geometry* in the former Soviet Union and called *hyperbolic geometry* in the rest of the world. It is important to keep in mind that one can have more than one geometry, and what is true in one geometry may or may not be true in another geometry.

RELATED READING FOR APPENDIX II

Heath, Sir T. *Euclid: The Thirteen Books of the Elements, Vol. 1.* 2nd ed. New York: Dover Publications, 1956.

APPENDIX III

Selected Geometry Formulas

Distance in R^2

$$PQ = \sqrt{(x_1 - x_2)^2 + (y_1 - y_2)^2}$$
$$P = (x_1, y_1) \text{ and } Q = (x_2, y_2)$$

This can be seen by applying the Pythagorean theorem to the triangle with vertices P, Q, and $S = (x_2, y_1)$.

Distance in R^3

$$PQ = \sqrt{\begin{array}{c}(x_1 - x_2)^2 + (y_1 - y_2)^2 \\ + (z_1 - z_2)^2\end{array}}$$
$$P = (x_1, y_1, z_1) \text{ and } Q = (x_2, y_2, z_2)$$

This can be seen using two applications of the Pythagorean theorem.

Triangles

Perimeter = sum of lengths of sides

Definition of perimeter.

$s = \frac{1}{2} \cdot (\text{perimeter}) = \text{semi-perimeter}$

Definition of semi-perimeter.

$A = \frac{1}{2} \cdot b \cdot h = \left(\frac{1}{2}\right) \cdot (\text{base}) \cdot (\text{height})$

See Chapter 2.

$A = \sqrt{s \cdot (s - a) \cdot (s - b) \cdot (s - c)}$

Heron's formula.

$c^2 = a^2 + b^2$

Pythagorean theorem where $m(\angle C) = 90°$. See Chapter 3.

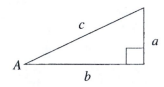

$\sin(A) = a/c = \text{opposite/hypotenuse}$

See Chapter 3.

$\cos(A) = b/c = \text{adjacent/hypotenuse}$

See Chapter 3.

$\tan(A) = a/b = \text{opposite/adjacent}$

See Chapter 3.

$c^2 = a^2 + b^2 - 2 \cdot a \cdot b \cdot \cos(C)$

Law of cosines (any triangle). See Chapter 3.

$$\frac{\sin(A)}{a} = \frac{\sin(B)}{b} = \frac{\sin(C)}{c}$$

Law of sines (any triangle). See Chapter 3.

Area

$A = b \cdot h = (\text{base}) \cdot (\text{height})$

For parallelograms (including rectangles). See Chapter 2.

$A = \frac{1}{2} \cdot b \cdot h = \frac{1}{2} \cdot (\text{base}) \cdot (\text{height})$

For triangles. See Chapter 2.

$A = \pi \cdot r^2 = (\pi) \cdot (\text{radius})^2$

For circles. See Chapter 2.

$A = \frac{\theta \cdot r^2}{2}$

Sector of a circle. See Chapter 2. Central angle θ in radians.

$A = \frac{1}{2} \cdot (b_1 + b_2) \cdot h = \frac{1}{2} \cdot$
(sum of bases) $\cdot$ (height)

For trapezoids. See Chapter 2.

Circles

$\pi = \frac{C}{d} = \frac{\text{Circumference}}{\text{Diameter}}$

Definition of π. See Chapter 2.

$C = \pi \cdot d = (\pi) \cdot (\text{diameter})$
$C = 2 \cdot \pi \cdot r = 2 \cdot (\pi) \cdot (\text{radius})$

From definition of π. Using definitions of π and of radius r. See Chapter 2.

$\theta = \frac{S}{r} = \frac{\text{arc subtended}}{\text{radius of circle}}$

Definition of angular measure using radians. See Chapter 1 and Chapter 2.

$S = r \cdot \theta = (\text{radius}) \cdot (\text{angle in radians})$

From definition of θ in radians.

Surface Area

$S_{\text{lateral}} = \pi \cdot r \cdot \ell = (\pi) \cdot (\text{radius}) \cdot$
(slant height)

Lateral surface area for a right circular cone. See Chapter 3.

$S_{\text{total}} = \pi \cdot r \cdot \ell + \pi \cdot r^2 =$
$S_{\text{lateral}} + (\text{area of base})$

Total surface area for a right circular cone. See Chapter 3.

$S_{\text{lateral}} = 2 \cdot \pi \cdot r \cdot h = 2 \cdot \pi \cdot$
(radius) $\cdot$ (height)

Lateral surface area for a right circular cylinder. See Chapter 2 and Chapter 3.

$S_{\text{total}} = 2 \cdot \pi \cdot r \cdot h + 2 \cdot \pi \cdot r^2 =$
$S_{\text{lateral}} + (2 \text{ bases})$

Total surface area for a right circular cylinder. See Chapter 2 and Chapter 3.

$S = 4 \cdot \pi \cdot r^2$

Surface area for a sphere of radius r. See Chapter 2.

Volume

$V = \ell \cdot w \cdot h = (\text{length}) \cdot (\text{width}) \cdot$
(height)

Volume of a box (also known as a *right rectangular prism*). See Chapter 2.

$V = B \cdot h = (\text{area of base}) \cdot (\text{height})$

Volume of a prism. See Chapter 2.

$V = \pi \cdot r^2 \cdot h = (\pi) \cdot (\text{radius})^2 \cdot$ (height)

Volume of a right circular cylinder. See Chapter 2 and Chapter 3.

$V = \frac{1}{3}B \cdot h = \frac{1}{3}(\text{area of base}) \cdot$ (height)

Volume of a cone or pyramid. See Chapter 2.

$V = \frac{1}{3} \cdot \pi \cdot r^2 \cdot h = \left(\frac{\pi}{3}\right) \cdot (\text{radius})^2 \cdot$ (height)

Volume of a right circular cone. See Chapter 2 and Chapter 3.

$V = \frac{4}{3} \cdot \pi \cdot r^3$

Volume of a sphere of radius r. See Chapter 2.

Complex Numbers

$i^2 = -1$

The number i is a square root of -1. See Chapter 5.

$(a + bi) + (c + di) = (a + c) + i(b + d)$

Addition of complex numbers. See Chapter 5.

$(a + bi) \cdot (c + di) = (ac - bd) + i(ad + bc)$

Multiplication of complex numbers. See Chapter 5.

$\overline{(a + bi)} = a - bi$

Complex conjugate. See Chapter 5.

$|(a + bi)| = \sqrt{a^2 + b^2}$

Absolute value of a complex number is its length, where the complex number is considered a vector. See Chapter 5.

$e^{i\theta} = \cos(\theta) + i\sin(\theta)$

Euler's identity. See Chapter 5.

Glossary

Acute angle

An acute angle is one with measure less than $90°$.

Adjacent angles

Two angles with the same vertex that share a common ray and that do not have any interior points in common are called adjacent angles.

Altitude

A line segment from one vertex of a triangle to the line containing the opposite side and perpendicular to that line is an altitude.

Angle

An angle is the union of two rays with the same endpoint and with no other points in common. The common endpoint is the angle's vertex and the rays are the angle's sides. Angles may be measured in degrees or radians. If the rays point in opposite directions, then the angle is called a **straight angle** and has measure $180°$ (π radians). If the rays are not opposite, then the interior of the angle is the set of points between the rays. If the rays are perpendicular, then the angle is $90°$ ($\pi/2$ radians).

Axiom

A statement that is assumed to be true without proof is an axiom (or postulate). The words axiom and postulate are used interchangeably in mathematics. However, for historical reasons, when teaching Euclidean geometry, one often makes an artificial distinction and reserves the word postulate for Euclid's first five assumptions and reserves the word axiom for Euclid's second set of assumptions (i.e., his "common notions").

Bisector

Given a line segment $\overline{AB}$, the **bisector** or **midpoint** of the line segment $\overline{AB}$ is the point D such that $AD = BD = \frac{1}{2}AB$. The **perpendicular bisector** of the segment $\overline{AB}$ is the line that passes through the midpoint D of $\overline{AB}$ and is perpendicular to $\overline{AB}$ at D. Given an angle $\angle ABC$, the **bisector of this angle** is a ray $\overrightarrow{BD}$ such that the point D lies interior to the angle $\angle ABC$, and $m(\angle ABD) = m(\angle DBC) = \frac{1}{2}m(\angle ABC)$. Given a triangle, the angle bisectors of a triangle meet at a point called the **incenter** of the triangle. The largest circle inside the triangle has a center at the incenter. The perpendicular bisectors of the triangle's sides meet at a point called the **circumcenter** of the triangle. The smallest circle that contains the triangle has a center at the circumcenter.

Cavalieri's principle

Cavalieri's principle in the plane states that given a plane figure and a line in the plane, if some or all line segments in this figure parallel to the given line are shifted in the plane and remain parallel to the given line, then the resulting figure has the same area as the original figure. Cavalieri's principle in space states that given a solid figure and a plane, if some or all plane sections in this figure parallel to the given plane are shifted in space parallel to the given plane, then the resulting figure has the same volume as the original figure.

Centimeter

A centimeter (denoted cm) is a measure of length. One hundred centimeters is one meter. One inch is equal to 2.54 centimeters. One centimeter is approximately the width of a piece of chalk. Area may be measured in square units. Thus, one cm^2 (i.e., one square centimeter) is the area of a square with sides of length 1 cm. Volume may be measured in cubic units. Thus, one cm^3 (i.e., one cubic centimeter) is the volume of a cube with edges of length 1 cm.

Central angle

Given a circle with center O, then an angle $\angle AOB$ with O as vertex is called a *central angle*. If the rays $\overrightarrow{OA}$ and $\overrightarrow{OB}$ intersect the circle at points A' and B', respectively, then let $\overarc{A'B'}$ denote the arc of the circle from A' to B', intersecting the interior and sides of the angle $\angle AOB$. The **angular measure** of $\overarc{A'B'}$ is the same as the measure of the angle $\angle AOB$. In other words, $m(\overarc{A'B'}) = m(\angle AOB)$. Since angles may be measured in both degrees and radians, the angular measure of an arc may also be measured in either degrees or radians.

Chord

Given a circle, a line segment joining two points of the circle is called a *chord*.

Circle

Given a point O and a positive number r, the circle with **center** O and **radius** r is defined to be the set of all points at distance r from O (i.e., $\{P \mid OP = r\}$). Notice that the circle itself is just the "edge." The **interior** of the circle is the set of points at a distance less than r from O and includes the center point O (i.e., interior $= \{Q \mid OQ < r\}$). The **circumference** of the circle is given by $C = 2\pi r$. The **area** of the circle is the area of its interior and is given by $A = \pi r^2$. The **diameter** of the circle is twice the radius $d = 2r$. Actually, the word radius may refer to the number r or may refer to a segment with one endpoint at the center O and the other endpoint on the circle. Likewise, the word diameter may refer to the number d or may refer to a chord that contains the center O. A **tangent** is a line that intersects the circle in exactly one point. A **secant** is a line that intersects the circle in exactly two points.

Collinear

A set of points in the Euclidean plane (or space) is said to be collinear if there is a line that contains all the points of the set. A set of points on a sphere are said to be collinear in spherical geometry if they all lie on some great circle of the sphere.

Column matrix

A matrix of the form $k \times 1$ is called a *column matrix* (also known as a **column vector**). For $k = 2$, any matrix of the form $\begin{bmatrix} x \\ y \end{bmatrix}$ where x and y are real numbers is a column matrix.

Column vector

A column matrix is also known as a column vector. Geometrically, one may interpret a column vector as an oriented line segment with the tail at the origin and the head at the point with corresponding components. For $k = 2$, the column vector $\begin{bmatrix} x \\ y \end{bmatrix}$ represents the oriented line segment with the tail at the origin $O = (0,0)$ and the head at the point $P = (x, y)$. One writes $\overrightarrow{OP} = \begin{bmatrix} x \\ y \end{bmatrix}$ and says that $\begin{bmatrix} x \\ y \end{bmatrix}$ is the coordinate representation of the vector.

Common notions

After stating his five postulates, Euclid stated five common notions. His common notions are often called **axioms**. In doing Euclidean geometry, one accepts Euclid's five postulates and five common notions as true without proof. The words *axiom* and *postulate* are generally used interchangeably in mathematics, but in geometry one often reserves the word *axiom* for Euclid's common notions.

Complementary

Two angles are said to be complementary if their measures add up to $90°$ ($\pi/2$ radians).

Complex number

A complex number is a number of the form $x + iy$, where x and y are real and i is the **unit imaginary number** ($i^2 = -1$). Since complex multiplication is commutative, the i may be placed on either side of the y. Thus, $x + iy = x + yi$. In general, the complex number $x + iy$ is identified with the point (x, y) in the xy-plane. If $x = 0$, the number is said to be **pure imaginary** and may be graphed as a point on the y-axis. If $y = 0$, then the number is said to be **real** and may be graphed as a point on the x-axis.

Congruent angles

Two angles are said to be **congruent** if they have the same measure. Thus, $\angle A \cong \angle B$ iff $m(\angle A) = m(\angle B)$.

Congruent figures

For figures that are not polygons, one defines congruence without reference to angles. Two figures (i.e., sets) T_1 and T_2 are congruent if there is a correspondence that preserves distance. In other words, for each pair of points A, B in T_1 that correspond, respectively, to A' and B' in T_2, one has $AB = A'B'$. One often writes $CPCF$ to note that one has congruent parts when one has congruent figures.

Congruent line segments

Two line segments are said to be congruent if they have the same length. Thus, $\overline{AB} \cong \overline{CD}$ iff $AB = CD$.

Congruent polygons

Let G_1 and G_2 be polygons with the same number of vertices and assume there is a correspondence between their vertices such that each pair of vertices joined by a side of G_1 corresponds to vertices joined by a side of G_2. The correspondence yields congruent polygons if each pair of corresponding sides are congruent and each pair of corresponding angles are congruent. For example, if G_1 has vertices $\{P_1, P_2, \ldots, P_n\}$ and if G_2 has vertices $\{Q_1, Q_2, \ldots, Q_n\}$, then the correspondence $F(P_i) = Q_i$, taking each P_i to Q_i, will show that the polygons are congruent iff (1) each pair of corresponding sides are congruent (i.e., $\overline{P_iP_{i+1}} \cong \overline{Q_iQ_{i+i}}$ for $i = 1, 2, \ldots, n$) and (2) each pair of corresponding angles are congruent (i.e., $\angle P_iP_{i+1}P_{i+2} \cong \angle Q_iQ_{i+1}Q_{i+2}$ for $i = 1, 2, \ldots, n$). Here, $P_{n+1} = P_1, P_{n+2} = P_2, Q_{n+1} = Q_1$ and $Q_{n+2} = Q_2$. The notation is $P_1P_2 \ldots P_n \cong Q_1Q_2 \ldots Q_n$. Just as with congruence of triangles, the order in which the vertices are listed is important. Two polygons may be congruent with one correspondence of vertices and yet fail to be congruent with a different correspondence of vertices.

Congruent triangles

Two triangles are said to be congruent if all three pairs of corresponding sides are congruent and if all three pairs of corresponding angles are congruent. Thus, $\triangle ABC \cong \triangle DEF$ iff $\overline{AB} \cong \overline{DE}, \overline{AC} \cong \overline{DF}, \overline{BC} \cong \overline{EF}, \angle A \cong \angle D, \angle B \cong \angle E$, and $\angle C \cong \angle F$.

Conjugate

If $z = x + iy$ is a given complex number, then $\bar{z} = x - iy$ is its conjugate. Geometrically, the conjugate $\bar{z}$ may be obtained by reflecting z across the x-axis.

Coordinates

The xy-plane is given by $R^2 = \{(x, y) \mid x, y \text{ are real}\}$. A **point** is an ordered pair (x, y) of real numbers, and a **line** is a set of the form $\{(x, y) \mid Ax + By + C = 0\}$, where at least one of A or B is not zero. A line in the xy-plane may also be given parametrically by

$$x = a_1 t + b_1$$

$$y = a_2 t + b_2$$

where at least one of a_1 or a_2 is not zero and the parameter t takes on all real values (i.e., $-\infty < t < +\infty$). The **Euclidean distance** from $P = (x_1, y_1)$ to $Q = (x_2, y_2)$ in the xy-plane is given by $d_E(P, Q) = PQ = \sqrt{(x_1 - x_2)^2 + (y_1 - y_2)^2}$.

Three-dimensional xyz-space is given by $R^3 = \{(x, y, z) \mid x, y, z \text{ are real}\}$. A point is an ordered triple (x, y, z) of real numbers and a **plane** in xyz-space is a set of the form $\{(x, y, z) \mid Ax + By + Cz + D = 0\}$, where at least one of $A, B,$ or C is not zero. A line is given parametrically in xyz-space by

$$x = a_1 t + b_1$$

$$y = a_2 t + b_2$$

$$z = a_3 t + b_3$$

where at least one of $a_1, a_2,$ or a_3 is not zero and the parameter t takes on all real values (i.e., $-\infty < t < +\infty$). The Euclidean distance from $S = (x_1, y_1, z_1)$ to $T = (x_2, y_2, z_2)$ in R^3 is given by

$$d_E(S, T) = ST = \sqrt{(x_1 - x_2)^2 + (y_1 - y_2)^2 + (z_1 - z_2)^2}.$$

Degrees

The two most common units for measuring angles are degrees and radians. If a regular polygon with exactly 360 sides is inscribed in a circle, then each side of this polygon will be a chord, and the (smaller) arc determined by a chord will subtend a central angle of $1°$ (i.e., 1 degree). Thus, a right angle has measure $90°$, and a straight angle has measure $180°$. Since a circle of radius 1 has circumference of 2π, it follows that a straight angle (corresponding to half a circle of radius 1) has measure π radians. Thus, one writes $180(\text{degrees}) = \pi(\text{radians})$. If a given angle has measure $A°$, then its measure in radians is $A \cdot (\pi/180)$ (radians). If an angle has measure B radians, then its measure in degrees is $B \cdot (180/\pi)$ (degrees). One degree is equal to 60 minutes ($1° = 60'$) and 1 minute is equal to 60 seconds ($1' = 60''$).

Diameter

Given a circle with center O and radius r, the diameter d is $2r$. Also, a chord that contains O is called a *diameter*. Thus, the word *diameter* may refer to either a chord passing through the center of the circle or the length of such a chord.

Dilation

If a positive number S and fixed point O are given, then a transformation known as a dilation is determined by mapping each point P to a point P' on the ray $\overrightarrow{OP}$ such that $OP' = S \cdot OP$. The point O is called the **center** and the number S is known as either the **dilation factor** or scale factor. In particular, if $S = 2$, then each point P, other then the center O, is moved radially away from O to a new point P' which is twice as far from O as P. Dilations are special cases of similarity transformations.

Distance

The distance PQ from point P to point Q is the length of the line segment $\overline{PQ}$ joining the two points.

For points $P = (x_1, y_1)$ and $Q = (x_2, y_2)$ in the xy-plane, the distance from P to Q is given by $PQ = \sqrt{(x_1 - x_2)^2 + (y_1 - y_2)^2}$.

For points $P = (x_1, y_1, z_1)$ and $Q = (x_2, y_2, z_2)$ in xyz-space, the distance from P to Q is given by $PQ = \sqrt{(x_1 - x_2)^2 + (y_1 - y_2)^2 + (z_1 - z_2)^2}$.

Elements, The

Euclid wrote thirteen books called *The Elements* in about 300 BC. These books systematized geometry known at that time. They contain a great deal of geometry and several other topics in mathematics.

Equiangular triangle

A triangle with all three angles of the same measure is **equiangular**. In Euclidean geometry, a triangle is equiangular iff it is equilateral.

Equilateral triangle

A triangle with all three sides of the same length is **equilateral**. In Euclidean geometry, a triangle is equilateral iff it is equiangular.

Equivalence relation

A relation $\sim$ on a set S is said to be an equivalence relation if it satisfies the reflexive, symmetric, and transitive laws. The **reflexive law** is that $a \sim a$ for all $a \,\varepsilon\, S$. The **symmetric law** is that $a \sim b$ implies $b \sim a$ for all $a, b \,\varepsilon\, S$. The **transitive law** is that $a \sim b$ and $b \sim c$ implies $a \sim c$ for all $a, b, c \,\varepsilon\, S$.

Euler number

Given a polyhedron, let the number of faces be denoted by F, let the number of edges be denoted by E, and let the number of vertices be denoted by V. The number $F - E + V$ is called the *Euler number*. This number is also sometimes called the **Euler characteristic** and can be expressed as either $F - E + V$ or as $F + V - E$. If the polyhedron can be topologically deformed into a sphere (i.e., can be deformed into a sphere in a continuous fashion with stretching and bending but without tearing), the Euler number is 2. Thus, the Euler number for a prism is

2, and the Euler number for a pyramid is also 2. The Euler number for a torus (i.e., a doughnut-shaped surface) is zero.

Euler's identity

Let i be the unit imaginary number. Thus, $i^2 = -1$, and let e be the base of the natural logarithms. The number e is irrational and is given approximately by $e \approx 2.71828$. Euler's identity is $e^{i\theta} = \cos(\theta) + i\sin(\theta)$, where θ is measured in radians.

Function

A **function** $f\colon A \to B$ is a way of relating elements of the set A to elements of the set B. Each $a \in A$ must correspond to exactly one element $b \in B$. The element b in B corresponding to a in A is denoted by $f(a)$. It may happen that some elements $b \in B$ may not be the image of any element $a \in A$. Also, it may happen that some elements $b \in B$ are the image of more than one element in A.

Hyperbolic geometry

The postulates for hyperbolic geometry are the same as for Euclidean geometry with one exception. The parallel postulate (i.e., Euclid's fifth postulate) is replaced with an alternative which is often called the **hyperbolic parallel postulate**. There are several models of two-dimensional hyperbolic geometry. In this book we have used the **Poincaré upper half-plane** model. The points are taken to be the points in the upper half of the xy-plane (i.e., $y > 0$). The hyperbolic lines in this model are defined as (1) the intersection of the upper half of the xy-plane with Euclidean circles having center on the x-axis and (2) the intersection of the upper half plane with Euclidean lines that have undefined slope in the xy-plane.

Hyperbolic parallel postulate

Given a point P off a line ℓ in the hyperbolic plane, the hyperbolic parallel postulate states that there are at least two lines containing P which are parallel to ℓ. This postulate actually implies that there are an infinite number of lines parallel to ℓ through each point off ℓ.

Inscribed angle

Given a circle, an inscribed angle is one with its vertex on the circle and such that each of the two rays making up the angle have nonempty intersection with the interior of the circle. If $\overline{BA}$ and $\overline{BC}$ are chords of the circle, then the angle $\angle ABC = \overrightarrow{BA} \cup \overrightarrow{BC}$ is an inscribed angle. In other words, an inscribed angle is one with its vertex on the circle and such that the angle points "into" the circle.

Interior

The **interior of a line segment** consists of the segment points that are not endpoints. Thus, given the segment $\overline{PQ}$, the interior consists of the points $S \in \overline{PQ}$ such that $S \neq P$ and $S \neq Q$. If a circle has center O and radius r, then the **interior of the**

circle consists of the points at distance less than r from O. Thus, the interior is the set $\{P \mid OP < r\}$. Given an angle $\angle ABC = \overrightarrow{BA} \cup \overrightarrow{BC}$, the **interior of the angle** consists of the points that lie interior to segments that do not have B as an endpoint but that do have one endpoint on the ray $\overrightarrow{BA}$ and one endpoint on the ray $\overrightarrow{BC}$. The **interior of a triangle** $\triangle ABC$ is the set of points trapped inside the union of the three sides. The interior of a triangle does not include points on the sides of the triangle.

Isometry

An isometry is a transformation that **preserves distance**. In other words, if P and Q correspond respectively to P' and Q', one must have $PQ = P'Q'$. Isometries take congruent figures to congruent figures. They preserve angles, lengths of curves and areas, and distance. Some isometries preserve orientation and some do not. An isometry that is orientation-reversing takes figures to their mirror images. Isometries are also called *rigid motions*.

Isosceles trapezoid

An isosceles trapezoid is a trapezoid that is not a parallelogram and that has the two nonparallel sides of equal length.

Isosceles triangle

A triangle is **isosceles** if it has at least two sides that are of the same length. The triangle's third side may or may not be the same length as the first two. A triangle with all three sides of the same length is equilateral. An equilateral triangle is a special case of an isosceles triangle.

Kilometer

A kilometer (denoted km) is a measure of length. One kilometer is 1,000 meters long. A kilometer is approximately .62 miles.

Lateral surface

Intuitively, the lateral surface of an object in three dimensions is its "side" surface. A **pyramid** has a base that is a polygon in a given plane and a (top) vertex off the plane of the polygonal base. The **lateral** or "side" surface of the pyramid is formed by line segments from the (top) vertex to the edges of the polygonal base. In other words, the lateral surface of a pyramid includes all but the pyramid's base. A **prism** has two congruent bases in parallel planes. The other faces of the prism form the **lateral** surface of the prism and are required to be parallelograms. A **cone** has a base in a given plane and a (top) vertex off the plane of the base. The lateral or "side" surface of the cone is that part of the total surface that does not include the base. The formula for the lateral surface area of a **right circular cone** of radius of base r and slant height ℓ is given by the formula $S_{\text{lateral}} = \pi \cdot r \cdot \ell$. The lateral surface of a cylinder includes all but the two bases of the cylinder. The lateral surface area of a **right circular cylinder** of height h and radius of base r is given by the formula $S_{\text{lateral}} = 2 \cdot \pi \cdot r \cdot h = 2 \cdot \pi \cdot (\text{radius}) \cdot (\text{height})$.

Law of cosines

Given any triangle $\triangle ABC$, the Law of cosines states that $c^2 = a^2 + b^2 - 2 \cdot a \cdot b \cdot \cos(C)$.

Law of sines

Given any triangle $\triangle ABC$, the Law of sines states that $\frac{\sin(A)}{a} = \frac{\sin(B)}{b} = \frac{\sin(C)}{c}$.

Line

The shortest curve joining two points on a line is a line segment lying on the line. Given two distinct points A and B, there is a unique line containing both of them. This line is denoted by $\overleftrightarrow{AB}$.

Linear pair

Two adjacent angles that are supplementary (i.e., have measures that sum to $180°$) form a linear pair. If $\angle ABC$ and $\angle CBD$ are adjacent angles, then they form a linear pair if $\angle ABD$ is a straight angle (i.e., $B \in \overleftrightarrow{AD}$).

Line segment

The line segment with endpoints A and B is denoted by $\overline{AB}$. The line segment $\overline{AB}$ is the shortest curve joining the two points A and B.

Matrix

A matrix is an array of numbers arranged as a rectangle. If the rectangle's height is k and the length is r, then the matrix is said to be $k \times r$. In general, one writes a_{ij} for the element of matrix A, which lies in the ith row and in the jth column.

Meter

A meter is a unit of length and is denoted by the single letter m. A meter is equal to 100 centimeters. Also, 1 meter ≈ 39.37 inches. Thus, a meter is slightly longer than 1 yard. Area may be measured in square meters (m^2), and volume may be measured in cubic meters (m^3). Notice that $1 \text{ m} = 10^2 \text{ cm} = 100 \text{ cm}$, $1 \text{ m}^2 = 10^4 \text{ cm}^2 = 10,000 \text{ cm}^2$, and $1 \text{ m}^3 = 10^6 \text{ cm}^3 = 1,000,000 \text{ cm}^3$.

Net

A flat diagram which can be folded to form a solid figure is known as a net. Nets are especially helpful in illustrating volume and surface area. They are also known as **flat patterns**.

Noncollinear

A set of points is noncollinear if no line contains all of the set's points. Thus, three points are noncollinear if any line containing two of them fails to contain the third.

Obtuse angle

An obtuse angle is one with measure greater than $90°$.

Oriented line segment

Given a line segment $\overline{PQ}$, it may be ordered with the tail at P and the head at Q. It can also be oriented with the tail at Q and the head at P. The line segment $\overline{PQ}$ oriented with the tail at P and the head at Q is denoted by $\overrightarrow{PQ}$. The line segment $\overline{PQ}$ oriented with the tail at Q and the head at P is denoted by $\overrightarrow{QP}$. An oriented line segment is called a **vector**, and one writes $\overrightarrow{QP} = -\overrightarrow{PQ}$. Notice that the same notation $\overrightarrow{PQ}$ is used both for the vector with the tail at P and the head at Q and for the ray with endpoint P containing Q. This rarely causes confusion since one rarely considers rays and vectors at the same time.

Parallel

Two lines in the same plane are said to be parallel if they don't intersect (i.e., if their intersection is the empty set). When studying plane geometry, one understands that lines under consideration are all lying in the same plane, and one defines parallel lines as nonintersecting lines. In higher-dimensional Euclidean spaces, two nonintersecting lines that fail to lie in a common plane are said to be skew.

Parallelogram

A quadrilateral with each pair of opposite sides parallel is called a parallelogram.

Parallel postulate

Both Euclid's fifth postulate and Playfair's postulate are referred to as *the* parallel postulate. These two postulates are equivalent given the other axioms and postulates of Euclidean geometry. **Playfair's postulate:** If a point P in a given plane lies off the line ℓ, then there is exactly one line through P that is parallel to ℓ in this plane. **Euclid's fifth postulate:** If a straight line falling on two straight lines makes the interior angles on the same side less than two right angles, the two straight lines, if extended indefinitely, meet on that side on which the angles are less than the two right angles.

Perimeter

Given a figure in a plane, the perimeter is the length of the figure's boundary. For example, the perimeter of a polygon with vertices $P_1, P_2, \ldots, P_n$ is the sum of the lengths of the sides. Thus, in this case, the perimeter p is given by $p = \sum_{i=1}^{i=n} P_i P_{i+1}$. In addition to using the word *perimeter* to denote the length of the boundary, one sometimes uses the word *perimeter* to refer to the boundary itself.

Perpendicular

Two intersecting lines are said to be perpendicular if they form four right angles at their point of intersection.

Perpendicular bisector

The perpendicular bisector of the segment $\overline{AB}$ in the Euclidean plane is the line ℓ which passes through the midpoint of the segment and is also perpendicular to the segment. The points of the line ℓ are the points that are equally distant from the endpoints A and B of the original segment.

Playfair's postulate

Given a point P off a line ℓ in the Euclidean plane, Playfair's postulate states that there is exactly one line containing P which is parallel to ℓ. Playfair's postulate may be substituted for Euclid's fifth postulate.

Polar coordinates

Polar coordinates (r, θ) are defined in the plane by letting r be the distance from the origin and letting θ be the angle made with the positive x-axis. Thus, $r = \sqrt{x^2 + y^2}, x = r\cos(\theta)$ and $y = r\sin(\theta)$.

Polygon

A polygon is a union of line segments $\overline{P_1P_2} \cup \overline{P_2P_3} \cup \ldots \cup \overline{P_{n-1}P_n} \cup \overline{P_nP_1}$ with no three consecutive vertices collinear and no two sides meeting except at a vertex and only when the sides are in consecutive order. The **vertices** are the points $P_1, P_2, \ldots, P_n$ and one sets $P_{n+1} = P_1$. The polygon is **convex** if it is contained by the sides and interior of each of its angles $\angle P_{i-1}P_iP_{i+1}$. The perimeter of a polygon with vertices $P_1, P_2, \ldots, P_n$ is the sum of the lengths of the sides.

Polyhedron

A polyhedron is the surface of a solid figure in three dimensions that has faces that are polygons. A tetrahedron has four faces, a pentahedron has five faces, a hexahedron has six faces, etc.

Postulate

A postulate is a statement that is assumed to be true without proof. The words *postulate* and *axiom* are used interchangeably in mathematics. However, for historical reasons, when teaching Euclidean geometry, one often makes an artificial distinction and reserves the word *postulate* for Euclid's first five assumptions and reserves the word *axiom* for Euclid's second set of assumptions (i.e., his "common notions").

Prism

A prism is a solid with two faces that are congruent polygons and lie in parallel planes. These two faces are called **bases**, and the other faces (**lateral faces**) are required to be parallelograms. When the lateral faces are rectangles, the prism is called a **right prism**. Otherwise, the prism is called an **oblique prism**. Prisms are named according to the polygonal shape of their bases. Thus, a triangular prism is one where the bases are triangles. The **height** of a prism is the (perpendicular) distance between the parallel planes containing the bases. The **volume** of a prism is the area of a prism's base times the prism's height.

Pyramid

A pyramid is a solid formed by taking a polygon in one plane, selecting a point off the polygon's plane, and then joining the polygon's vertices to the point selected off the plane. The point off the plane is called the (top) **vertex**, and the height of the pyramid is the (perpendicular) distance from the vertex's top to the polygon's plane. The polygon is called the base and the lateral faces of the pyramid are triangles with one vertex equal to the top vertex of the pyramid. Pyramids are named by the shape of the base.

Quadrilateral

A polygon with exactly four sides is a quadrilateral.

Radians

The two most common units for measuring angles are degrees and radians. If a circle of radius r has a central angle that subtends an arc of length S, then the radian measure of this angle is $\theta = S/r$. Since a circle of radius one has circumference of 2π, it follows that a straight angle (corresponding to half a circle of radius one) has measure π radians.

Radius

Given a circle with point O as center, the radius r is the distance from O to any point on the circle. The word *radius* may refer to the number r or it may refer to a segment with one endpoint at the center O and the other endpoint on the circle. In three dimensions, the object corresponding to a circle is a sphere. One defines center and radius for a sphere in three dimensions in basically the same way one defines these for a circle in two dimensions.

Ray

A **half line** with endpoint P is said to be a ray with endpoint P. The ray with endpoint P that contains the point Q is denoted by $\overrightarrow{PQ}$. The same notation $\overrightarrow{PQ}$ is used both for the vector with the tail at P and the head at Q and for the ray with endpoint P containing Q.

Rectangle

A rectangle is defined to be a parallelogram with at least one right angle. It can be proven that a rectangle has right angles for all four of its angles.

Reflection across a line

Given a line m in the Euclidean plane and a point P off m, draw a perpendicular from P to the point F on m. The point F is the closest point on m to P and is said to be the **foot** of P on m. Extend the segment $\overline{PF}$ to get P' on the other side of m from P and such that $PF = P'F$. The reflection across the line m takes P to P' and takes P' to P. It leaves each point of m fixed.

Rhombus

A rhombus is a parallelogram with two adjacent sides of equal length. It can be proven that all four sides of a rhombus are of equal length.

Right angle

A right angle is one with measure equal to $90°$. When two lines are perpendicular, then they form four right angles at the point of intersection.

Right triangle

A triangle with a right angle is called a *right triangle*. If $\angle C$ in $\triangle ABC$ is a right angle, then the Pythagorean theorem states that $a^2 + b^2 = c^2$.

Rotation

If C is a point in the Euclidean plane, the rotation by angle α about C is denoted by $R_{C,\alpha}$. Angles are considered positive when measured counterclockwise and negative when measured clockwise. Using degrees, a clockwise rotation by α is equivalent to a counterclockwise rotation by $360° - \alpha$.

Row matrix

A matrix that is $1 \times k$ is called either a *row matrix* or a **row vector**.

Scale factor

If two figures T_1 and T_2 are similar (denoted $T_1 \sim T_2$), the scale factor S is the ratio $A'B'/AB$, where A and B are distinct points of T_1 and A' and B' are their respective images in T_2. If $T_1 \sim T_2$ with scale factor S, then $T_2 \sim T_1$ with scale factor $1/S$.

Scalene triangle

A triangle with all three sides of different lengths is **scalene**.

Sector

If a circle is given with center O and a central angle $\angle AOB$, the corresponding sector consists of the intersection of the circle and its interior with the set formed by the interior of the angle and its sides. Intuitively, the sector circle is a pie-shaped wedge with the angle $\angle AOB$ as the "point" of the wedge.

Similar figures

Two figures T_1 and T_2 are said to be **similar** with scale factor S if there is a correspondence between them such that whenever A and B in T_1 correspond respectively to A' and B' in T_2, then the ratio $A'B'/AB$ is equal to the fixed number S. The general definition of *similar* does not explicitly require corresponding angles in the figures to be congruent; however, when defining *similar* for triangles and other polygons, one uses the requirement that corresponding angles be congruent.

Similar polygons

Two polygons are defined to be similar if their corresponding angles are congruent and if the corresponding sides are in the same ratio. More explicitly, let Q_1 and Q_2 be polygons and assume there is a correspondence between their vertices such that each pair of vertices joined by a side of Q_1 corresponds to vertices joined by a side of Q_2. If all angles at corresponding vertices are congruent and if the ratio S of corresponding sides (of polygon Q_2 to polygon Q_1) is the same for all pairs of corresponding sides, then the polygons are similar. The ratio S is called the *scale factor*.

Similar triangles

Two triangles are defined to be similar if their corresponding angles are congruent and if the corresponding sides are in the same ratio. The ratio S is called the *scale factor*. More explicitly, $\triangle ABC \sim \triangle A'B'C''$ with scale factor S means that $\angle A \cong \angle A', \angle B \cong \angle B', \angle C \cong \angle C'$, and $S = \frac{A'B'}{AB} = \frac{A'C'}{AC} = \frac{B'C'}{BC}$.

Sphere

Let O be a point in three-dimensional Euclidean space and let r be a positive number. The sphere with center O and radius r is defined to be the points in the three-dimensional space that are at exactly distance r from O. The surface area of a sphere of radius r is given by $S = 4 \cdot \pi \cdot r^2$. The volume of a sphere of radius r is given by $V = \frac{4}{3} \cdot \pi \cdot r^3$.

Spherical coordinates

Spherical coordinates (ρ, θ, ϕ) have ρ representing the distance of the point (x, y, z) from the origin $O = (0, 0, 0)$. The angle ϕ is the angle that the positive z-axis makes with the vector from the origin to the point (x, y, z). The angle coordinate θ is an angle measured in the xy-plane and is basically the same as the angle θ used in polar coordinates.

Spherical distance

Given two points P and Q on a sphere, the spherical distance between them is the length of a shortest arc of a great circle joining them. If they are distinct ($P \neq Q$) and are not diametrically opposed, there is exactly one great circle containing both of them, and in this case, the great circle containing both of them will have one shortest arc joining them. The length of this shortest arc will be the spherical distance between them. If they are diametrically opposed, then there are an infinite number of great circles joining them. The distance between P and Q when they are diametrically opposed is half the radius of a great circle. Thus, $PQ = \pi \cdot r$, where r is the sphere's radius and of each great circle on the sphere.

Spherical excess

Given a spherical triangle, the spherical excess is the angle sum less one straight angle. In other words, it is the excess amount over what a triangle would have in the Euclidean plane. If $\triangle ABC$ is a spherical triangle, then using radians, one has that the excess is given by $\varepsilon(\triangle ABC) = m(\angle A) + m(\angle B) + m(\angle C) - \pi$. If the sphere has radius r, then the **area of the spherical triangle** is its excess (in radians) times r^2. Area$(\triangle ABC) = \varepsilon(\triangle ABC) \cdot r^2$.

Spherical geometry

The points of spherical geometry are the points on the surface of a sphere and the lines of the geometry are the great circles of the sphere. **Great circles** are the largest circles on the sphere. They are formed by the intersection of the sphere with a plane though the center of the sphere. A small circle on the sphere is defined as a circle on the sphere formed by the intersection of sphere with a plane which does not pass through the center of the sphere.

Spherical triangle

On a sphere, a great circle is formed by intersecting the sphere with a plane through the sphere's center. The great circles are the "lines" of spherical geometry. Three points on a sphere are noncollinear if they all fail to lie on one great circle. Given three noncollinear points, then no two are diametrically opposed. Let A, B, C be such a noncollinear triple. The spherical triangle $\triangle ABC$ they determine is found by taking the shortest arcs of the great circles joining them in pairs. Thus, one takes the shortest great circle arc joining A to B, the shortest joining B to C, and the shortest joining C to A. These three arcs form the spherical triangle. The spherical triangle divides the sphere into two separate pieces. The smaller of these is said to be the **interior** of the spherical triangle $\triangle ABC$, and the larger of these is said to be the **exterior**.

Square

A square is a rectangle that is also a rhombus. Thus a square is a parallelogram with all four angles equal to 90° (i.e., $\pi/2$ radians) and all four sides of the same length.

Supplementary angles

Two angles are supplementary if the sum of the measures of their angles equals 180° (i.e., π radians).

Surface area

The total surface area of a polyhedron is the sum of the areas of the faces of the polyhedron. A solid that has one or two bases such as a pyramid, cone, prism or cylinder has both a total surface area and a **lateral** (that is side) surface area. For such a solid, the total surface area is the sum of the lateral surface area and the area of the base(s).

Symmetry

A symmetry of a given set is an isometry which takes the set onto itself. If a set is mapped onto itself by reflection across some line, then the line is a **line of symmetry**. If a set is mapped onto itself by a rotation about some given point, then the set is said to have a **rotational symmetry**.

Taxicab distance

The taxicab distance between the two points (x_1, y_1) and (x_2, y_2) in the xy-plane is given by $d_T((x_1, y_1), (x_2, y_2)) = |x_1 - x_2| + |y_1 - y_2|$.

Taxicab geometry

In taxicab geometry, angles and lines are the same as in the Euclidean plane. However, there is a fundamental difference in how distance is measured. In order to find the taxicab distance between the two points (x_1, y_1) and (x_2, y_2) in the xy-plane one adds the distance $|x_1 - x_2|$ between their x-components to the distance $|y_1 - y_2|$ between their y-components. Taxicab geometry shows the SAS axiom does not follow from the other axioms and postulates of Euclidean geometry.

Torus

A torus is a doughnut-shaped surface in three-dimensional space. If one has a circle and a line in the circle's plane that does not intersect the circle, then a torus is obtained when the circle is rotated in space about the line. The Euler number is zero for any surface that can be topologically deformed to a torus.

Transversal

A transversal is a line that "cuts" two other lines. More precisely, given two lines ℓ_1 and ℓ_2, a third line ℓ_3 is said to be a transversal if (1) it has nonempty intersection with each of ℓ_1 and ℓ_2, and if (2) $\ell_1 \cap \ell_2 \notin \ell_3$.

Trapezoid

A trapezoid is a quadrilateral that has at least one pair of opposite sides parallel. Two opposite sides that are parallel sides are called *bases*. A parallelogram is a special case of a trapezoid and either pair of opposite sides may be taken as the bases. The interior of a trapezoid is the area trapped inside. The area of a trapezoid is given by $A = \frac{1}{2} \cdot (b_1 + b_2) \cdot h = (\frac{1}{2})$ (sum of bases) $\cdot$ (height), where b_1 and b_2 are the lengths of the bases and h is the height (perpendicular distance between the bases).

Triangle

A triangle is defined to be the union of the three line segments joining three noncollinear points. Thus, if the three points A, B, C fail to be collinear, $\triangle ABC = \overline{AB} \cup \overline{AC} \cup \overline{BC}$. An equivalent definition is that a triangle is a polygon with exactly three sides. The triangle's perimeter is the sum of the lengths of its sides $p = AB + AC + BC$. The interior of the triangle is the set trapped inside. The area of the triangle is the area of its interior and is given by $A = (1/2) \cdot b \cdot h = (1/2) \cdot$ (base) $\cdot$ (height). A scalene triangle has all three sides of different lengths. An **isosceles** triangle has at least two sides of the same length. An **equilateral** triangle has all three sides of the same length. A **right triangle** has a right angle. A **median** of a triangle is a segment from the midpoint of a side to the opposite vertex. The three medians of a triangle meet at the **centroid**. An **altitude** of a triangle is a segment with one endpoint at a vertex and the other endpoint on the line containing the opposite side and is such that it is perpendicular to the opposite side. The three altitudes meet at a point called the **orthocenter**. The three perpendicular bisectors of the sides of a triangle meet at the circumcenter. The three angle bisectors of a triangle meet at the incenter. Two triangles are defined to be congruent if they have respective angles congruent and respective sides congruent.

Vector

An oriented line segment is called a *vector*. The starting endpoint is called the **tail** and the other endpoint is called the **head**. Vectors have **the movement property**. Thus, a vector may be moved so as to locate its tail at any desired point. When moving a vector, one must keep its length unchanged and must keep it pointing in the same direction. Using parallelograms, one may obtain a more precise description of how one moves a vector so that its tail has a new starting position.

Vertical angles

Two different angles that share a common vertex and whose sides form lines are said to be vertical. Thus, when two lines intersect, they will form two sets of vertical angles. It can be proven that vertical angles have the same measure.

Answers and Hints for Odd Exercises

EXERCISES 1.1

1. The intersection $\ell_1 \cap \ell_2$ has zero, one, or an infinite number of points.

EXERCISES 1.2

1. a. F
   b. T
   c. F
   d. T

3. If, $n \neq 40$, then $n = 30$ (False)

5. a. If $m(\angle A) < 90°$, then $\angle C$ in $\triangle ABC$ is a right angle. (False)
   b. If $m(\angle A) \geq 90°$, then $\angle C$ in $\triangle ABC$ is not a right angle. (True)
   c. If $\angle C$ in $\triangle ABC$ is not a right angle, then $m(\angle A) \geq 90°$. (False)

7. One example is the following: If a triangle has all three sides of the same length, then it has all three angles of the same measure.

EXERCISES 1.3

1. 5.58 feet (approx.) (≈ 5 ft. 7 in.). One approach to this problem is to use 1 meter $\approx$ 39.37 inches to obtain $1 \approx \dfrac{39.37 \text{ inches}}{1 \text{ meter}}$ and to use 1 foot $= 12$ inches to obtain $1 = \dfrac{1 \text{ foot}}{12 \text{ inches}}$. Multiplying 1.7 meters by the two fractions yields 1.7 meters $\approx$

$$(1.7 \text{ meters}) \cdot \left(\frac{39.37 \text{ inches}}{1 \text{ meter}}\right) \cdot \left(\frac{1 \text{ foot}}{12 \text{ inches}}\right) \approx 5.58 \text{ feet.}$$

3. $3\sqrt{2}$ units (≈ 4.243 units)

5. a. $x = 2, y = 3$
   b. $\sqrt{13}$ units

7. a. $2x - y + 0 = 0$
   b. $y = 2x$
   c. The graph is the straight line of slope 2, which passes through the origin.
   d. The line intersects the x-axis at $(0, 0)$.

9. 30 units.

EXERCISES 1.4

1. 14 units
3.

Polyhedron	Type of Faces	F (#faces)	E (#edges)	V (#vertices)	F − E + V
Tetrahedron	Equilateral triangles	4	6	4	2
Cube	Squares	6	12	8	2
Octahedron	Equilateral triangles	8	12	6	2
Dodecahedron	Regular pentagons	12	30	20	2
Icosahedron	Equilateral triangles	20	30	12	2

5. $F = 7, E = 15, V = 10,$ and $F - E + V = 2$

EXERCISES 1.5

1. 6,440 km (approx.)
3. a. 25,130 mi (approx.)
   b. 40,450 km (approx.)
5. 21,330 km (approx.)

CHAPTER 1 REVIEW EXERCISES

1. a. If the quadrilateral $ABCD$ has all of its sides of the same length, then it is a square. (False)
   b. If the sides of the quadrilateral $ABCD$ are not all of the same length, then it is not a square. (True)
   c. If the quadrilateral $ABCD$ is not a square, then its sides are not all of the same length. (False)
3. $\sqrt{2}$ units ≈ 1.414 units
5. 13 units
7. $8 + 2\sqrt{2}$ units
9. $F = 7, E = 12, V = 7,$ and $F - E + V = 2$
11. 1.6 million mi (approx.)

EXERCISES 2.1

1. $m(\angle A) < m(\angle B) < m(\angle C)$. The measure of $\angle C$ must be the largest since it lies across from the longest side, and the measure of $\angle B$ must be the second largest since it lies across from the second longest side. The measure of $\angle A$ must be the smallest since it lies across from the shortest side.
3. $m(\angle B) < m(\angle A) < m(\angle C)$

5. **a.** $GH = HK$
   b. $m(\angle G) = m(\angle K)$
   c. $m(\angle H) > m(\angle G) = m(\angle K)$

7. **a.** Hint: One may prove $\triangle HGE \cong \triangle HGF$ using SAS.
   b. $HE = HF$ [CPCF (corresponding parts of congruent figures are congruent) shows that $\overline{HE} \cong \overline{HF}$].
   c. Isosceles

9. This relation is transitive. It is not reflective and not symmetric.

11. **a.** This is an equivalence relation. It is reflective, symmetric, and transitive.
    b. This is an equivalence relation. It is reflective, symmetric, and transitive.

13. Hint: This may be proven using the exterior angle inequality (i.e., Theorem 2.1.5).

EXERCISES 2.2

1. $m(\angle x) = 60°, m(\angle y) = 60°, (\angle u) = 120°$
   $m(\angle w) = 60°, m(\angle k) = 60°, m(\angle h) = 120°$

3. **a.** $360°$
   b. $540°$
   c. $720°$

5. 7

7. $160°$

9. Draw the segment $\overline{CB}$. Since ℓ_1 and ℓ_2 are parallel, alternate interior angles must be congruent. Thus, $\angle ACB$ and $\angle DBC$ are congruent. After noting that $\angle A$ and $\angle D$ must be congruent and that $\overline{BC} \cong \overline{BC}$, it follows that $\triangle ACB \cong \triangle DBC$ using AAS. Then CPCF yields $\overline{AB} \cong \overline{CD}$.

EXERCISES 2.3

1. **a.** $120°$
   b. Yes. This must be a parallelogram since each set of consecutive angles forms a supplementary pair.

3. $m(\angle K) = 130°, m(\angle L) = 50°$, and $m(\angle M) = 130°$

5. Let $ABCD$ be the parallelogram and draw diagonal $\overline{AC}$ (see diagram here). Using the alternate interior angle criteria (see Theorem 2.2.1), show that $m(\angle CAB) = m(\angle ACD)$ and that $m(\angle CAD) = m(\angle ACB)$. Using $m(\angle A) = m(\angle CAB) + m(\angle CAD)$ and $m(\angle C) = m(\angle ACD) + m(\angle ACB)$, show that $m(\angle A) = m(\angle C)$. Also, one may prove that $m(\angle B) = m(\angle D)$ using similar reasoning.

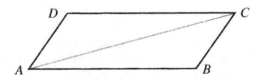

EXERCISES 2.4

1. 2

3. a.

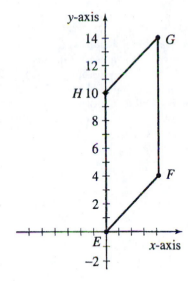

b. 40 units^2

5. a. 1.75 in^2

   b. 1.25 in^2

   c. 3 in^2

7. 30$\sqrt{2}$ cm$^2 \approx$ 42.43 cm^2

9. 12 units^2

EXERCISES 2.5

1. $\dfrac{\pi}{6}$ units

3. 6π units^2

5. The circle O has chords $\overline{AB}$ and $\overline{CD}$ subtending central angles of measures θ_1 and θ_2, respectively.

($\Longrightarrow$) (Assume that $AB = CD$ and then prove this implies that $\theta_1 = \theta_2$.)

Proof: Since all radii of circle O have the same length, it follows that $OA = OB = OC = OD$. Using $AB = CD$, one obtains $\triangle AOB \cong \triangle COD$ by SSS. It follows that $\theta_1 = \theta_2$ using CPCF.

($\Longleftarrow$) (Assume that $\theta_1 = \theta_2$ and prove this implies $\overline{AB} \cong \overline{CD}$, hence $AB = CD$.)

Proof: Since all radii of circle O have the same length, it follows that $OA = OB = OC = OD$. Using $\theta_1 = \theta_2$, one obtains $\triangle AOB \cong \triangle COD$ by SAS. It follows that $\overline{AB} \cong \overline{CD}$ using CPCF.

EXERCISES 2.6

1. (18) C

   (19) B

   (20) A

(21) A is a right hexagonal prism and C is a cube.

(22) Figure D is *not* a prism. Given any two opposite faces lying in parallel planes, there are lateral faces that are not parallelograms.

3. **a.** 27 ft^3
   b. 54 ft^2

5. There are several possible answers to this exercise. One possible net is an equilateral triangle with sides of 2 inches with this triangle subdivided into four equilateral subtriangles by joining the midpoints of the sides. The net for a (regular) tetrahedron that is shown with the other nets for the platonic solids illustrates this possibility. Another possible net looks like a parallelogram with two angles of 60° and two angles of 120°. The parallelogram is subdivided into four equilateral triangles. This second possibility is illustrated in #19 of Figure 2.6.2.

7. **a.** $5\pi \text{ in}^2$
   b. $6\pi \text{ in}^2$

9. **a.** $V = 36,000 \text{ cm}^3$
      $A = 7,200 \text{ cm}^2$
   b. $V \approx 2,197 \text{ in}^3$ (approx.)
      $A \approx 1,116 \text{ in}^2$ (approx.)

11. $96\pi \text{ ft}^3$

13. **a.** 4 units^3
    b. 18 units^2

CHAPTER 2 REVIEW EXERCISES

1. $m(\angle B) < m(\angle C) < m(\angle A)$

3. 7 inches

5. 4 sides

7. 140°

9. $m(\angle A) = 60°, m(\angle ABC) = 120°, m(\angle BCD) = 60°, m(\angle CBF) = 60°,$
   $m(\angle BCF) = 95°$

11. 32 units^2

13. 16 units^2

15. $\dfrac{2\pi}{3}$ feet ≈ 2.094 feet

17. **a.** $12,000 \text{ in}^3$
    b. $3,160 \text{ in}^2$

19. A triangle with two obtuse angles would have to have an angle sum of more than 180°, which would contradict the fact that the angle sum of a triangle is 180°.

21. Let ℓ_1, ℓ_2, and ℓ_3 be the respective perpendicular bisectors of $\overleftrightarrow{AB}$, $\overleftrightarrow{AC}$, and $\overleftrightarrow{BC}$ as shown here.

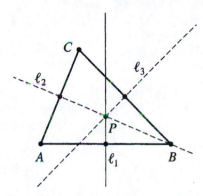

Let $P = \ell_1 \cap \ell_2$. Then it must be shown that P is also on the line ℓ_3. Since P is on the perpendicular bisector of $\overline{AB}$, it follows that $AP = BP$. Since P is on the perpendicular bisector of $\overline{AC}$, it follows that $AP = CP$. These equalities yield that $BP = CP$. Thus, P is on the equidistant locus of the points B and C. But this equidistant locus coincides with the perpendicular bisector of the segment $\overline{BC}$. Hence P is on the line ℓ_3.

23. Let parallelogram $ABCD$ have diagonals intersecting at point P as shown here.

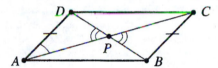

Notice that $AD = BC$ since opposite sides of a parallelogram have equal lengths. Also, $\angle DAC \cong \angle BCA$ since they are interior angles on opposite sides of a transversal to two parallel lines. Furthermore, $\angle DPA \cong \angle BPC$ since they are vertical angles. It follows that $\triangle APD \cong \triangle CPB$ using AAS. Thus, $AP = CP$ and $DP = BP$ using CPCF.

25. Let circle O have chord $\overline{AB}$ as shown here.

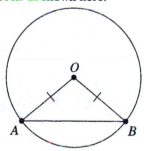

Notice that $AO = BO$ since all radii of a circle have the same length. Thus, O is on the equidistance locus of the points A and B. It follows that O must be on the perpendicular bisector of the segment $\overline{AB}$, as desired.

EXERCISES 3.1

1. a. $S = 3$. Using $S = \dfrac{A'B'}{AB} = \dfrac{B'C'}{BC} = \dfrac{C'A'}{CA}$ as well as $A'B' = 12$ and $AB = 4$, it follows that $S = \dfrac{A'B'}{AB} = \dfrac{12}{4} = 3$.

    b. $B'C' = 9$. Using $B'C' = S \cdot BC$ as well as $S = 3$ and $BC = 3$, it follows that
        $B'C' = 3 \cdot 3 = 9$.

    c. $C'A' = 12$. Using $C'A' = S \cdot CA$ as well as $S = 3$ and $CA = 4$, it follows that
        $C'A' = 3 \cdot 4 = 12$.

3. 30

5. Using the fact that any two consecutive angles of a parallelogram are supplementary, it is easy to see that four copies of a parallelogram may always be joined together to form a larger similar parallelogram as illustrated here.

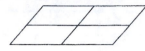

7. a. $3\sqrt{2}$

   b. $(0,6)$

9. a. Middle ring

   b. Outer ring

   c. Innermost circle

   d. Outer ring

11. Let G_1 be a square and let G_2 be a rectangle that is not a square.

13. Let $ABCD$ and $XYZQ$ be squares. Consider the vertex correspondence of $A \leftrightarrow X, B \leftrightarrow Y, C \leftrightarrow Z$, and $D \leftrightarrow Q$. Since all squares have all angles of measure $90°$, this correspondence matches up angles with angles of equal measure. Let $S = AB/XY$. Using $XY = YZ = ZQ = QX$ and $AB = BC = CD = DA$, it follows that $S = AB/XY = BC/YZ = CD/ZQ = DA/QX$. Thus, they are similar.

EXERCISES 3.2

1. 21

3. 80 ft

5. 12 ft

7. 4 ft

EXERCISES 3.3

1. a. 3/5

   b. 4/5

   c. 3/4

   d. 4/5

   e. 3/5

   f. 4/3

3. 52 ft

5. 12 in

7. $\sqrt{28}$ ft ≈ 5.29 ft

9. Yes. Let $\ell = MN = \sqrt{2}, n = LM = \sqrt{8}$, and $m = LN = \sqrt{10}$. Since $m^2 = n^2 + \ell^2$, there is a right angle at M.

11. $AC = BD = \sqrt{145}$ units ≈ 12.04 units

13. There are several ways to do this exercise. One approach is to use the Pythagorean theorem.
    Proof: Given the rectangle $ABCD$, apply the Pythagorean theorem to $\triangle ABC$ and then to $\triangle BAD$ to obtain $(AC)^2 = (AB)^2 + (BC)^2$ and $(BD)^2 = (AB)^2 + (AD)^2$. Using $BC = AD$, one finds $(AC)^2 = (BD)^2$ and hence, $AC = BD$, as desired.

EXERCISES 3.4

1. **a.** $(1,3), (5,3), (6,5), (2,5)$
   b. Yes
3. **a.** $(2,2), (6,8), (2,12)$
   b. $AB = \sqrt{13}, AC = 5, BC = 2\sqrt{2}, A'B' = 2\sqrt{13}, A'C' = 10, B'C' = 4\sqrt{2}$
   c. Yes. The SSS similarity theorem guarantees they are similar.
5. **a.** $8\sqrt{3}$ cm
   b. $24\sqrt{3}$ cm
   c. $48\sqrt{3}$ cm^2
7. **a.** .05
   b. 300 in

EXERCISES 3.5

1. **a.** 4
   b. 8
3. 50 ft
5. 1,000,000 liters
7. 2 lb. The scale factor S will be the ratio of the tall statue's height to the short statue's height. Thus, $S = \dfrac{6 \text{ feet}}{6 \text{ inches}} = \dfrac{6 \text{ feet}}{.5 \text{ feet}} = 12$. Since the statues are made of the same material, the weights will be proportional to the volumes. Hence, one has that (volume of tall statue) $= S^3 \cdot$ (volume of short statue) implies that (weight of tall statue) $= S^3 \cdot$ (weight of short statue). Using 3,456 lb. for the weight of the tall statue and $S = 12$, yields $(3{,}456 \text{ lb.}) = (12)^3 \cdot$ (weight shorter statue). Solving this last equation yields that the weight of the short statue is $\dfrac{3{,}456 \text{ lb.}}{(12)^3} = \dfrac{3{,}456 \text{ lb.}}{1{,}728} = 2$ lb.
9. **a.** .00008 cm
   b. 2.68×10^{-13} cm^3 (approx.)
11. The figure T_2 is congruent to T_1, but has the opposite orientation. The left hand of T_2 is touching her head and the right hand of T_2 is by her right side. The figurine T_2 is the mirror image of the original T_1 resulting from reflection across the yz-plane, hence "right" and "left" are interchanged.

CHAPTER 3 REVIEW EXERCISES

1. 9 cm
3. 1.25 cm^2
5. $NL = 12$ in and $LM = 10.5$ in
7. **a.** False
   b. False
   c. True
   d. False
9. $3\sqrt{2}$ cm
11. $\sqrt{5}$
13. **a.** 26 feet
    b. $\dfrac{5}{13}$

15. a. No. The sides do not satisfy the Pythagorean relation.

   b. $\cos(H) = \dfrac{2}{\sqrt{13}} \approx .5547$

17. a. Yes. The sides satisfy the Pythagorean relation.

   b. $\cos(A) = \dfrac{3}{\sqrt{13}} \approx .832$

19. $\sqrt{13}$ feet

21. a. Both have measure $150°$

   b. $18\ \text{cm}^2$

23. $6\sqrt{2} + \sqrt{3} \approx 11.59\ \text{cm}$

25. a. Yes

   b. No

   c. Yes

27. a. No

   b. Yes

   c. Yes

   d. Yes

29. a. $E' = (2,3)$, $F' = (3,3)$, $G' = (3,5)$, $H' = (2,5)$.

   b. The graph of $EFGH$ is a square with sides having length 1 and with one vertex at the origin. The graph of $E'F'G'H'$ is a rectangle with height 2 and length 1. The sides are parallel to the coordinate axes, and the lower-left vertex of $E'F'G'H'$ is at $E' = (2,3)$.

   c. No. The quadrilateral $E'F'G'H'$ is a rectangle that is not a square.

31. 500 cm

33. 70 inches

EXERCISES 4.1

1. a. $2\ \text{units}^2$

   b. $A' = (4,0)$, $B' = (2,0)$, $C' = (3,2)$

   c. $2\ \text{units}^2$

3. Let m and k be the perpendicular bisectors of $\overline{AB}$ and $\overline{AD}$.

5. a. $(2,-3)$

   b. $(-2,-3)$

   c. $(-2,-3)$

   d. $(-x,-y)$

   e. One possible answer is rotation about the origin by $180°$. Another valid answer is that it is a reflection across the origin.

7. a. The graph is exponential. It lies in the top half of the xy-plane, it is asymptotic to the negative x-axis, passes through the point $(0,1)$, and "quickly" becomes very high as x increases along the positive x-axis.

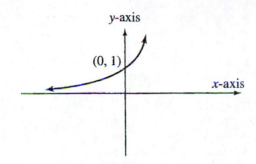

b. The graph is in the right half of the *xy*-plane. It is asymptotic to the negative *y*-axis as *x* approaches zero from positive values. The graph passes through the point $(1,0)$ on the *x*-axis and rises higher at slower and slower rates as *x* increases along the positive *x*-axis.

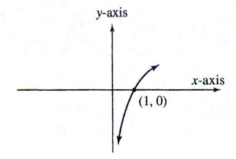

c. $y = \log_{10} x$

EXERCISES 4.2

1. $h = 5, k = 7$

3. $F(x,y) = (x + 4, y)$, $h = 4$, $k = 0$. This is a horizontal translation by 4 units in the positive *x*-direction.

5. $\alpha = 90°$

7. a. $(-1, 2)$
  b. $(-1, -2)$

EXERCISES 4.3

1. Let the equilateral triangle be $\triangle ABC$. Under an isometry, the image A' of A may be any of the three vertices. Once A' is determined, then the image B' of B may be any of the remaining two vertices. The images A' and B' fully determine the isometry since the image C' of C must be the only remaining vertex in the triangle. It follows that there are $3 \cdot 2 = 6$ isometries in this group.
The group consists of rotations by angles of $0°$, $120°$, and $240°$ about the "center" of the triangle and reflections across each of the three perpendicular bisectors of the sides. More precisely, let v, m, and k be the three perpendicular bisectors of the triangle's sides and let O be the point of intersection of these three lines. The group G is given by $G = \{I, R_{O,120}, R_{O,240}, r_v, r_m, r_k\}$.

3. There are many possible answers to this question. It is only necessary to find two isometries F, G and a single point P with $F \circ G(P) \neq G \circ F(P)$. For example, let $F = R_{(0,0)90}$ and $G = r_h$, where h is the x-axis. Let $P = (1, 0)$. Then $F \circ G(1, 0) = F(1, 0) = (0, 1)$ and $G \circ F(1, 0) = G(0, 1) = (0, -1)$. Hence $F \circ G(P) \neq G \circ F(P)$, as desired.

EXERCISES 4.4

1. $a = 3, b = -6, c = 2$
3. **a.** The condition is that n should be even.
   b. The condition is that n should be odd.
5. Left front = 3,   left back = 5,   top = 6,   right back = 4,   right front = 2, bottom = 1.

CHAPTER 4 REVIEW EXERCISES

1. **a.** $P' = (5, -1)$.
   b. Your graph should have the point $P = (5, 1)$ 1 unit above the point $x = 5$ on the x-axis, and the point $P' = (5, -1)$ should be 1 unit below the point $x = 5$ on the x-axis.
3. **a.** $S' = (5, 0), T' = (4, 1)$
   b. $S'' = (7, 0), T'' = (8, 1)$
   c. Your graph should have the segments $\overline{ST}$ and $\overline{S''T''}$ of length $\sqrt{2}$ and have slope $+1$. Segment $\overline{S'T'}$ has length $\sqrt{2}$ and has slope -1.
5. **a.** $A' = (2, -3), B' = (6, -3), C' = (4, -4)$
   b. Your graph should have sides $\overline{AB}$ and $\overline{A'B'}$ parallel to the x-axis. Triangle $\triangle ABC$ is in the first quadrant and $\triangle A'B'C'$ is in the fourth.
   c. The triangle $\triangle A'B'C'$ is traversed clockwise, and the triangle $\triangle ABC$ is traversed counterclockwise.
7. **a.** $L' = (1, -7), M' = (2, -7), N' = (2, -8)$
   b. Your graph should have both $\angle M$ and $\angle M'$ as right angles. Triangle $\triangle LMN$ is in the first quadrant and $\triangle L'M'N'$ is in the fourth.
9. **a.** $W' = (0, -1), X' = (0, -5), Y' = (3, -5), Z' = (3, -1)$
   b. Your graph should have both rectangles with sides parallel to the coordinate axes. Rectangle $WXYZ$ is in the first quadrant. It has height 3 and length 4. Rectangle $W'X'Y'Z'$ is in the fourth quadrant. It has height 4 and length 3.
11. **a.** $S' = (0, -1), T' = (-4, 0)$
    b. Your graph should have segment $\overline{ST}$ in the first quadrant with slope -4. The segment $\overline{S'T'}$ is in the third quadrant and has slope $-.25$.
13. **a.** $(5, -7)$
    b. $(-5, -7)$
    c. $(-5, 7)$
    d. $(-5, 7)$
    e. $(-x, y)$
    f. Reflection across the y-axis (i.e., r_y)
15. $y = 5$
17. **a.** ℓ_1 and ℓ_2 are lines of symmetry.
    b. Rotation about P by $180°$
    c. 4 (counting the trivial isometry)

19. **a.** Your graph should have the set T shaped like a Y that lies on its side. The three segments are each of length 1 and make angles of $120°$ with each other. The set T looks like three spokes of a wheel of radius 1 centered at the origin.

   b. The six symmetries of the set T are as follows: (1) reflection across $\overleftrightarrow{OA}$, (2) reflection across $\overleftrightarrow{OB}$, (3) reflection across $\overleftrightarrow{OC}$, (4) the identity map, (5) rotation about O by $120°$, and (6) rotation about O by $240°$.

EXERCISES 5.1

1. Using $\vec{0} = \overrightarrow{BB}$, and the definition of vector addition, one has $\overrightarrow{AB} + \vec{0} = \overrightarrow{AB} + \overrightarrow{BB} = \overrightarrow{AB}$. Using $\vec{0} = \overrightarrow{AA}$ and the definition of vector addition, one has $\vec{0} + \overrightarrow{AB} = \overrightarrow{AA} + \overrightarrow{AB} = \overrightarrow{AB}$. Thus, $\overrightarrow{AB} + \vec{0} = \vec{0} + \overrightarrow{AB} = \overrightarrow{AB}$.

3. $(3,3)$

5. $\sqrt{2}$

7. $(2,1)$

EXERCISES 5.2

1. $\begin{bmatrix} 3 \\ 4 \end{bmatrix}$

3. $\begin{bmatrix} 5 \\ 0 \end{bmatrix}$

5. $\begin{bmatrix} 5 \\ -24 \end{bmatrix}$

7. **a.** $\begin{bmatrix} 0 \\ 4 \end{bmatrix}$

   b. $\begin{bmatrix} 0 \\ -4 \end{bmatrix}$

EXERCISES 5.3

1. $\begin{bmatrix} 5 & 6 \\ 3 & 2 \end{bmatrix}$

3. $\begin{bmatrix} 6 & 11 \\ -5 & -1 \end{bmatrix}$

5. $\begin{bmatrix} 24 & 23 \\ 7 & 4 \end{bmatrix}$

7. $\begin{bmatrix} -1 & \sqrt{2} \\ 3 & -7 \end{bmatrix}$

9. **a.** $\begin{bmatrix} -1 & 7 \\ 16 & 4 \end{bmatrix}$

   b. $\begin{bmatrix} -1 & 7 \\ 16 & 4 \end{bmatrix}$

11. a. $\begin{bmatrix} 0 & 6 \\ 2 & 14 \end{bmatrix}$

b. $\begin{bmatrix} 0 & 6 \\ 2 & 14 \end{bmatrix}$

13. $\begin{bmatrix} 2 & 7 & 2 \\ 7 & 3 & 6 \\ 0 & -5 & 10 \end{bmatrix}$

15. $\begin{bmatrix} 21 \\ -48 \end{bmatrix}$

17. $\begin{bmatrix} x + 2y \\ 5x - 7y \end{bmatrix}$

19. $x = 1, y = 2$

21. $\begin{bmatrix} ae + bg & af + bh \\ ce + dg & cf + dh \end{bmatrix}$

23. Calculating $k(AB), (kA)B$, and $A(kB)$ independently, one obtains

$$k(AB) = k\left(\begin{bmatrix} a & b \\ c & d \end{bmatrix} \begin{bmatrix} e & f \\ g & h \end{bmatrix} \right) = k\left(\begin{bmatrix} ae + bg & af + bh \\ ce + dg & cf + dh \end{bmatrix} \right)$$

$$= \begin{bmatrix} k(ae + bg) & k(af + bh) \\ k(ce + dg) & k(cf + dh) \end{bmatrix},$$

$$(kA)B = \begin{bmatrix} ka & kb \\ kc & kd \end{bmatrix} \begin{bmatrix} e & f \\ g & h \end{bmatrix} = \begin{bmatrix} kae + kbg & kaf + kbh \\ kce + kdg & kcf + kdh \end{bmatrix}$$

$$= \begin{bmatrix} k(ae + bg) & k(af + bh) \\ k(ce + dg) & k(cf + dh) \end{bmatrix},$$

and

$$A(kB) = \begin{bmatrix} a & b \\ c & d \end{bmatrix} \begin{bmatrix} ke & kf \\ kg & kh \end{bmatrix} = \begin{bmatrix} ake + bkg & akf + bkh \\ cke + dkg & ckf + dkh \end{bmatrix}$$

$$= \begin{bmatrix} k(ae + bg) & k(af + bh) \\ k(ce + dg) & k(cf + dh) \end{bmatrix}.$$

Since all three of these calculations yield the same result, one finds $k(AB) = (kA)B = A(kB)$, as desired.

EXERCISES 5.4

1. 3

3. $\begin{bmatrix} -9 & 2 \\ \dfrac{5}{2} & -\dfrac{1}{2} \end{bmatrix}$

5. $\begin{bmatrix} 1 & 0 & 0 \\ -\dfrac{2}{3} & \dfrac{1}{3} & 0 \\ \dfrac{2}{3} & -\dfrac{4}{3} & 1 \end{bmatrix}$

7. a. $\begin{bmatrix} 16 \\ 9 \end{bmatrix}$

b. $(16, 9)$

c. $(7, 5)$

9. a. The matrix equation $X' = AX + H$ becomes

$$\begin{bmatrix} x' \\ y' \end{bmatrix} = \begin{bmatrix} 2 & 0 \\ 1 & 1 \end{bmatrix}\begin{bmatrix} x \\ y \end{bmatrix} + \begin{bmatrix} 5 \\ 5 \end{bmatrix} = \begin{bmatrix} 2x + 5 \\ x + y + 5 \end{bmatrix}.$$

Hence, $\begin{cases} x' = 2x + 0 \cdot y + 5 \\ y' = x + y + 5 \end{cases}$, which yields the following:

$a = 2, b = 0, c = 1, d = 1, h = 5, k = 5$.

b. Since $F(0,0) = (5,5)$, $F(2,0) = (9,7)$, $F(2,2) = (9,9)$, $F(0,2) = (5,7)$, the image of the square R is a figure with vertices at the four points $(5,5)$, $(9,7)$, $(9,9)$, $(5,7)$.

c. Parallelogram

d. 2

e. 2

11. a. 7

b. 7. Hence $\det(B) = \det(B^T)$.

c. $B^2 = \begin{bmatrix} 2 & 1 \\ 3 & 5 \end{bmatrix}\begin{bmatrix} 2 & 1 \\ 3 & 5 \end{bmatrix} = \begin{bmatrix} 7 & 7 \\ 21 & 28 \end{bmatrix}$.

d. $\det(B^2) = 49 = 7^2 = [\det(B)]^2$

e. $B^{-1} = \begin{bmatrix} 5/7 & -1/7 \\ -3/7 & 2/7 \end{bmatrix}$

f. $\det(B^{-1}) = 1/7 = [\det(B)]^{-1}$

g. Some possible (and correct) conjectures:

$$\det(B^T) = \det(B), \ \det(B^2) = [\det(B)]^2, \ \det(B^{-1}) = [\det(B)]^{-1}.$$

The first two equations hold for any square matrix B, and the last equation holds for any square matrix B with nonzero determinant.

13. Calculating $\det(AB)$ and $\det(A) \cdot \det(B)$ independently, one obtains

$$\det(AB) = \det\left(\begin{bmatrix} a & b \\ c & d \end{bmatrix}\begin{bmatrix} e & f \\ g & h \end{bmatrix}\right) = \det\left(\begin{bmatrix} ae + bg & af + bh \\ ce + dg & cf + dh \end{bmatrix}\right)$$

$$= (ae + bg)(cf + dh) - (ce + dg)(af + bh)$$

$$= aecf + aedh + bgcf + bgdh - ceaf - cebh - dgaf - dgbh$$

$$= aedh + bgcf - cebh - dgaf$$

and

$$\det(A) \cdot \det(B) = \det\left(\begin{bmatrix} a & b \\ c & d \end{bmatrix}\right) \cdot \det\left(\begin{bmatrix} e & f \\ g & h \end{bmatrix}\right) = (ad - bc)(eh - gf)$$

$$= adeh - adgf - bceh + bcgf.$$

Since $aedh + bgcf - cebh - dgaf = adeh - adgf - bceh + bcgf$, one finds $\det(AB) = \det(A) \cdot \det(B)$.

EXERCISES 5.5

1. $A = \begin{bmatrix} -1 & 0 \\ 0 & -1 \end{bmatrix}$

3. $C = \begin{bmatrix} 0 & 1 \\ 1 & 0 \end{bmatrix}$

EXERCISES 5.6

1. a. $6 + 8i$
   b. $-7 + 22i$
   c. $23 - 2i$

3. a. Use $\theta = \pi$ to find that $A = \begin{bmatrix} \cos(\pi) & -\sin(\pi) \\ \sin(\pi) & \cos(\pi) \end{bmatrix} = \begin{bmatrix} -1 & 0 \\ 0 & -1 \end{bmatrix}$.

   b. Since $e^{i(3\pi/2)} = -i$, it follows that $\theta = 3\pi/2$. Hence,

$$A = \begin{bmatrix} \cos(3\pi/2) & -\sin(3\pi/2) \\ \sin(3\pi/2) & \cos(3\pi/2) \end{bmatrix} = \begin{bmatrix} 0 & 1 \\ -1 & 0 \end{bmatrix}.$$

   c. Use $\theta = \dfrac{5\pi}{2} = 2\pi + \dfrac{\pi}{2}$ to find that

$$A = \begin{bmatrix} \cos(5\pi/2) & -\sin(5\pi/2) \\ \sin(5\pi/2) & \cos(5\pi/2) \end{bmatrix}$$

$$= \begin{bmatrix} \cos(\pi/2) & -\sin(\pi/2) \\ \sin(\pi/2) & \cos(\pi/2) \end{bmatrix} = \begin{bmatrix} 0 & -1 \\ 1 & 0 \end{bmatrix}.$$

CHAPTER 5 REVIEW EXERCISES

1. $(10, 15)$
3. $(-3\pi, -3)$
5. $2\sqrt{2}$
7. $\sqrt{x^2 + y^2}$
9. a. $(0, 0)$
   b. $(1, 2)$
11. a. 23
    b. 3
13. a. 10
    b. 10

c. 0

d. 0

15. $\begin{bmatrix} 18 \\ -3 \end{bmatrix}$

17. $\begin{bmatrix} -1 \\ -1 \end{bmatrix}$

19. $\begin{bmatrix} 2 \\ -7 \end{bmatrix}$

21. $\begin{bmatrix} 4 & 5 \\ 6 & 12 \end{bmatrix}$

23. $\begin{bmatrix} 2 & 8 \\ 22 & 8 \end{bmatrix}$

25. $\begin{bmatrix} -3 & -2 \\ 5 & 12 \end{bmatrix}$

27. $\begin{bmatrix} 11 & 17 \\ -2 & 23 \end{bmatrix}$

29. $\begin{bmatrix} 1 & 1 & 0 \\ -3 & 2 & 4 \\ 1 & 3 & 2 \end{bmatrix}$

31. $\begin{bmatrix} 2x + 3y \\ 4x - 5y \end{bmatrix}$

33. 5

35. $\begin{bmatrix} 3 & -\dfrac{2}{3} \\ 1 & -\dfrac{1}{3} \end{bmatrix}$

37. Calculating $(AB)^T$ and $B^T A^T$ independently, one obtains

$$(AB)^T = \left(\begin{bmatrix} a & b \\ c & d \end{bmatrix} \begin{bmatrix} e & f \\ g & h \end{bmatrix} \right)^T = \left(\begin{bmatrix} ae + bg & af + bh \\ ce + dg & cf + dh \end{bmatrix} \right)^T$$

$$= \begin{bmatrix} ae + bg & ce + dg \\ af + bh & cf + dh \end{bmatrix}$$

and

$$B^T A^T = \begin{bmatrix} e & f \\ g & h \end{bmatrix}^T \begin{bmatrix} a & b \\ c & d \end{bmatrix}^T = \begin{bmatrix} e & g \\ f & h \end{bmatrix} \begin{bmatrix} a & c \\ b & d \end{bmatrix}$$

$$= \begin{bmatrix} ea + gb & ec + gd \\ fa + hb & fc + hd \end{bmatrix}.$$

Since $\begin{bmatrix} ae + bg & ce + dg \\ af + bh & cf + dh \end{bmatrix} = \begin{bmatrix} ea + gb & ec + gd \\ fa + hb & fc + hd \end{bmatrix}$, one finds $(AB)^T = B^T A^T$.

39. Calculating $A(B + C)$ and $AB + AC$ independently, one obtains

$$A(B + C) = \begin{bmatrix} a & b \\ c & d \end{bmatrix}\left(\begin{bmatrix} e + i & f + j \\ g + k & h + l \end{bmatrix}\right)$$

$$= \begin{bmatrix} a(e + i) + b(g + k) & a(f + j) + b(h + l) \\ c(e + i) + d(g + k) & c(f + j) + d(h + l) \end{bmatrix}$$

and

$$AB + AC = \begin{bmatrix} a & b \\ c & d \end{bmatrix}\begin{bmatrix} e & f \\ g & h \end{bmatrix} + \begin{bmatrix} a & b \\ c & d \end{bmatrix}\begin{bmatrix} i & j \\ k & l \end{bmatrix}$$

$$= \begin{bmatrix} ae + bg & af + bh \\ ce + dg & cf + dh \end{bmatrix} + \begin{bmatrix} ai + bk & aj + bl \\ ci + dk & cj + dl \end{bmatrix}$$

$$= \begin{bmatrix} ae + bg + ai + bk & af + bh + aj + bl \\ ce + dg + ci + dk & cf + dh + cj + dl \end{bmatrix}.$$

This shows $A(B + C) = AB + AC$ since clearly

$$\begin{bmatrix} a(e + i) + b(g + k) & a(f + j) + b(h + l) \\ c(e + i) + d(g + k) & c(f + j) + d(h + l) \end{bmatrix} \text{equals}$$

$$\begin{bmatrix} ae + bg + ai + bk & af + bh + aj + bl \\ ce + dg + ci + dk & cf + dh + cj + dl \end{bmatrix}.$$

41. Calculating $\det(hA)$ and $h^2\det(A)$ independently, one obtains

$$\det(hA) = \det\left(h\begin{bmatrix} a & b \\ c & d \end{bmatrix}\right) = \det\left(\begin{bmatrix} ha & hb \\ hc & hd \end{bmatrix}\right) = (ha)(hd) - (hb)(hc)$$

and

$$h^2\det(A) = h^2\det\left(\begin{bmatrix} a & b \\ c & d \end{bmatrix}\right) = h^2(ad - bc).$$

This shows $\det(hA) = h^2\det(A)$ since clearly
$(ha)(hd) - (hb)(hc) = h^2(ad - bc)$.

43. **a.** $E' = (-3, 4)$, $F' = (1, 6)$, $G' = (21, 4)$
  b. 3
  c. 3

45. **a.** $1 - i$
  b. $7 - 12i$
  c. $4 + 7i$
  d. $-\dfrac{8}{5} - \dfrac{i}{5}$

47. **a.** $\sqrt{10}$
  b. $\sqrt{5}$

    c. $5\sqrt{2}$

    d. $\sqrt{2}$

49. a. $-1(= -1 + 0 \cdot i)$

    b. $i(= 0 + 1 \cdot i)$

51. a. $90°$

    b. $90°$

    c. $45°$

EXERCISES 6.1

1. a. $d_E(R,S) = 5$

    b. $d_T(R,S) = 7$

3. a. This is an isosceles right triangle for Euclidean geometry with a right angle at C. Thus, $m(\angle C) = 90°, m(\angle A) = m(\angle B) = 45°$.

    b. $d_E(A,B) = 2, d_E(B,C) = d_E(C,A) = \sqrt{2}$.

    c. $d_T(A,B) = d_T(B,C) = d_T(C,A) = 2$.

    d. Yes. All of the sides have the same taxicab length.

    e. No. Using taxicab angular measure (which is the same as Euclidean angular measure), the measure of $\angle C$ is larger than that of the other two angles.

5. No. Consider the triangles $\triangle ABC$ and $\triangle A'B'C'$ of problems 3 and 4, respectively. These triangles have SSS for taxicab geometry (all sides of each have taxicab length equal to 2), yet the triangles are not congruent. They are not congruent since they cannot have corresponding angles, that are congruent (i.e., $\triangle ABC$ is a right triangle, and $\triangle A'B'C'$ is not a right triangle).

EXERCISES 6.2

1. a. $8,000\pi$ mi $\approx 25,130$ mi

    b. $4,000\pi$ mi $\approx 12,565$ mi

3. a. $\dfrac{4,000\pi}{3}$ mi $\approx 4,190$ mi

    b. $\dfrac{8,000\pi}{3}$ mi $\approx 8,380$ mi

5. a. $181°$

    b. $\dfrac{\pi}{180}$ radians $\approx .01745$ radians.

    c. $\dfrac{800,000\pi}{9}$ mi^2 $\approx 279,200$ mi^2

EXERCISES 6.3

1. Let $\triangle ABC$ be a hyperbolic triangle in the Poincaré upper half-plane. The defect of this triangle is given by $\text{def}(\triangle ABC) = \pi - m(\angle A) - m(\angle B) - m(\angle C)$. Clearly, $\text{def}(\triangle ABC) < \pi$. Hence, $A_{\text{hyperbolic}}(\triangle ABC) = \text{def}(\triangle ABC) < \pi$.

3. $(x - 8)^2 + y^2 = 80$

CHAPTER 6 REVIEW EXERCISES

1. **a.** 5 units
   b. 7 units
3. **a.** $m(\angle A) = m(\angle A') = m(\angle B) = m(\angle B') = 45°$ and $m(\angle C) = m(\angle C') = 90°$
   b. $d_E(A, C) = d_E(B, C) = 2$, $d_E(A, B) = 2\sqrt{2}$, $d_E(A', C') = d_E(B', C') = \sqrt{2}$ and $d_E(A', B') = 2$.
   c. $d_T(A, C) = d_T(B, C) = 2$, $d_T(A, B) = 4$, $d_T(A', C') = d_T(B', C') = 2$ and $d_T(A', B') = 2$.
   d. Yes
   e. No. (This shows SAS is not valid for taxicab geometry.)
5. **a.** .125 radians

   b. $\dfrac{2,000,000}{(1,080)^2}$ radians ≈ 1.715 radians

   c. .0729
   d. .0729

7. $\left(\dfrac{r_2}{r_1}\right)^2$

9. $1,000\pi$ mi $\approx 3,142$ mi. This spherical distance is larger than the Euclidean distance.
11. **a.** $\ln(3)$ units
    b. $\ln(3)$ units
    c. $2 \cdot \ln(3)$ units
13. $x^2 + y^2 = 25$

15. $\ln\left(\dfrac{\sqrt{2} + 1}{\sqrt{2} - 1}\right)$ units ≈ 1.763 units

Photo Credits

Page 9:   G. Kleiman, *Mathscape: Seeing and Thinking Mathematically, Grades 6, 7, 8,* 1999, McGraw-Hill. Reprinted by permission of McGraw-Hill.

Page 19:   Excerpts from *Math Thematics: Book 1* by Rick Billstein and Jim Williamson. Copyright © 1999 by McDougal Littell. Reprinted by permission of McDougal Littell, a division of Houghton Mifflin Company.

Page 21:   From *Connected Mathematics: Filling and Wrapping* by Glenda Lappan, James T. Fey, William M. Fitzgerald, Susan N. Friel, and Elizabeth Defanis Phillips. © 2004 by Michigan State University. Published by Pearson Education, Inc., publishing as Pearson Prentice Hall. Used by permission.

Page 22:   G. Kleiman, *Mathscape: Seeing and Thinking Mathematically, Grades 6, 7, 8,* 1999, McGraw-Hill. Reprinted by permission of McGraw-Hill.

Page 25:   From *Mathematics in Context: Packages and Polygons.* Copyright © 2003 by Encyclopaedia Britannica, Inc. Reproduced by permission of Holt, Rinehart and Winston, under an exclusive agreement with Encyclopaedia Britannica, Inc.

Page 25:   Used by permission of North Wind Picture Archives.

Page 30:   G. Kleiman, *Mathscape: Seeing and Thinking Mathematically, Grades 6, 7, 8,* 1999, McGraw-Hill. Reprinted by permission of McGraw-Hill.

Page 43:   From *Mathematics in Context: Triangles and Beyond.*Copyright © 2003 by Encyclopaedia Britannica, Inc. Reproduced by permission of Holt, Rinehart and Winston, under an exclusive agreement with Encyclopaedia Britannica, Inc.

Page 48:   From *Mathematics in Context: Triangles and Beyond.*Copyright © 2003 by Encyclopaedia Britannica, Inc. Reproduced by permission of Holt, Rinehart and Winston, under an exclusive agreement with Encyclopaedia Britannica, Inc.

Page 55:   G. Kleiman, *Mathscape: Seeing and Thinking Mathematically, Grades 6, 7, 8,* 1999, McGraw-Hill. Reprinted by permission of McGraw-Hill.

Page 61:   G. Kleiman, *Mathscape: Seeing and Thinking Mathematically, Grades 6, 7, 8,* 1999, McGraw-Hill. Reprinted by permission of McGraw-Hill.

Page 66:   G. Kleiman, *Mathscape: Seeing and Thinking Mathematically, Grades 6, 7, 8,* 1999, McGraw-Hill. Reprinted by permission of McGraw-Hill.

Page 67:   Billstein, Rick and Jim Williamson, *Math Thematics: Book 1,* McDougal Littell, 1999. Reprinted by permission of McDougal Littell.

Page 69:   G. Kleiman, *Mathscape: Seeing and Thinking Mathematically, Grades 6, 7, 8,* 1999, McGraw-Hill. Reprinted by permission of McGraw-Hill.

Page 77:   From *Mathematics in Context: Reallotment.* Copyright © 2003 by Encyclopaedia Britannica, Inc. Reproduced by permission of Holt, Rinehart and Winston, under an exclusive agreement with Encyclopaedia Britannica, Inc.

Page 78:   Billstein, Rick and Jim Williamson, *Math Thematics: Book 1,* McDougal Littell, 1999. Reprinted by permission of McDougal Littell.

Page 79:   Billstein, Rick and Jim Williamson, *Math Thematics: Book 1,* McDougal Littell, 1999. Reprinted by permission of McDougal Littell.

Page 79:   From *Connected Mathematics: Filling and Wrapping* by Glenda Lappan, James T. Fey, William M. Fitzgerald, Susan N. Fiel, and Elizabeth Defanis Phillips. © 2004 by Michigan State University. Published by Pearson Education, Inc., publishing as Pearson Prentice Hall. Used by permission.

Page 80:   G. Kleiman, *Mathscape: Seeing and Thinking Mathematically, Grades 6, 7, 8,* 1999, McGraw-Hill. Reprinted by permission of McGraw-Hill.

Page 81:	From *Connected Mathematics: Filling and Wrapping* by Glenda Lappan, James T. Fey, William M. Fitzgerald, Susan N. Fiel, and Elizabeth Defanis Phillips. © 2004 by Michigan State University. Published by Pearson Education, Inc., publishing as Pearson Prentice Hall. Used by permission.
Page 82:	From *Connected Mathematics: Filling and Wrapping* by Glenda Lappan, James T. Fey, William M. Fitzgerald, Susan N. Fiel, and Elizabeth Defanis Phillips. © 2004 by Michigan State University. Published by Pearson Education, Inc., publishing as Pearson Prentice Hall. Used by permission.
Page 93:	From *Connected Mathematics: Stretching and Shrinking* by Glenda Lappan, James T. Fey, William M. Fitzgerald, Susan N. Fiel, and Elizabeth Defanis Phillips. © 2004 by Michigan State University. Published by Pearson Education, Inc., publishing as Pearson Prentice Hall. Used by permission.
Page 99:	From *Connected Mathematics: Stretching and Shrinking* by Glenda Lappan, James T. Fey, William M. Fitzgerald, Susan N. Fiel, and Elizabeth Defanis Phillips. © 2004 by Michigan State University. Published by Pearson Education, Inc., publishing as Pearson Prentice Hall. Used by permission.
Page 100:	From *Connected Mathematics: Stretching and Shrinking* by Glenda Lappan, James T. Fey, William M. Fitzgerald, Susan N. Fiel, and Elizabeth Defanis Phillips. © 2004 by Michigan State University. Published by Pearson Education, Inc., publishing as Pearson Prentice Hall. Used by permission.
Page 111:	From *Mathematics in Context: Looking at an Angle.* Copyright © 2003 by Encyclopaedia Britannica, Inc. Reproduced by permission of Holt, Rinehart and Winston, under an exclusive agreement with Encyclopaedia Britannica.
Page 120:	From *Connected Mathematics: Stretching and Shrinking* by Glenda Lappan, James T. Fey, William M. Fitzgerald, Susan N. Fiel, and Elizabeth Defanis Phillips. © 2004 by Michigan State University. Published by Pearson Education, Inc., publishing as Pearson Prentice Hall. Used by permission.
Page 124:	Billstein, Rick and Jim Williamson, *Math Thematics: Book 1,* McDougal Littell, 1999. Reprinted by permission of McDougal Littell.
Page 124:	Used by permission of Crazy Horse Memorial.
Page 125:	G. Kleiman, *Mathscape: Seeing and Thinking Mathematically, Grades 6, 7, 8,* 1999 McGraw-Hill. Reprinted by permission of McGraw-Hill.
Page 142:	Billstein, Rick and Jim Williamson, *Math Thematics: Book 2,* McDougal Littell, 1999. Reprinted by permission of McDougal Littell.
Page 143:	From *Mathematics in Context: Triangles and Beyond.* Copyright © 2003 by Encyclopaedia Britannica, Inc. Reproduced by permission of Holt, Rinehart and Winston, under an exclusive agreement with Encyclopaedia Britannica, Inc.
Page 145:	Billstein, Rick and Jim Williamson, *Math Thematics: Book 2,* McDougal Littell, 1999. Reprinted by permission of McDougal Littell.
Page 148:	Billstein, Rick and Jim Williamson, *Math Thematics: Book 2,* McDougal Littell, 1999. Reprinted by permission of McDougal Littell.
Page 156:	G. Kleiman, *Mathscape: Seeing and Thinking Mathematically, Grades 6, 7, 8,* 1999, McGraw-Hill. Reprinted by permission of McGraw-Hill.
Page 181:	Billstein, Rick and Jim Williamson, *Math Thematics: Book 2,* McDougal Littell, 1999. Reprinted by permission of McDougal Littell.
Pages 239–244:	*The Geometer's Sketchpad*®, Key Curriculum Press, 1150 65th Street, Emeryville, CA 94608, 1-800-895-MATH, www.keypress.com/sketchpad.

Index